biology
of
aging

biology of aging

Roger B. McDonald

 **Garland Science**
Taylor & Francis Group

NEW YORK AND LONDON

Vice President: Denise Schanck
Senior Editor: Janet Foltin
Assistant Editor: Allie Bochicchio
Production Editor: Natasha Wolfe
Typesetter and Senior Production Editor: Georgina Lucas
Text Editor: Margy Kuntz
Copy Editor: Linda Strange
Proofreader: Sally Livitt
Illustrations and Design: Matthew McClements, Blink Studio, Ltd.
Permissions Coordinator: Sheri Gilbert
Indexer: Indexing Specialists (UK) Ltd

Front Cover:
Photograph of an African bush woman,
courtesy of Dietmar Temps.

ISBN 978-0-8153-4213-7

Library of Congress Cataloging-in-Publication Data

McDonald, Roger B.
  Biology of aging / Roger B. McDonald.
    pages cm
  ISBN 978-0-8153-4425-4
1. Cells--Aging. 2. Aging--Molecular aspects. 3. Longevity. 4.
Physiology, Pathological. I. Title.
  QH608.M33 2013
  571.6--dc23

                              2013012208

Published by Garland Science, Taylor & Francis Group, LLC,
an informa business,
711 Third Avenue, New York, NY, 10017, USA,
and 2 Park Square, Milton Park, Abingdon, OX14 4RN, UK.

Printed in the United States of America

15 14 13 12 11 10 9 8 7 6 5 4 3 2 1

Dedication

Christoffer and Jacob, the lights of
my life, you make it all worthwhile.

Rodney and Jennifer, my best
friends, you have always been there
to put the neurons back in order.

Jack, you shall never know just
how much you did. You made it all
possible.

Taylor & Francis Group

Visit our web site at http://www.garlandscience.com

PREFACE

Why do we age? And what determines how long we—or any other species—will live? Sir Peter Medawar, biologist and Nobel laureate, suggested in 1952 that biological aging was one of the great unsolved mysteries of the time—and it would take another 40 years for this baffling mystery to be solved. We now know the underlying cause of aging and why a species lives as long as it does. In brief, *aging* is a random process; it is not a process that evolved. *Longevity* did evolve, through the selection of genes that promote reproductive success. With a better understanding of the differences between aging and longevity came a different approach to the study of biological aging: the science of biogerontology. This led to exciting new research that is providing a greater understanding of the cellular and molecular mechanisms underlying aging and longevity. *Biology of Aging* is the first textbook on the subject of biological aging since the causes of aging and longevity were elucidated.

Over the past 15–20 years, a general consensus has arisen among scientists on the evolutionary explanation of why we age and why we live as long as we do, resulting in an explosion of research on the basic mechanisms underlying aging and longevity. The results of this research have rapidly transformed biogerontology from a primarily observational and biomedical science into a more experimental and rigorous discipline within the general biological sciences. This new and exciting direction for biogerontology created a need for courses on the biology of aging. However, the pace of course development has been hampered by the lack of a textbook that approaches biogerontology from the biological rather than the biomedical perspective. *Biology of Aging* fills that gap. It is a biology textbook written for the biology student.

This book adheres to the fundamental principle that the key to fully understanding the biological process of aging is to first understand the basic concepts in biochemistry and physiology that apply to all life and all life stages. Each chapter begins with an overview of basic biological principles of non-aged systems. With this knowledge, the student is more fully prepared to understand how time-dependent alterations in molecular and cellular functions lead to aging. Working within a framework of basic biology also provides students with the background necessary to consider what interventions might lead to a slowdown in the rate of aging and an extension of life span.

The sequence of chapters follows a standard biology textbook organization, with each chapter providing the concepts and principles needed for understanding subsequent chapters. The 10 chapters are divided into three broad organizational groups. Chapters 1 and 2 present the basic concepts of biogerontology that are common to both experimental and clinical applications. Chapters 3 through 6 explore findings in evolutionary, cellular, and genetic biogerontology that have led to our current understanding of why and how we age. And Chapters 7 through 10 focus on how the basic sciences described in earlier chapters apply to human aging and longevity. We also establish the relationship between aging and disease and discuss what can be done now to slow the rate of human aging.

Chapter 1, Basic Concepts in the Biology of Aging, lays the foundation for all the chapters that follow. It introduces and explains the terminology of biogerontology, discusses the rise of biogerontology as a subdiscipline of biology, and describes the types of model systems that researchers use to study aging and longevity. In Chapter 2, Measuring Biological Aging, we take a closer look at the basic methods used by biogerontologists to measure the rate of aging in the individual and in populations. The principles of measuring mortality rate, concepts important for understanding the evolution of longevity, are thoroughly discussed in the context of life tables. Here, the student is introduced to the field of demography and learns how studies in demography help to predict trajectories in the rate of aging and longevity and how this prediction can be used to better understand how and why we age.

Chapter 3, Evolutionary Theories of Longevity and Aging, is the heart of the book. Only through an understanding of *why* we age—that is, an understanding of aging and evolution—have biogerontologists been able to more precisely construct and test hypotheses for *how* we age. Chapter 3 traces the development of evolutionary theories on longevity and aging from the early observational postulations to the mathematics and laboratory experimentation of the contemporary evolutionist.

Chapters 4 and 5 present the basic cellular and genetic findings that reveal how we age. Chapter 4, Cellular Aging, elucidates how the basic forces of the universe that affect all matter also provide the explanation for the underlying principle of aging. Here, we explore how the laws of thermodynamics explain what has become the fundamental principle underlying cellular aging: cellular aging reflects the accumulation of damaged proteins. We describe two mechanisms, oxidative damage and telomere shortening, in demonstrating the biochemistry and physiology behind the accumulation of damaged proteins and the finite life span of the cell.

Chapter 5, Genetics of Longevity, expands on the concepts presented in Chapter 3—that the longevity of a species is related to genes selected for reproductive success. We discuss the results of elegant laboratory research on yeast and worms that has identified specific genes affecting longevity. Because these genes are known to have similar effects in higher-order animals such as mice and rats, the student will get a sense of how biogerontology is on the verge of major discoveries that will allow the genetic manipulation of the rate of aging and longevity.

Chapter 6, Plant Senescence, contains material unique to this textbook: this is the first book on the biology of aging to discuss plant senescence and its importance to aging in humans and other animals. Plant sciences are an important part of biology, and plant senescence is an important part of the biology of aging.

The exploration of human aging begins in Chapter 7, Human Longevity, with a look into an exciting new area of aging and longevity research: gerontological biodemography. Biodemography is a computational science combining biology with demography. Results from this emerging field are providing evidence that the origin of aging and longevity in humans may differ significantly from that in other species, including nonhuman primates. In the second half of the chapter, we explore the reasons behind the unprecedented increase in human life expectancy during the twentieth century.

Chapter 8, The Physiology of Human Aging, and Chapter 9, Age-Related Disease In Humans, provide a detailed look at time-dependent changes in the major physiological systems: age-related changes in physiological systems that do not, in general, increase the risk of disease or mortality (Chapter 8) and those that are more likely to develop into diseases that lead to an increased mortality rate or morbidity (Chapter 9). As with other chapters, we cover basic physiology before describing age-related alterations.

The book concludes with Chapter 10, Modulating Human Aging and Longevity, a brief discussion of the current state of scientific knowledge about the modulation of aging and longevity. We begin by considering why modulation of biological aging might not be possible, then discuss the only two interventions that have been scientifically established as capable of modulating the rate of aging or life span: (1) reducing caloric intake and (2) maintaining physical activity throughout life. The chapter ends with a discussion of the possible implications of halting aging and/or extending the human life span.

This textbook provides an accessible introduction to biogerontology. The material is presented as an engaging, story-like narrative, with no loss of the accuracy essential to a biological science text. The illustrations are easy to follow and are accompanied by expanded legends that supplement the text description, rather than just repeating information. All chapters contain key terms (in bold type throughout the text) that are included in the book's extensive Glossary and available as flash-cards online; boxed text, providing further detail or interesting asides on a chapter topic; Essential Concepts, summarizing the main points of the chapter; Discussion Questions, to aid with studying (with answers provided on the student and instructor's site at www.garlandscience. com); and Further Reading, a list of sources and references, grouped by the chapter's section titles. An appendix explaining the mathematical derivations of life tables will assist the student in understanding Chapters 2 and 7. The appendix is an excerpt from the article E. Arias, United States life tables, 2006, Natl Vital Stat. Rep. 58:1-40, 2010. The full report can be accessed on the US Centers for Disease Control website (www.cdc.gov).

Biogerontology is a relatively young science. And it is an area of science that is of intense interest to us all—whether planning a career in biological research and teaching or in other areas of science, health, or medicine, or simply planning, while growing older, to understand as much about the process as possible. We hope this textbook serves the biology student well, and we look forward to feedback from both students and teachers.

Online Resources

While the book is the centerpiece of the course, we provide students and instructors with access to online resources to aid in the teaching and learning process. Accessible from http://garlandscience.com/aging, Student and Instructor Resource websites provide learning and teaching tools created for *Biology of Aging*. The Student Resources site is open to everyone, and users have the option to register in order to use book-marking and note-taking tools. The Instructor's Resource site requires registration; access is available to instructors who have assigned the book to their course. To access the instructor's resources, please contact your local sales representative or email science@garland.com. Below is an overview of the resources available for this book. Resources may be browsed by individual chapters and there is a search engine.

For students:

- A selection of animations and movies illustrating relevant concepts from the book.
- Answers to end-of-chapter problems.
- Glossary terms available as online flashcards.

For instructors:

- All of the images from the book are available in two convenient formats: Microsoft PowerPoint® and JPEG. Figures are searchable by figure number, figure name, or by keywords used in the figure legend from the book.
- The animations and movies that are available to students are also available on the Instructor's Resource site in two formats. The WMV-formatted movies are created for instructors who wish to use the movies in PowerPoint presentations on computers running Windows; the QuickTime-formatted movies are for use in PowerPoint for Apple computers or Keynote presentations. The movies can easily be downloaded to your personal computer using the "download" button on the movie preview page.
- Answers to end-of-chapter problems.
- Qualified instructors will be able to access a Question Bank comprising of multiple choice and open-ended questions.
- Lecture outlines are available to aid in framing a course around *Biology of Aging*.
- Access to resources from all of the Garland Science textbooks.

ACKNOWLEDGMENTS

I begin by acknowledging the three people whose tireless efforts on my behalf made this book a reality. Margy Kuntz, you were the right person at the right time, taking my writing from a meandering two-lane country road and making it into a four-lane expressway. You taught me so much in so little time. Janet Foltin, your leadership on this project, the confidence you gave me, and your knowledge of publishing were simply remarkable. Allie Bochicchio, wow! The unknown player drafted in the late rounds only to come off the bench and dazzle the crowd. It has been an honor and privilege to work with you.

There are many others at Garland Science who may not have been on the frontlines daily, but nonetheless had a major impact on this book. Denise Schanck, what an amazing job of management. You put together an outstanding team of professionals that accepted nothing less than excellence. Matthew McClements, an astonishing talent for graphics. Thanks for taking my finger paintings and making them into Rembrandts. Linda Strange, I don't know how you did it, but you transformed my writing into real English. Natasha Wolfe, few people have ever had the ability to keep me on time and in line the way you did. And, you did it with such grace and elegance. Thanks also go to Georgina Lucas, Sally Livitt, and Sheri Gilbert for doing all the detail work that never gets enough credit. Thank you to Adam Sendoff and Lucy Brodie for making sure that the right people know that this book exists. Finally, I wish to give my thanks to Michael Morales for walking into my office.

A huge thanks goes out to the individuals who took the time to review the chapters: Steven Bloomer, Ashok Upadhyaya, Olav Rueppell, Deborah Roach, Kenneth M. Crawford, Susheng Gan, Carol Itatani, Claudio Franceschi, Joel Parker, and Suresh Rattan. Your insight and exceptional knowledge of the subject improved the text vastly. I hope that you can see your influence in the finished product.

There are people who did not edit a single page or help write a single sentence that still need to be recognized for the pivotal role they have played in the development of this book. Jessica Coppola, you never cease to make my day just a little brighter. No matter how difficult it may get, I can count on your unconditional love to put the smile back on my face. Lisa Martinez, whether we are biking a 100 miles or just walking our dogs, you always seem to know the right thing to say to make the day better.

My graduate students, Kristin, Lisa, Maria, Annette, Cynthia, Mary, Michelle, Carol and David. Thanks for the wonderful times we have had together. My journey with *Biology of Aging* could not have been done without you.

Finally to the people who are the foundation of all that I do. Lois McDonald, you always told me that I could do it no matter the obstacles. Your baby boy done good. Mike Muirhead, you gave me a chance and showed me that everybody has something to offer. Paul Saltman, you may be physically gone, but your spirit is alive and well in me. To the thousands of undergraduate students who are and have been my daily inspiration—you give me great hope for the future. Party on, Garth.

DETAILED CONTENTS

CHAPTER 9: AGE-RELATED DISEASE IN HUMANS   253

BASIC CONCEPTS IN THE BIOLOGY OF AGING

1

"HOW OLD WOULD YOU BE IF YOU DIDN'T KNOW HOW OLD YOU WAS?"

-SATCHEL PAIGE, BASEBALL PLAYER (1906-1982)

Four billion years ago, two amino acids collided, bonded, and formed the first bioorganic molecule, a molecule that would one day lead to life. At precisely the same instant of its creation, the molecule began to interact with its environment, and a time-dependent history of those interactions was recorded in the form of changes in its chemistry. From that moment on, the molecules of life would always be linked to the process of biological change. Aging had begun.

So what does it mean to age? How and why do we and other organisms age? How do we measure aging? Are the causes of aging the same in different species? What are the consequences of aging? And what can we do, if anything, about it? The answers to these and many other questions are the subject of this text. In this chapter, we focus on general principles and concepts used in the study of **biogerontology**, the scientific investigation of the biological mechanisms of how and why we age. We begin by tracing the brief history of biogerontology, from its origins to its rise as an independent subfield within general biology. We then explore the underlying cause of aging and how biogerontologists define aging. The final two sections examine how biogerontologists study aging through the use of laboratory animals as models for human aging and through observations in wild animals.

BIOGERONTOLOGY: THE STUDY OF BIOLOGICAL AGING

Research in the biological sciences is all about searching for answers to the "how" and "why" of life. Biogerontology focuses on the how and why of aging. This relatively new field explores the biological processes that occur inside living things as they age and integrates research from many different fields, including biophysics, physical chemistry, molecular biology, neurobiology, biochemistry, genetics, evolutionary biology, medicine, and **gerontology** (the study of human aging and the problems of the aged). The scope of the field is broad—it can cover everything from molecular protein damage occurring inside the smallest cells to arterial atherosclerosis in a full-grown human adult.

Biologists began studying aging when human life spans increased

In the opening paragraph of this chapter, we suggested that life and aging arose simultaneously. However, although serious research in the life sciences can be traced back 400 years, the mechanisms of aging have been investigated rigorously for only the past 50–60 years. Why have the life sciences paid so little attention to the mechanisms of biological aging and **longevity**, the potential maximum age that an individual of a particular species can attain?

Until the beginning of the twentieth century, aging was an unimportant problem for biologists, because humans had relatively short **life spans** (the length of life of an individual organism). Between 1500 and 1900 c.e., the average life span for people in Western Europe and the United States hovered between 35 and 45 years **(Figure 1.1)**. For most of the population during this time, death commonly occurred at birth and, for women, in childbirth; childhood diseases killed millions of children under the age of 10; and infectious disease, such as influenza and tuberculosis, affected all age groups **(TABLE 1.1)**. There were no compelling reasons to investigate a phenomenon—aging—that affected so few humans. Instead, biologists were focused on studying and curing the diseases that killed the majority of people before they had a chance to grow old. Thoughts about growing old were left to philosophers and theologians.

Biogerontology became an independent field of research during the 1940s

Beginning around 1900, scientific and technological advances occurred that significantly increased life span. However, research on biological aging and longevity remained in the hands of only a few scientists. As a result, knowledge about the biological bases of aging and potential

Figure 1.1 Average life expectancy at birth for humans in Western Europe and the United States from 1500 to 2000 c.e. The numbers above the graph line are the percentage increases in life expectancy from one century to the next. The inset table shows life expectancy by decade for the United States since 1910. Note that the average life span did not rise above 50 years until after the 1900s. (Data from Gy Acsádi and J. Nemeskéri, History of human life span and mortality. Translated by K. Balas. Budapest: Akadémiai Kiadó, 1970. With permission from the University of Chicago Press; E. Arias, United States life tables, 2006, Natl. Vital Stat. Rep. 58:1–40, 2010. With permission from the National Center for Health Statistics.)

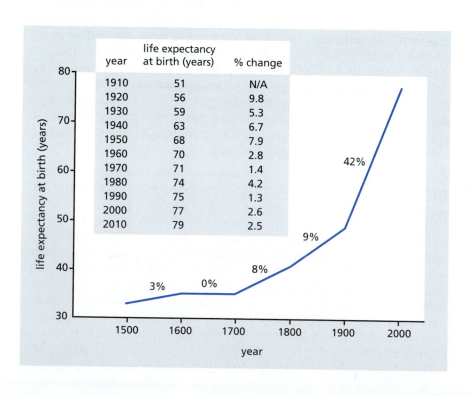

year	life expectancy at birth (years)	% change
1910	51	N/A
1920	56	9.8
1930	59	5.3
1940	63	6.7
1950	68	7.9
1960	70	2.8
1970	71	1.4
1980	74	4.2
1990	75	1.3
2000	77	2.6
2010	79	2.5

TABLE 1.1
LEADING CAUSES OF DEATH IN THE UNITED STATES FOR THE YEARS 1900 AND 2005

1900	% of deaths	2005	% of deaths
Influenza and pneumonia	12	Heart disease	31
Tuberculosis	11	Cancer	26
Diarrheal disease	8	Stroke	8
Heart disease	8	COPD[1]	6
Stroke	6	Influenza and pneumonia	3
Kidney disease	5	Alzheimer's disease	3
Accidents	4	Diabetes mellitus	3
Cancer	4	Kidney disease	2
Senility[2]	3	Accidents	2
Diphtheria	2	Septicemia	1

[1] COPD = chronic obstructive pulmonary disease.
[2] All dementias were referred to as senility. Alzheimer's disease had not yet been characterized.

treatments for age-related dysfunction did not keep pace with the increase in life span. The slow pace of aging research between 1900 and the mid-1930s was due, at least in part, to the lack of national organizations that promoted aging research and provided a mechanism for scientists to exchange ideas and findings. Other biological fields, such as physiology, chemistry, and anatomy, had strong professional societies that helped to attract funding for their members and had been in place for as long as 150 years, holding annual meetings and publishing scientific journals.

It was not until 1937 that a group of scientists held the first meeting of the Club for Aging Research at Woods Hole, Massachusetts. The Club for Aging Research later became the Gerontological Society of America. In 1946, this professional organization published the first scientific journal that focused exclusively on research in aging—the *Journal of Gerontology*. At about the same time, physicians also recognized that the diseases of aging were increasing with the increase in life span and, in 1942, established the American Geriatrics Society (**geriatrics** is the branch of medicine that deals with the problems and diseases of old age and aging people). The creation of these two professional societies marked the beginning of organized aging research.

The Gerontological Society of America and the American Geriatrics Society were instrumental in increasing the awareness that the biological and medical problems of aging needed a highly focused and organized research program. Without such a program, the United States and other economically developed countries would be facing a health crisis in the decades to come. To this end, the National Institutes of Health (NIH), the primary source of research funding in medicine and biology in the United States, established the Center for Aging Research in 1957. The program grew substantially over the next decade and a half, and in 1974 the National Institute on Aging (NIA) was established as an independent granting arm of the NIH. Today, the NIA has an annual budget of just over $1.1 billion and funds research in the biological, medical, and behavioral sciences.

Current aging research considers the health of the total person

The research programs at the Center for Aging Research and then the NIA initially focused largely on biological and biomedical research as a mechanism to improve the health of the ever increasing older population. As soon became apparent, the growth of the older population was outpacing the advances in research, and a significant number of people were experiencing age-related dysfunction without remedy. Quality, rather than quantity, of life was becoming a significant health issue for the older population. In response, the NIA began programs that included research in psychology, sociology, nursing, hospice care, and other fields that centered on the care and overall well-being of the older individual.

The inclusion of the behavioral and palliative care sciences in the overall research agenda in gerontology and geriatrics points to the uniqueness of aging research with respect to other health-related research. That is, aging and death have no cure, and thus gerontological and geriatric research, more than any other organized research field, must take a holistic approach. Biogerontological research that leads to improved health and extended life must be reconciled with the facts that aging will occur no matter how successful the remedy for a specific age-related dysfunction and that death will be the endpoint for the individual. Thus, biogerontologists are required not only to be experts in their particular field but also to be active participants in discussions on the psychological, social, and economic consequences of improved health and well-being in the older population.

Biological aging in nonhuman species shares many of the traits observed in human aging

Until recently, biological aging in other organisms received even less attention than that given to humans. The primary reason was that most scientists accepted the premise that, due to predation, few wild animals could reach an advanced age. Today, scientists recognize that nature provides many examples of aging in the wild. Moreover, all **eukaryotes**, organisms whose cells contain their genetic material inside a nucleus, from the simplest single-cell yeast to the most complex organism, *Homo sapiens*, share some aspects of the aging process. We are now at a stage in which discoveries in a nematode worm, *Caenorhabditis elegans*, concerning the process of aging and longevity can be directly applied to studies in mice or other complex life forms. How aging in the wild is providing clues to human aging is discussed later in the chapter.

The study of aging is a complex process

Recall that organized research in aging has been in existence for only 50–60 years, a very short time in the history of biological research. Although biogerontologists have learned a great deal about the causes of human aging and longevity, they have also found that the study of aging is complex and often influenced by factors that are difficult to control. For example, the outcomes of aging are the result, in large part, of a lifetime of interactions with our environment. No two humans have the same interactions with their environment. Because of this, the rate of aging, as you will learn in the next chapter, is highly individualized and cannot be determined by investigations that compare mean data across populations. Although environmental factors can be controlled through the use of animals to model human aging, variation in the rate of aging within a species remains. Thus, differences between individuals' specific genomes can also cause significant variation in the rate of aging among individuals, and researchers are finding that this variation cannot be controlled easily, even when they use sophisticated genetic engineering technology designed to create genetically identical animals.

Differences between species in the rate of aging also make research in aging and longevity challenging and have been an obstacle to defining aging precisely. For example, humans living in economically developed countries can expect to live, on average, 70–80 years. Some may even live to see 120 **(Figure 1.2A)**. The adult female of the mayfly species *Dolania americana* emerges from its nymph stage, lays eggs, and dies within just 5 minutes—provided that a trout does not have it for dinner first (Figure 1.2B). Examples of aging diversity within the plant kingdom are no less spectacular. Common sweet corn (*Zea saccharata*) germinates, matures, and dies over a four-month season (Figure 1.2C). Travel to the White Mountains of eastern California and you can touch a bristlecone pine tree (*Pinus aristata*) that has been in existence for more than 5000 years (Figure 1.2D).

Figure 1.2 Examples of the diversity of animal and plant life spans.
(A) Jeanne Calment, the oldest person on record, died on August 4, 1997, at the age of 122. (B) Some species of the mayfly die within 5 minutes of emerging from the nymph stage. (C) The life cycle of sweet corn is only four months long. (D) Bristlecone pine trees may live for more than 5000 years. (A, courtesy of G. Gobet/AFP/Getty Images; B–D, courtesy of Thinkstock.)

DEFINITIONS OF BIOLOGICAL AGING

How is "biological aging" defined? This has proved difficult because, until recently, the cause of aging was unknown—or, at least, controversial. As a result, hundreds of definitions have been proposed over the years. We now know the cause of aging and can construct a more precise definition of biological aging. Nonetheless, biogerontology is a diverse field that includes researchers from many different disciplines. The definition used for this text and intended for the biologist may not have relevance for other sectors within the broad field of biogerontology (although the process of aging will be identical). This may be particularly true for fields that deal exclusively with human aging.

We begin this section by tracing the history and development of definitions of aging and considering why these definitions remain relevant to specific areas within the general scope of biogerontology. The section ends with a definition of aging that will serve as your guide throughout this text.

The first definitions of biological aging were based on mortality

Many scientists defined biological aging as an increased risk of **mortality**, or death. For example, "Biological aging is characterized by an increase in the mortality rate," and is "an increased susceptibility to die, or increasing loss of vigor, with increasing chronological age, or with the passage of the life cycle." Mortality-based definitions are particularly useful in the research field of gerontological demography, a statistical science that studies size and mortality characteristics within populations. The usefulness of mortality as a descriptor for aging is discussed in detail in Chapter 7 when we explore biodemography, a subfield of demography that combines classical demography with evolutionary theory to study aging patterns in populations.

Mortality-based definitions of biological aging are less useful to researchers who correlate biological events to aging outcomes in individuals rather than in populations. For example, in humans, the leathery skin and gray hair of an 80-year-old can be shown to be a result of changes in the biochemistry of these tissues, making them less functional than their 10-year-old counterparts. These are clear signs of biological aging. However, it is unlikely that changes in the skin and hair of an aging human significantly increase mortality risk. That is, aging of these organs does not equate with death. In a similar way, the fruit of the apple tree develops, reaches maturity, and dies without significantly affecting the mortality risk of the entire tree. Using the risk of death as a measure of aging also fails to distinguish between longevity and aging. As you will discover in Chapter 3, "longevity" refers to only a single point in time on a scale established by the observer, whereas "aging" reflects changes that occur over a period of time.

For some species, death and aging *are* the same, and mortality-based definitions are appropriate. In the mayfly described earlier, death occurs so rapidly after the completion of adult development that measuring the rate of aging can be difficult. The Pacific sockeye salmon (*Oncorhynchus nerka*) provides another good example for which death equals aging. This salmon spends 99% of its life span in the open ocean and does not show measurable signs of aging during that time. However, when the fish returns to fresh water for spawning, it undergoes immediate deterioration and observable signs of aging **(Figure 1.3)**. Death occurs almost immediately after the spawning phase is complete.

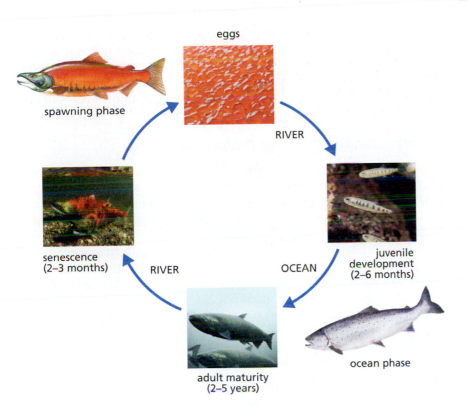

eggs

RIVER

spawning phase

senescence
(2–3 months)

RIVER

OCEAN

juvenile
development
(2–6 months)

adult maturity
(2–5 years)

ocean phase

Figure 1.3 The relationship between life cycle and aging in the Pacific sockeye salmon (*Oncorhynchus nerka*). The Pacific salmon begins its life in freshwater streams, grows into the juvenile fish, and then migrates to the ocean. Once in the ocean, the fish grows into an adult but does not reproduce. Two to five years after hatching, the Pacific salmon migrates back to the freshwater stream of its birth and begins spawning. At this point, the fish ages rapidly, developing a characteristic humped back and hook jaw. It typically dies within two weeks of spawning. (clockwise from top, courtesy of A. Nakazawa/Getty Images; courtesy of Thinkstock; courtesy of Visual Photos; courtesy of Thinkstock; courtesy of Ocean/Corbis; courtesy of Ocean/Corbis.)

For other forms of life, however, death does not equate with biological aging.

Functional-based definitions help describe biological aging over specific time periods

Scientists who correlate specific biological events with the rate of aging find functional-based definitions of aging—how well something works—more useful than mortality-based definitions. Two widely accepted definitions of this type include (1) "[Aging is the] deteriorative changes with time during post maturational life that underlie an increasing vulnerability to challenges, thereby decreasing the ability of the organism to survive" (Masoro, 1995); and (2) "Senescence [aging] is mainly used to describe age-related changes in an organism that adversely affect its vitality and functions but most importantly, increase the mortality rate as a function of time. Senility represents the end stage of senescence, when mortality risk is approaching 100%" (Finch, 1990).

The strength of these definitions is that they identify processes associated with advanced age, "increasing vulnerability to challenges" and "changes . . . that adversely affect . . . vitality and functions," which can be measured and followed over time. For example, muscle function can easily be evaluated by measuring the mass that a specific muscle group can move or lift. Indeed, numerous studies have shown that muscle strength, as well as many other physiological functions, declines after maturation. These definitions also specify a particular time period in which to "look" for aging: **postmaturation**—that is, the period after the organism has reached full growth.

There are limitations to both of these definitions, however. Both are organism-based; that is, they address the aging of the whole organism rather than aging at a lower level of organization, such as cellular function. Also, neither addresses possible events that occur during development and growth that might have a direct impact on postmaturational life. In addition, functional-based definitions make it difficult to determine when "aging" starts. It is possible that some physiological functions begin to decline while others are still developing. The human thymus gland, for example, begins to atrophy at around 14 years of age, at which point the rate of bone growth may be at its greatest.

A definition of aging for *Biology of Aging*

Several factors entered into developing the definition of aging that will be your guide throughout this textbook. Primary among these is that we now know the cause of cellular aging. Cellular aging reflects the random and **stochastic** (a process that has a probability distribution or pattern that may be analyzed statistically but may not be predicted precisely) accumulation of damaged proteins resulting from an organism's interaction with the environment. This means that our cells accumulate proteins that either do not function at their optimal level or do not function at all. The random nature of aging also means that aging did not evolve and thus there are no genes that regulate aging. The mechanism underlying the random and stochastic accumulation of damage and the reasons that aging could not have evolved are thoroughly discussed in Chapters 3, 4, and 5.

Three other factors also were important in developing the definition of aging, all of which are explained in detail throughout this text. (1) Biological aging occurs at many levels of biological organization that may not be directly applicable to the whole organism. (2) Factors that influence the biochemical and physiological decline that leads to the deterioration of old age may begin early in biological development. (3) Longevity and aging are related but distinct processes. Based on these considerations, I define biological aging as follows:

> Aging is the random change in the structure and function of molecules, cells, and organisms that is caused by the passage of time and by one's interaction with the environment. Aging increases the probability of death.

Random change to the structural and functional relationships of molecules is fundamental to this definition. As you will learn throughout this text, alterations in the structure and function of molecules that arise randomly are the result of environmental conditions and have a significant impact on the aging process.

Our definition does not include a specific point in time as the start of aging; it only suggests that aging occurs over time. This approach is taken because significant evidence has begun to accumulate suggesting that an individual's trajectory in the rate of aging may be influenced by environmental factors operating as early as fetal development.

Development, maturity, and senescence are event-related stages used to describe aging

Our definition of aging does not specify a time period during which aging is most likely to occur. The definition implies that biological aging is a continuum that starts at birth and ends at death. Although this description has theoretical value, in practice it makes comparisons of changes across the life span difficult. Therefore, we need to establish specific event-related points within the entire life span that describe distinct periods of biological aging. In this text, biological aging is discussed in terms of development, maturity, and senescence.

Development refers to the stage of the life span during which functional change is generally positive. This stage includes events such as the transition from larva to pupa, the expression of sexual characteristics, and the progression of protein synthesis from mRNA transcription to formation of quaternary structure. The developmental period ends when the organism achieves its maximal growth, a period in which many organisms experience their optimal reproductive fitness. In terms of bioactive molecules, cells, and organs, development ends when optimal functionality is reached. **Maturity** is the period during which function remains at optimal levels or slowly declines. The end of maturity occurs when the capacity to resist the force of **entropy** (the degradation of matter and energy in the universe to an ultimate state of inert uniformity) within the organism or molecule begins to wane. The capacity to resist the force of entropy as a factor in aging is discussed in detail in Chapter 4. **Senescence**, or the process of postreproductive aging, generally manifests as declines in vitality and function. Death is the end stage of senescence.

As shown in **Figure 1.4**, the duration and percentage of total life span for each of these stages varies greatly across different life forms. The life-stage curve of a human (Figure 1.4A), for example, illustrates the curve of organisms in which development and maturity take up the majority of the life span. In general, plants and animals that fit this pattern of aging grow to a fixed size and are **iteroparous**, organisms capable of reproducing more than once in a lifetime. Another characteristic of these organisms is that they have a significant amount of life after reproduction has ended. Senescence in these organisms tends to be gradual.

The cicada (*Magicicada septendecim*) shown in Figure 1.4B is an example of organisms that have an extended period of development (these cicadas live 16.5 years underground as nymphs). The developmental period is followed by a very short maturity stage during which all of the animal's energy is focused on reproduction. A rapid senescence

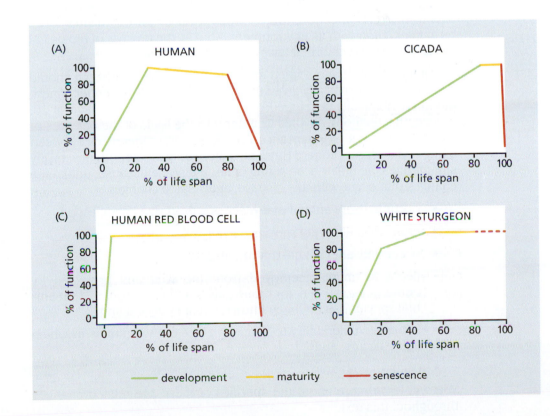

Figure 1.4 Life-stage curves show patterns of the three stages of biological aging in different types of organisms. The development stage for each organism is shown in green; the maturity stage, in yellow; the senescence stage, in red. (A) The curve for humans is typical of organisms in which development and maturity take up the majority of the life span. (B) The curve for the 17-year periodical cicada (*Magicicada septendecim*) represents organisms that have an extended developmental period. (C) The curve for human red blood cells illustrates aging in organic molecules. (D) The curve for the white sturgeon (*Acipenser transmontanus*) represents the life stages in organisms that do not seem to senesce.

stage occurs immediately following the maturity stage. Animals and plants with this type of life-stage curve do not have a postreproduction life span following maturity and usually produce offspring in a single season.

The shape of the life-stage curve for the human red blood cell (Figure 1.4C) characterizes aging in organic molecules. The short and rapid developmental period equates with the synthesis of the molecule, protein, or cell and can be measured in seconds or days. Maturity represents total functionality of the protein—in this case, the oxygen-carrying and carbon dioxide-carrying capacity of the hemoglobin in the cell. Senescence is denoted by **catabolism** (degradative metabolism that breaks down complex materials into simpler compounds).

Finally, Figure 1.4D shows the life-stage curve of the white sturgeon (*Acipenser transmontanus*) and illustrates the general pattern for organisms that either do not seem to senesce or demonstrate negligible senescence. This life-stage pattern can be the most difficult of the four to describe, due in part to the limited amount of accurate life span data for these organisms. Nonetheless, there are several commonalities among organisms that seem to have escaped senescence, including continuous growth, reproduction that is delayed until late in the developmental stage, and being iteroparous.

Biological aging is distinct from the diseases of old age

You may have noticed that the description of biological aging does not include any mention of diseases of old age. This is because, in our view, using the diseases of old age as a model for the underlying mechanisms of biological aging is not useful to an understanding of the process of biological aging—just as examining the results from research on chickenpox would not add to our understanding of developmental biology. Disease is a process within the animal or plant that impairs normal function. In contrast, as you will discover in Chapter 4, the functional changes and physical deterioration of biological aging are due to a loss in the resistance to entropy, brought on by an organism's long-term interaction with its environment. That is, the process of biological aging abides by the normal laws of physics and biology.

The importance of age-related diseases to an aging individual should be self-evident, especially given the dramatic rise in the number of people in economically developed countries who are over the age of 70. Indeed, this textbook devotes Chapter 9 to the topic of the diseases of aging. Even so, it is important to recognize the differences between aging and disease. Leonard Hayflick, a pioneer in biogerontology, nicely summarizes the differences: aging is not a disease because, unlike the changes that occur with any disease, age-related changes

- occur in every animal that reaches a fixed size in adulthood;
- cross virtually every species barrier;
- occur only after sexual maturation;
- occur in animals removed from the wild and protected by humans, even when, for thousands or millions of years, that animal species has not been known to experience aging;
- increase vulnerability to death in 100% of the animals in which aging occurs; and
- occur in both animate and inanimate objects.

You will learn more about the specifics of these age-related changes throughout this text.

HOW BIOGERONTOLOGISTS STUDY AGING: THE USE OF LABORATORY ORGANISMS IN HUMAN AGING RESEARCH

Ethical and practical considerations limit the type of research that can be done in humans. Therefore, biogerontologists use a variety of organisms, including single-cell organisms, invertebrates including insects, a range of mammals and fish, birds, nonhuman primates, as well as a few genetic disorders in humans, to investigate the basic nature of human aging. This section contains a brief survey of **eukaryotic** cells and organisms (that is, cells and organisms with membrane-enclosed nuclei) that serve as laboratory models for the investigation of mechanisms underlying human aging and longevity (**prokaryotes**—single-cell organisms that lack nuclei—have yet to secure their place in the study of aging). In the following section, "How Biogerontologists Study Aging: Comparative Biogerontology," we explore the use of wild animals as models for human aging and longevity.

Regardless of the type of organism being used, the research will have relevance to human aging because of the phylogenic relationships among all eukaryotes **(BOX 1.1)**. **Phylogenetics** describes the relatedness among organisms that is based on gene similarities.

BOX 1.1 THE PHYLOGENETIC TREE OF LIFE

Before the twentieth century, the classification of the diversity of living organisms and their relationships to each other was greatly influenced by centuries of philosophical and theological teachings. The founders of **taxonomy**—John Ray (1627–1705) and Carolus (Carl) Linnaeus (1707–1778)—developed their classification of organisms to reflect the Divine Order of creation, with "order" being the key word. For almost two hundred years, Linnaeus's system of classification used **morphology** (the form and structure of an organism) exclusively to suggest that evolution marches toward greater complexity—bacteria being the simplest and earliest life form, and humans being the most complex and most recently evolved **(Figure 1.5)**.

After the rediscovery and ultimate understanding of Mendel's principles of genetics, biologists began questioning whether evolution actually reflected an orderly procession of life from low to high complexity. Pressure to use an alternative form of classification grew even stronger with the discovery that the structure of DNA was identical in all organisms and that many genes found in "lower" life forms were identical to those found in "higher" animals. These findings provided solid and conclusive evidence that all life descended

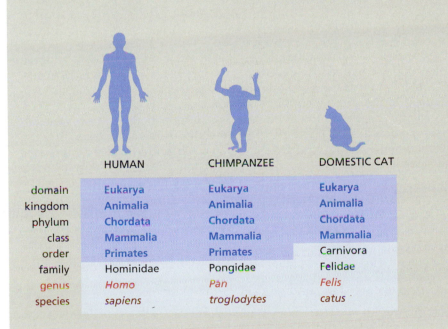

	HUMAN	CHIMPANZEE	DOMESTIC CAT
domain	Eukarya	Eukarya	Eukarya
kingdom	Animalia	Animalia	Animalia
phylum	Chordata	Chordata	Chordata
class	Mammalia	Mammalia	Mammalia
order	Primates	Primates	Carnivora
family	Hominidae	Pongidae	Felidae
genus	*Homo*	*Pan*	*Felis*
species	*sapiens*	*troglodytes*	*catus*

Figure 1.5 The Linnaean classification system. Carolus Linnaeus's system of classification is a series of hierarchically arranged categories based on an organism's resemblance to other life forms. Although morphological classification systems are being replaced by phylogenetic systems, the taxonomic names developed by Linnaeus are still widely used.

BOX 1.1 THE PHYLOGENETIC TREE OF LIFE

from a single common ancestor or, at most, a few common ancestors. In addition, evolutionists were finding that morphological complexity was a poor descriptor of evolutionary history for a species. Species were only as complex as they needed to be for survival in their environment. Thus, complexity was related more closely to the species' ability to survive in its environment than to a human definition of purpose.

Advancement in the biological sciences during the mid- to late-twentieth century led to the development of a classification system based on phylogeny rather than on morphology. **Phylogeny** is the evolutionary sequence of events involved in the development of a species or of groups of organisms. Modern phylogenetics uses a combination of factors and techniques to establish the evolutionary relationships among species. These include morphological characteristics, DNA sequences, ecological data, and mathematical algorithms to predict likely gene relationships. Phylogenetics does not consider one species more advanced than another. It simply holds that a species evolved from a previous group due to a genetic adaptation to its proximal environment.

Phylogenetic relationships can be visualized using a **phylogenetic tree**, a branching diagram showing the inferred evolutionary relationships among various species **(Figure 1.6)**. The branches of the tree define the ancestry and the descendant relationships among monophyletic groups. A **monophyletic group** contains all the descendants of a common ancestor. The nodes of the tree represent taxonomic units—such as an organism, a species, or a population—connected by a single branch. The topology, or branching pattern, of the tree can be scaled or unscaled. A scaled tree uses branch lengths proportional to the number of evolutionary changes that have occurred between taxonomic units. An unscaled tree uses the branches

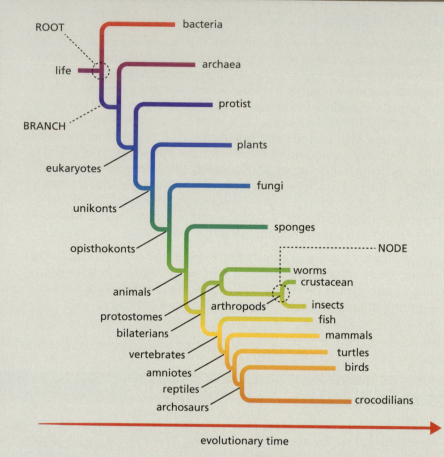

Figure 1.6 A rooted, unscaled phylogenetic tree. A phylogenetic tree is composed of nodes, each representing a taxonomic unit (species, populations, individuals), and branches, which define the relationship between the taxonomic units in terms of descent and ancestry. For example, arthropods are a monophyletic group that contains insects and crustaceans. In the tree, arthropods are represented by a node, and insects and crustaceans are represented by branches off that node.

only to connect relationships. Trees may also be rooted or unrooted. A rooted tree, such as the one in Figure 1.6, has a common ancestor for all other species or groups on the tree. Unrooted trees illustrate only relationships, without reference to common ancestors.

Phylogenetics is more than just a system for classification, however. It is a useful tool. For example, gene sequence comparisons made by molecular phylogeneticists have established a close evolutionary relationship between humans and the domestic pig, indicating a close

relationship in physiology. Indeed, the pig heart is very similar to the human heart in both structure and function. Medical researchers used this information to test whether healthy heart valves taken from a pig could be transplanted into a failing human heart. As it turns out, pig valves are an almost perfect match for human valves. Today, many people are alive and well with heart valves from pigs that replaced their own defective valves—due, in part, to the science of phylogenetics.

The discussion here serves as an introduction to groups of species generally used in laboratory experimentation in the science of biogerontology. Later chapters explore in much greater detail the specific use of these model systems. Plants are not discussed here; Chapter 6 describes plant biogerontology in detail. It is important to remember that no single animal or plant model is a "perfect" system for studying biological aging. The choice of an organism, instead, depends on such things as the question being asked, the rate of aging and longevity of the organism, the organism's reproductive type and success, and the cost for the care and keeping of the organism.

Isolated cell systems can be studied to describe the basic biochemistry of aging and longevity

Humans are the most complex organisms on Earth, containing sophisticated neural, vascular, and endocrine systems that have allowed us to be evolutionarily successful. Nonetheless, to work efficiently, these advanced systems depend on the proper functioning of their intracellular biochemistry. This is why cell function and its changes over time will ultimately describe how humans age. Investigating aging from a cellular perspective has a long history in gerontology, dating back to 1912, when the first cell culture was successfully developed (see Chapter 4). From these early studies arose the four basic cell systems that are used in the study of biogerontology: primary cell cultures, replicating cell cultures, cell lines, and stem cells.

Primary cell cultures are **differentiated cells** (highly specialized cells) removed directly from their *in vivo* location and maintained in an *in vitro* environment **(Figure 1.7)**. In biogerontology work, primary cell cultures typically contain post-mitotic cells, or cells with limited proliferation capacity, and remain viable for a very short time, usually only a few days. Primary cell cultures allow researchers to compare differences among particular types of differentiated cells. For example, techniques are available for determining the contractile properties of smooth muscle cells. To this end, smooth muscle cells for young and old animals can be removed, placed in culture, and then evaluated for age-related differences.

Replicating cell cultures are the most widely used type of cell culture systems in biogerontology. Replicating cell cultures are nondifferentiated mitotic cells, such as fibroblasts, that have been removed from tissue and allowed to divide until they reach **confluence** (maximum capacity within the confines of the culture dish). They are then separated into another flask and allowed to grow again, a process known a **population doubling**. Mammalian cells can be doubled about 30–50 times before the population dies. By sampling cells within the culture at different times, the biogerontologist can compare intracellular factors in young versus old cells. These systems are generally used to evaluate factors that lead to cell senescence and death, since mitotic cells have a finite replicative life span *in vitro*.

Cell lines are mitotic cells that do not have a finite life span. These cell populations either are derived from cancerous tumors or are normal cells that have had their internal biochemistry altered to make them immortal. Cell lines are a staple in general cell biology research, but they have not found wide use in biogerontology, most likely because these cells do not age and do not display the age-related functional loss observed in normal cells. However, some researchers use cell lines to investigate the pathways common to aging and cancer.

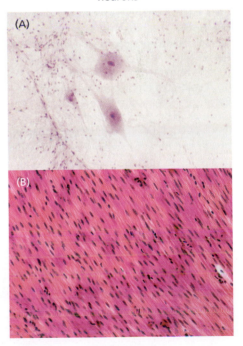

neurons

(A)

(B)

smooth muscle

Figure 1.7 Primary cell cultures. (A) Human neurons. (B) Smooth muscle cells. (A, courtesy of Thinkstock; B, courtesy of S. Gschmeissner/Getty Images.)

(A) totipotent stem cells (blastocyst)

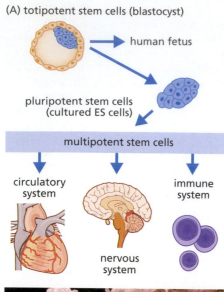

Figure 1.8 **Embryonic stem cells can give rise to the different cell types of the body.** (A) Embryonic stem (ES) cells are harvested from the inner cell mass of a blastocyst. These cells can be encouraged to differentiate into specific cell types. (B) A colony of embryonic cells in culture. Because of their unlimited capacity for self-renewal, embryonic stem cells have been proposed as a mechanism to regenerate tissues and organs damaged by aging or age-related disease. (B, courtesy of S. Gschmeissner/Science Photo Library/Getty.)

Stem cells are undifferentiated cells that have the ability to renew themselves indefinitely; they divide and create a differentiated cell. Stem cells exist in two forms, embryonic and adult. Embryonic stem cells are either **totipotent**, having the ability to generate an entire organism, including the placenta, or **pluripotent**, having the ability to generate cells and tissues from the three types of germ layers—endoderm, ectoderm, and mesoderm **(Figure 1.8)**. Adult stem cells are **multipotent** and form the type of tissue from which they were extracted: liver stem cells produce liver cells, muscle stem cells produce muscle cells, and so on. Stem cells have value in biogerontological research because of their ability to rejuvenate or replace aging tissue. For example, hematopoietic stem cells, which produce blood cells, are being implanted into the bone marrow of elderly patients following chemotherapy, to reduce the risk of infection by speeding up the regenerative process.

Fungi are good models for studying environmental factors that affect aging and longevity

Fungi, in yeast and mycelial form, do not have a sophisticated vascular, neural, or endocrine system, making intercellular signaling difficult (cell-to-cell communication occurs through gaps or pores in the cell wall). They must rely on direct cellular contact with the environment for sensing the world around them. This property makes fungi an excellent subject for the study of environmental factors that affect aging. In addition, the study of aging in fungi provides researchers with some practical advantages. First, fungi survive in virtually every environment on Earth **(Figure 1.9)**. A fungal species can be selected to match the environmental condition that the investigator hypothesizes has an effect on aging. Second, the nuclear and mitochondrial genomes of fungi have a compact, high coding–to–regulatory sequence ratio and are comparatively easy to sequence. As you will see in Chapter 5, the high coding–regulatory sequence ratio allows investigators to determine more precisely which gene has which function. Third, fungi have a wide range of life spans, ranging from a few days to 8000 years. And fourth, large quantities of individual fungi can be grown quickly in the laboratory at very low cost.

Primitive invertebrates may provide clues to extended cellular life, cell signaling, and whole-body aging

Primitive invertebrates are a diverse group that includes sponges, jellyfish, sea anemones, coral, worms, rotifers, and mollusks **(Figure 1.10)**. Many aquatic invertebrates have extreme life spans and have only recently received significant attention in aging research (see the next section, "How Biogerontologists Study Aging: Comparative Biogerontology").

Worms and rotifers are easy to raise in the laboratory, and most have relatively short life spans. Although cell and tissue specialization in these organisms is primitive compared with that of higher-order animals, these species have sophisticated intercellular communication through cell junctions. Because these animals also have a compact genome, they are excellent models for investigating how cellular events may be linked to whole-body aging. Chapter 5 provides a detailed description of how the genetic manipulation of cell-signaling pathways connecting the environment to the start of reproduction in *C. elegans* led to the discovery of genes that may regulate longevity. Moreover, these organisms are **eutelic**; that is, they have a fixed number of cells when they reach maturity. Because they cannot renew tissue, species within

Figure 1.9 The diversity of fungi. Fungi live in diverse environments and have a wide range of life spans. (A) Budding yeast (*Saccharomyces cerevisiae*) is easily grown in culture. (B) Honey mushroom (*Armillaria ostoyae*; also called shoe-string fungus) may be the oldest living organism on Earth. A single *A. ostoyae* discovered in the Malheur National Forest of northeastern Oregon may be as old as 8000 years. Both (C) goblet fungus (*Cookeina sulcipes*), found in tropical rain forests, and (D) reindeer moss (*Rangifera*) which grows on the tundra, are examples of long-lived fungi that can survive in harsh conditions. (A, courtesy of S. Gschmeissner/Science Photo Library/Corbis; B, courtesy of M. Watson/moodboard/Corbis; C, courtesy of M. Read/123RF; D, courtesy of A. Romanov/123RF.)

this group may provide biogerontologists with a model to investigate stochastic aging, a principle of whole-body aging that you will learn about in Chapters 3 and 4.

Insects can be used to investigate how whole-body and intracellular signaling affect life history

Insects are the single largest class of animals on the planet, with three million known species and several times that number of undiscovered species. The short life span and extremely high rate of reproduction of many insects provide the opportunity to study and manipulate the genetics of several generations in a short time. In addition, the **life history** (the sum of all biological events occurring in an organism throughout its life span) of insects is more easily modulated through manipulation

Figure 1.10 Long-lived anemones. Sea anemones, such as this giant green anemone (*Anthopleura sola*), are reported to have extreme life spans and unlimited growth. (Courtesy of altrendo nature/Thinkstock)

of the environment than are those of many other, more complex animals. For example, the reproductive activity of insects, and thus their life spans, can be altered through changes in temperature, food availability, and amount of light in the day. This type of modulation is often associated with changes in signals from the neuroendocrine system. Thus, researchers can use insects to investigate how whole-body and intracellular signaling affect an organism's life history.

Although the benefits of insects as models for human aging are clear, only a handful of species have been studied to any great extent. The fruit fly, *Drosophila melanogaster*, has been used extensively in aging research and was the first animal to have its life span precisely determined. *D. melanogaster*'s niche in aging research lies primarily in the study of the genetics of life span and is discussed in great detail in subsequent chapters.

Mice and rats are common research subjects in the investigation of nutritional, genetic, and physiological questions

The vast majority of biogerontological research has been conducted on rats or mice as model organisms. Rodents are particularly useful in research because of the similarity between rodent and human physiology and cellular function. Rats and mice are relatively inexpensive to house compared with other animals with similar life spans. And unlike for human subjects, the diet and environment of rodents can be strictly controlled. In addition, rodents can easily be genetically manipulated, which allows for the testing of gene products and age-related changes. Because so much of the current research has been conducted on these animals, their specific use in research is described in detail in later chapters.

Nonhuman primates display many of the same time-dependent changes as humans

Nonhuman primates are the genetically closest relatives to humans and, as such, are the ultimate model for investigating the biological basis of human aging. Several species of nonhuman primates, such as lemurs, marmosets, monkeys, and great apes, have been used to study the biology of aging, but the majority of well-controlled laboratory studies involve the rhesus macaque (*Macaca mulatta*) monkey. Aging rhesus macaques display many of the time-dependent physiological declines that are also observed in humans and not often observed in other species **(Figure 1.11)**. These include visual and auditory deficits, motor function decline, loss in bone mineral content, a true menopause in females and decreasing testosterone levels in males, muscle mass decreases, and a general decline in metabolic function.

Rhesus macaques are also susceptible to many time-dependent human diseases, such as type 2 diabetes, cardiovascular disease, and pseudo forms of both Alzheimer's and Parkinson's disease. The etiology of type 2 diabetes and cardiovascular disease seems to be identical in rhesus macaques and humans.

The similarity between nonhuman primates and humans in time-dependent functional loss and disease provides scientists with highly controlled populations in which to perform repeated noninvasive or low-risk invasive procedures, testing drugs and other physical therapies. For example, many of the osteoporosis and anti-bone-loss prescription drugs currently available to humans were tested on rhesus macaques.

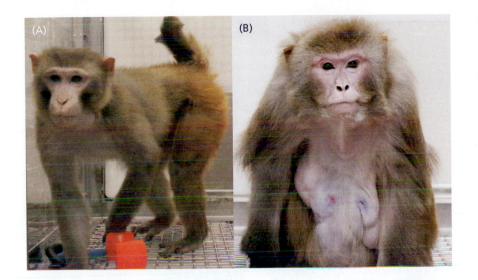

Figure 1.11 Rhesus macaque monkeys. Nonhuman primates such as rhesus macaque monkeys are invaluable models for human aging because of their genetic similarity to *Homo sapiens.* As shown here, rhesus macaque monkeys show visible signs of aging: (A) an 18-month-old monkey; (B) a 25-year-old. They are also susceptible to many age-related diseases and dysfunctions common in humans. (A, courtesy of J. Lenon; B, courtesy of J. Miller.)

However, the same physiological similarities that make these species attractive as models for aging also present a major limitation. The genetic similarity to humans has led to questions about the ethics of conducting invasive experiments on such close relatives. Regulatory agencies charged with insuring the humane treatment of research animals have responded by limiting the type of research that can be conducted in nonhuman primates. In general, investigative procedures are limited to those that are also permitted in humans, although the safety standards may be lower (for example, X-ray exposure limits are higher for monkeys than humans).

Another major limitation of using nonhuman primates as models for human aging is cost. Animals must be maintained in highly controlled conditions throughout their life span. The average life span of a rhesus macaque housed in a certified animal facility is approximately 35 years, with a maximum life span approaching 45 years. The average cost for maintaining a rhesus monkey varies with the research institution, but generally ranges from $9 to $10 per day. Thus, maintaining a single rhesus macaque for 35 years would cost approximately $121,000. It is unlikely that an individual investigator could receive sufficient funding to conduct an investigation with the numbers of rhesus monkeys needed to complete a well-controlled study in aging. The NIA, for example, currently supports research in rhesus monkeys at only two locations in the country.

Human progerias can be used to model normal human aging

Werner syndrome and Hutchinson-Gilford progeria syndrome are diseases that many believe are associated with premature aging **(Figure 1.12)**. **Progerias** are rare genetic conditions marked by slowed physical growth and characteristic signs of rapid aging. Hutchinson-Gilford progeria syndrome affects people at birth and during young childhood, whereas Werner syndrome normally begins to express itself in the second or third decade of life. While both syndromes place the individual at greater risk for developing age-related diseases, patients who have Werner syndrome tend to die of cancer and atherosclerosis, whereas those with Hutchinson-Gilford progeria syndrome are more prone to cardiovascular and neurological disorders.

Figure 1.12 Woman who has Werner syndrome. The woman is shown (A) at 13 years of age and (B) at 56 years of age. (A, from F.M. Hisama, V.A. Bohr, and J. Oshima, *Sci. Aging Knowl. Environ.* 10:18, 2006. With permission from the American Association for the Advancement of Science. B, courtesy of J. Oshima)

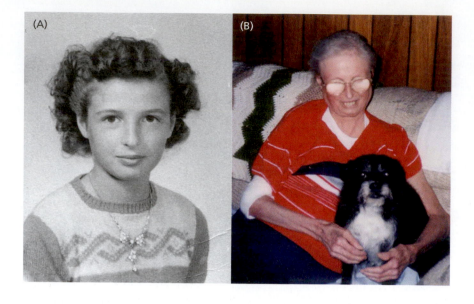

Because people with Werner syndrome tend to have longer life spans than those with Hutchinson-Gilford (45–50 years vs. 12–15 years), Werner syndrome is considered a better model for aging. There are four features common to all individuals affected with Werner syndrome: short stature, early graying and loss of hair, cataracts in both eyes, and scleroderma-like skin changes. Many affected individuals also have flat feet, changes to their voice, and hypogonadism. People with Werner syndrome have a high risk for developing type 2 diabetes, atherosclerosis, coronary heart disease, hypertension, and osteoporosis. These changes in appearance and increased risk for age-related diseases are clearly similar to those that happen during normal aging.

Werner syndrome is caused by mutations in the *WRN* gene. This gene is responsible for production of the WRN protein, which plays a role in the maintenance and repair of DNA. The protein may also assist in DNA replication. The lack of or decreased function of the WRN protein mimics the outcome predicted in some theories of aging that are discussed in Chapter 4.

HOW BIOGERONTOLOGISTS STUDY AGING: COMPARATIVE BIOGERONTOLOGY

Although research on laboratory species has provided, and continues to provide, significant insight into the basic biological mechanism of aging, the short life spans of these species make them less effective as models for investigating the mechanisms that underlie the exceptional longevity observed in humans. Some biogerontologists study longevity by observing wild animals that have long life spans, a subfield of biogerontology called **comparative biogerontology**. Comparative biogerontology identifies wild species that show resistance to aging and thus have extended longevity in environments that are otherwise conducive to short life spans. The long-lived species can then be bred in captivity, where possible genetic and biochemical mechanisms underlying their extended longevity can be evaluated. Identification of the mechanisms that have allowed the evolution of resistance to early death and thus lengthened life spans provides clues as to how extended longevity evolved in humans.

Our focus here is on a general overview of comparative biogerontology, describing a few relationships suggested to extend longevity in wild animals. We also describe a few specific animals with extended longevity and the evolutionary adaptations that have provided these animals with long life spans in the wild. The evolutionary, genetic, and biochemical mechanisms accounting for extended longevity are covered thoroughly in Chapters 3, 4, and 5.

Species' body size is related to maximum life span

A field mouse's maximum life span is much shorter than a rabbit's; a rabbit's much shorter than an elephant's. The casual observation that large mammals live longer than smaller ones was first noted in the scientific literature more than a hundred years ago and was confirmed in more recent studies **(Figure 1.13)**. The relationship between body size—here meaning overall dimensions, independent of underweight or overweight—and longevity also holds when warm-blooded animals are separated into their taxonomic groups, such as primates, ungulates, carnivores, rodents, and so on, and within the avian class. Humans, interestingly, do not fit within this pattern, having the longest life span of any mammal, yet clearly not the largest mammal. Nonhuman primates also fall outside the range in the body size–versus–life span curve established for other mammals. The unique position of the primates, including *Homo sapiens*, within the body size–life span relationship most likely reflects factors involving intelligence; this is discussed more thoroughly in Chapter 7.

Humans' large brains, combined with their exceptional longevity among mammals, suggested to many scientists that brain size might be the factor that accounts for the body size–life span relationship: larger animals tend to have larger and more complex brains. Such a hypothesis would seem reasonable, given that larger brains tend to provide more intelligence and greater regulation of the physiological functions that help to maintain **homeostasis** (the ability to maintain internal stability). Intelligence would help animals escape predators and enhance their success at finding food. Superior control of homeostasis would

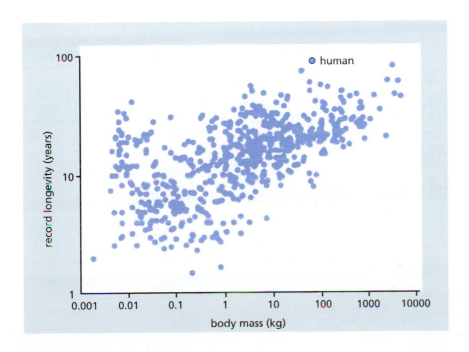

Figure 1.13 Relationship between life span and body mass in 605 mammalian species. Note that humans have the longest life span among the mammals, despite having a smaller body mass than some other mammals. (Adapted from S. Austad, in The Comparative Biology of Aging [N.S. Wolf, ed.], New York: Springer Science, 2006.)

allow animals to exist over a larger range of temperatures and to survive in many different types of environmental conditions. These animals would then have a bigger range over which to find food and thus increase their chances of survival.

Although the brain–weight hypothesis has innate appeal and has been supported in some investigations, the vast majority of studies do not find that brain weight significantly accounts for the body size–life span relationship. Indeed, the sizes of organs such as the liver, spleen, and heart are better predictors of longevity in most mammals than is brain size. Since the sizes of most internal organs other than the brain are determined, in large part, by the overall size of the body, it is not surprising that the size of internal organs would be as good a predictor of life span as body size.

The physiological complexity of the warm-blooded animals (also known as homeothermic and endothermic animals) has led biogerontologists to suggest that some factor (or factors) other than the simple measure of size may more closely reflect the relationship between longevity and body size. For example, scientists exploring the relationship between body size and longevity in the first half of the twentieth century noted that smaller mammals have significantly faster rates of metabolism than larger mammals, as measured by daily energy expenditure per total body mass. This led to a general theory suggesting that the greater the rate of energy expenditure, the shorter the life span—known scientifically as the "rate-of-living theory" and more popularly as "live fast, die young."

The rate-of-living theory as a general explanation for the body size–longevity relationship has not held up under more extensive and rigorous scientific experimentation, although it remains popular in the nonscientific literature. Animals of the avian class have metabolic rates twice those of mammals with a similar body mass **(Figure 1.14)**. However,

Figure 1.14 Relationship between basal metabolic rate (BMR) and maximum life span potential (MLSP) in birds and mammals. Birds have a longer life span and greater metabolic rate than mammals of similar body size. (From A.J. Hulbert, R. Pamplona, R. Buffenstein, and W.A. Buttemer, *Physiol. Rev.* 87:1175–1213, 2006. With permission from the American Psychological Society.)

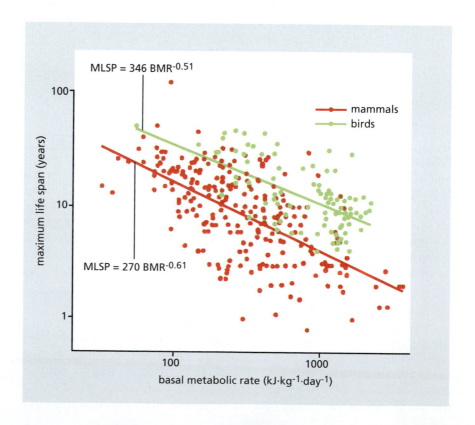

$$MLSP = 346\ BMR^{-0.51}$$

$$MLSP = 270\ BMR^{-0.61}$$

maximum life span (years)

basal metabolic rate (kJ·kg⁻¹·day⁻¹)

— mammals
— birds

birds live two to three times longer than their body size–matched mammalian counterparts. **Marsupials,** nonplacental mammals (for example, kangaroos, opossums), have shorter life spans and slower metabolic rates than comparably sized **eutherians**, the placental mammals.

Reduced vulnerability to extrinsic dangers explains extended longevity

Although observation and statistical analyses suggest that body size may be related to extended longevity in the wild, sufficient numbers of exceptions have been found among warm-blooded animals to suggest that this relationship is casual, not causal. The causal factor (or factors) that allows for extended longevity in wild animals, although not yet fully described, most likely reflects species' ability to adapt to their environment. Large animals may have extended longevity simply because their size and strength make them better at defending themselves against predators. Large animals also tend to have much larger foraging areas than small animals. This would reduce population density and decrease competition for food. Over evolutionary time, the ability to defend against predation and to decrease the risk of starvation would enhance survival and lead to the evolution of extended longevity. The underlying mechanism for the evolution of extended longevity is related to delayed reproduction, as discussed in detail in Chapter 3.

An ability to escape predation and to forage over a larger area as the mechanism for the evolution of long life spans in the wild can also be illustrated by animals that fly. Nonflying birds or weak-flying birds (chickens, for example) have significantly shorter life spans than strong-flying birds. Flight is an extremely effective way to escape predators and a more efficient mechanism for traveling long distances than the use of legs—thus providing a greater foraging area. Moreover, bats are the longest-lived mammals for their size.

Nonflying mammals and cold-blooded animals with expanded longevity share an ability to protect themselves from predation. The quills of the porcupine are most likely the reason that these mammals are thought to be the second longest lived (for body size), after the bats. Some species of turtles are known to live more than 150 years.

A highly organized social structure also extends longevity in the wild

Safety in numbers is another aspect of the evolution of extended life span in wild species. Social animals such as primates and herd animals consistently have longer life spans for their body size than nonsocial animals. An excellent example of societal influence on longevity occurs in the naked mole rat of Equatorial Africa **(Figure 1.15)**. The naked mole rat is the size of a mouse and lives out its life completely underground. That is, naked mole rats are never exposed to the dangers of the terrestrial world, and they live within large colonies. As a result, these animals live for 20–30 years, two to three times as long as other mammals of similar size.

Social insects, such as termites, ants, wasp, and bees, are another group of animals that have developed extended longevity. The social aspects of these insects involve a division of labor based on reproductive function, a type of social organization known as **eusociality**. The reproductive function also determines the life span of the insects. For example, each hive of honeybees has only one reproductively active queen, which can live 5–7 years. The queen is genetically identical to the thousands of functionally sterile female workers that attend to the

Figure 1.15 The naked mole rat (*Heterocephalus glaber*). Naked mole rats spend their entire lives underground, reducing their exposure to the harsh environment. This may be a factor in their relatively high longevity for their body size. (From R. Buffenstein *J. Gerontol. A. Biol. Sci. Med. Sci.* 60:1369-1377, 2005. With permission from Oxford University Press.)

Figure 1.16 Morphology of honeybees. The morphology and life spans of honeybees appear to be related to differences in nutrition during the larval phase. (A) A queen bee (the large bee in the middle) and workers (the smaller bees surrounding the queen). (B) A male drone. (A, courtesy of angelshot/Shutterstock; B, courtesy of alle/Shutterstock.)

larvae and pupae and live for just a few months. Some of the female workers will transition into foragers, the bees that collect pollen and make honey. These female foragers have life spans of less than 30 days. Finally, the male drones, whose only job is to mate with the queen, live through only one seasonal cycle. Thus, one colony of bees contains genetically identical females with three different phenotypes and life spans and males whose life span is directly related to reproduction **(Figure 1.16)**. Moreover, the castes into which the females are assigned appear to be controlled by nutritional status during development. The best-nourished larvae become queens. This unique feature of the eusocial insects provides biogerontologists with a method to easily manipulate life span and evaluate the interaction between nutrition and longevity.

A few aquatic animals have extreme longevity

Figure 1.17 Extreme longevity in aquatic animals. (A) The maximum life span of the white sturgeon (*Acipenser transmontanus*) is unknown but is estimated to be close to 200 years. (B) A clam of this *Arctica islandica* species was carbon-dated as 400 years of age. (A, courtesy of Shutterstock; B, courtesy of Z. Ungvari, Z. Ungvari et al., *J. Gerontol. A Biol. Sci. Med. Sci.* 66:741–750, 2011. With permission from Oxford University Press.)

Sponges, jellyfish, sea anemones, clams, and some fish, as noted earlier in the chapter, are thought to have extreme longevity, although precise age estimates are difficult to make, as most of these species have not been maintained in the laboratory. White sturgeon (*Acipenser transmontanus*), a freshwater fish found on the west coast of North America, has been estimated to live upward of 200 years. Unpublished accounts of sea anemones living more than 150 years in a fish tank are often cited. Only recently has confirmation of extreme longevity in a clam, *Arctica islandica*, been documented. Carbon-dating of the shell places one clam at 400 years of age and many others at 100 years **(Figure 1.17)**.

The biochemical and genetic mechanisms underlying the extreme longevity in these aquatic species have yet to be determined. However, this extreme longevity seems to be associated with continual growth. As you will learn in Chapter 3, growth and development are associated with many biological functions that enhance survival. The long-lived clam shows a heightened resistance to cellular damage, the basic mechanism underlying aging. In sponges and jellyfish, a unique mechanism has evolved that may account for their extreme longevity. The cells of sponges and jellyfish can transition between **somatic cells** (cells not involved in sexual reproduction) and **germ cells** (reproductive cells, produced by the sex organs or tissues of multicellular organisms, that transmit hereditary information). You will learn in Chapter 4 that germ cells are extremely efficient at protecting themselves from damage and are thought to have infinite life spans.

ESSENTIAL CONCEPTS

- Biogerontology explores the biological processes that occur inside living things as they age and integrates research from many different scientific fields.

- Biogerontology emerged as an independent field of study in the 1940s, after scientific and technological advances led to a significant increase in the average human life span.

- Aging research is unlike any other health-related research in that aging and death have no cure.

- Aging reflects the result of a lifetime of interactions with our environment, and no two humans have the same interactions with their environment.

- The differences in the rate of aging within and between species make research in aging and longevity challenging and have presented an obstacle to a precise definition of aging.

- There are many different definitions of aging, each having a particular place in the study of aging. For this textbook, aging is defined as random change in the structure and function of molecules, cells, and organisms caused by the passage of time and one's interaction with the environment. Aging increases the probability of death.

- Development, maturity, and senescence are specific event-related points, or stages, within the life span that describe distinct periods of biological aging that can be used to make comparisons of changes across the life span.

- Aging and disease are separate entities. A disease is a process within the animal or plant that impairs normal function. Aging occurs within the normal bounds of biology.

- Because ethical and practical considerations limit the type of research that can be done in humans, biogerontologists use a variety of laboratory organisms, from single-cell organisms, invertebrates including insects, a range of mammals and fish, birds, nonhuman primates, to understand the basic nature of human aging. However, no single animal or plant model can be seen as a "perfect" system for studying biological aging.

- Investigations of exceptional longevity of some wild animals are used to suggest evolutionary mechanisms that may apply to human longevity. The study of aging in the wild is called comparative biogerontology.

DISCUSSION QUESTIONS

Q1.1   Explain why biogerontological research did not become organized until the 1940s.

Q1.2   Discuss briefly the role played by the Gerontological Society of America and the American Geriatrics Society in establishing organized biogerontological research programs.

Q1.3   Consider the following statement: "No two humans have the same interactions with their environment." Why is it important for biogerontologists to understand this statement when studying aging?

Q1.4   Consider the following definition of aging: "Aging, the process of growing old, is defined as the gradual biological impairment of normal function resulting from changes made to cells and structural components." What are this definition's strengths and weaknesses?

Q1.5   Describe why a single definition of aging may not apply to all areas of biogerontology.

Q1.6   Identify the life stages of development, maturity, and senescence on the graph in **Figure 1.18**. Describe briefly the characteristics of an organism that has a similar pattern of life stages.

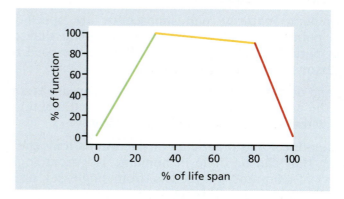

Figure 1.18

Q1.7   Many biogerontologists agree that aging is not disease. What evidence can you give to support this belief?

Q1.8   As a biogerontologist, you wish to study a particular gene's contribution to life span and whether changing the environment affects the outcome. However, the type of study you perform must be done on a budget that will not support long-term housing for mice. What other species might offer you the best chance of answering your research question? Why?

Q1.9   Discuss why the rhesus macaque may be a good model in which to investigate the etiology of age-related human disease.

Q1.10   Why would one expect a flying species to have a longer life span than a species that moves on the ground?

FURTHER READING

BIOGERONTOLOGY: THE STUDY OF BIOLOGICAL AGING

American Aging Association. www.americanaging.org

American Federation for Aging Research (AFAR). www.afar.org

American Geriatrics Society. www.americangeriatrics.org

Gerontological Society of America. www.geron.org

National Institute on Aging. www.nia.nih.gov

Park HW (2008) Edmund Vincent Cowdry and the making of gerontology as a multidisciplinary scientific field in the United States. *J Hist Biol* 41: 529–572.

Settersten RA, Flatt MA & Ponsaran RS (2008) From the lab to the front line: how individual biogerontologists navigate their contested field. *J Aging Stud* 22: 304–312.

DEFINITIONS OF BIOLOGICAL AGING

Finch CE (1990) Longevity, Senescence, and the Genome. Chicago: University of Chicago Press.

Hayflick L (2004) The not-so-close relationship between biological aging and age-associated pathologies in humans. *J Gerontol A Biol Sci Med Sci* 59:B547–550; discussion 551–553.

Hayflick L (2007) Entropy explains aging, genetic determinism explains longevity, and undefined terminology explains misunderstanding both. *PLoS Genet* 3:e220.

Masoro EJ (1995) Aging: current concepts. In Handbook of Physiology: Aging (EJ Masoro ed), pp 3–21. Oxford: Oxford University Press.

Masoro EJ (2006) Are age-associated diseases an integral part of aging? In Handbook of the Biology of Aging, 6th ed (EJ Masoro, SN Austad eds), pp 43–62. New York: Elsevier.

HOW BIOGERONTOLOGISTS STUDY AGING: THE USE OF LABORATORY ORGANISMS IN HUMAN AGING RESEARCH

Conn PM (2006) Handbook of Models for Human Aging, p 1075. New York: Elsevier.

Cristofalo VJ, Beck J & Allen RG (2003) Cell senescence: an evaluation of replicative senescence in culture as a model for cell aging in situ. *J Gerontol A Biol Sci Med Sci* 58:B776–779; discussion 779–781.

Erwin JM & Hof PR (2002) Aging in Nonhuman Primates, vol 31, p 239. Basel: Karger.

Martin GM & Oshima J (2000) Lessons from human progeroid syndromes. *Nature* 408:263–266.

Nystrom T & Osiewacz HD (2004) Model Systems in Aging, p 301. New York: Springer.

HOW BIOGERONTOLOGISTS STUDY AGING: COMPARATIVE BIOGERONTOLOGY

Austad SN (2009) Comparative biology of aging. *J Gerontol A Biol Sci Med Sci* 64:199–201.

Austad SN (2010) Methuselah's zoo: how nature provides us with clues for extending human health span. *J Comp Pathol* 142:S10–21.

Buffenstein R (2008) Negligible senescence in the longest living rodent, the naked mole-rat: insights from a successfully aging species. *J Comp Physiol B* 178:439–445.

Furness LJ & Speakman JR (2008) Energetics and longevity in birds. *Age (Dordr.)* 30:75–87.

Sacher GA (1959) Relation of lifespan to brain weight and body weight in mammals. In Life Span of Animals (GEO Wolstenholme & M O'Connor eds), pp 115–141. London: J. A. Churchill.

MEASURING BIOLOGICAL AGING

2

"IF YOU LIVE TO BE ONE HUNDRED, YOU'VE GOT IT MADE. VERY FEW PEOPLE DIE PAST THAT AGE."

-GEORGE BURNS, ACTOR (1896-1996)

Describing any event in the physical world requires a method for expressing observations. Biologists spend a great deal of effort establishing, for any given process, standard units of measurement that will be understood by the entire discipline. For better or worse, biogerontologists have traditionally used the passage of time as the standard unit of measurement for describing age-related change.

MEASURING BIOLOGICAL AGING IN THE INDIVIDUAL

Describing changes in a cell or organism as a function of seconds, hours, days, or years is easy. Simply measure something (a variable), wait the appropriate amount of time, and measure it again. Any change in the variable is then expressed as a difference over time. The hard part arises in trying to determine whether time-related differences observed at the chosen points in time have any meaning for biological aging. For example, we observe that a mouse reaches sexual maturity at about 2–3 months of age and remains reproductively active for the next 12–15 months. When mice reach the end of their reproductive life, only slight declines in physiological functions occur over the next 4–6 months. Then, between 22 and 28 months of age, most mice experience 1–2 months of senescence before death. These time-related markers give biogerontologists a general idea of the mouse's life history and convenient points for comparison in research **(Figure 2.1)**.

As valuable as these markers may be, many questions are left unanswered. For example, do the lengths of the development and maturity stages or the total length of life tell us anything about the physiological function of the mouse when half its life is over? Do all mice live 22–28 months, or is there some variation in length of life? If sexual development were to occur over 4 months instead of 3 months, would the mouse live longer? In this chapter, we examine similar questions and describe how biogerontologists measure the rate of biological aging, both in individuals and in populations.

Figure 2.1 Body weight provides a convenient marker for expected physiological functions in the mouse. The rapid rate of body weight gain between 50 and 100 days of age reflects the growth of tissues through cell division and the development of optimal physiological function. The slower rate of weight gain between 200 and 500 days of age results from an increase in fat deposits; physiological functions stabilize or show only slight decrements. From approximately 600 days of age to the end of the life span, the animal's body weight declines, as does physiological function. Note, however, the variable pattern in weight loss of the older animals—that is, large peaks and valleys—compared with the steady weight gain during development and maturity. This variability occurs because some animals are entering senescence while others remain in the maturity phase.

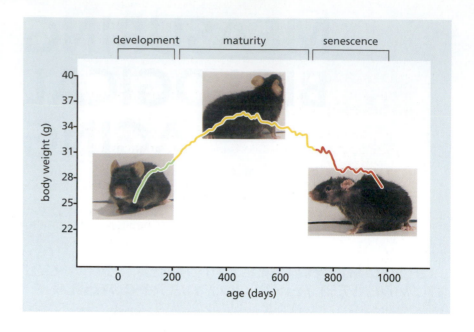

Most of us can think of examples of someone who, at 80 years of age, is running marathons, while another individual, at 70 years of age, is confined to a long-term care facility **(Figure 2.2)**. These examples demonstrate that the rate of biological aging is highly individualized and cannot be measured simply by consulting a calendar. What falls apart in one person may not be affected by aging in another. This means that population averages describing age-related changes can be very interesting as a general overview but may not have any specific relevance to us as individuals. We all want to know exactly what will happen to *us* as we grow old, rather than what may happen to the "average" person. If we knew which physiological system was likely to decline over time, we could take steps to prevent or treat that age-related dysfunction. This is why measuring the rate of aging precisely in each individual is the ultimate goal of biogerontological research.

Measuring the rate of aging in an individual requires that we identify a biological event that changes at a known rate over time. According to the American Federation for Aging Research, the following criteria must be met for a useful and accurate age-related biological marker, or **biomarker**, of aging:

Figure 2.2 Which of these people, of similar age, represent the rate of aging for humans? (A, courtesy of K. Chernus/Getty Images; B, courtesy of S. Hix/Somos Images/Corbis.)

- It must predict the rate of aging. In other words, the marker should reveal exactly where a person is within his or her total life span. It must be a better predictor of aging than chronological age.

- It must monitor a basic process that underlies the aging process, not the effects of disease.

- It must be something that can be tested repeatedly without harming the person; for example, a blood test or an imaging test.

- It must be a process that works both in humans and in laboratory animals, such as mice, so that it can be tested in lab animals before being validated in humans.

In this section, we explore issues dealing with the measurement of aging at the individual level and the identification of a biomarker for aging. We begin by discussing how our interaction with the environment and our lifestyle choices make measuring aging at the individual level so challenging. Then we look at past attempts to identify biomarkers and discuss why these methods have not produced reliable indicators of biological age. Finally, we discuss new directions in biomarker research that may have great potential for providing a mechanism by which we can reliably measure the rate of aging in each individual.

Differences in the age-related phenotype affect the measurement of aging in individuals

We are born with a genetic pattern that results in physical attributes specific to our species. We have hands and fingers; birds have wings and claws. The genetic constitution of an organism is known as the **genotype**. As shown throughout this text, the rate of aging caused by the genotype, known as the **intrinsic rate of aging**, can be measured with a fair amount of accuracy. During fetal development and throughout the life span, the genotype interacts with the environment to produce the **phenotype**. Biogerontologists refer to the impact of the environment on the genotype in producing the age-related phenotype as the **extrinsic rate of aging**. No two humans (genotypes) are ever exposed to identical environments for the same length of time so as to produce identical phenotypes; the age-related phenotype and the extrinsic rate of aging are infinitely diverse. As a result, measuring the impact of the extrinsic rate of aging on individual aging can be very difficult.

How the environment affects the genotype to produce the phenotype varies with age. From fetal development to about 3 years of age, our immune system, thermoregulatory ability, and brain are still developing. During this time, we are rather defenseless against the environment and must rely on the judgment and care of others, such as our parents, to survive. Geneticists have long known that the environment has the greatest impact on shaping our phenotype during this period. Human fetal development can be divided into two periods: one that is influenced primarily by the genotype, and one in which the phenotype begins to form. The first 8 weeks of fetal development are devoted exclusively to forming the basic tissues, organs, and anthropometric characteristics that distinguish us as a species—that is, to expressing the genotype. Alteration to the *in utero* environment during this phase can result in the death of the fetus or genetic abnormalities at birth. The period from 9 weeks of gestation until after birth is a time of tremendous growth, characterized by rapid and sustained cell division. The rate of cell division and growth depends on many nongenetic factors, such as nutrition, oxygen supply, and removal of metabolic waste products. It is at this stage that the unique person each human will become—that is, the phenotype—takes shape.

Recent studies have shown that fetal development and the shaping of the phenotype can have a significant impact on physiological function and disease during adulthood. In the 1990s, a group of scientists in England, led by Dr. David Barker, compared the birth weights of 16,000 men and women born between 1910 and 1930 with the incidence of adult disease. These scientists found that low birth weight at full term was highly correlated with the incidence of diabetes in adulthood **(Figure 2.3)**. The relationship remained even after the researchers removed, statistically, the effects of adult lifestyle choices on this disease. Two decades of animal experimentation have now confirmed that poor nutrition during fetal and early life development establishes a phenotype that leads to poor health outcomes in adulthood. These facts strongly suggest that determining a biomarker of aging and, thus, the individual rate of aging will include teasing out factors occurring during development that have an influence on the extrinsic rate of aging.

Once the individual has developed to the point of having enough physical and cognitive ability to survive on his or her own, the genotype that determines survival to reproductive age and the passing on of the germ line becomes dominant. Biologically speaking, we become very good at resisting the hazards of the environment. The incidence of cancer, heart disease, diabetes, and other potentially lethal diseases that have an environmental component is very low during the run-up to puberty. Even nonlethal hazards are less traumatic in the young. Think about how much better a 10-year-old recovers from a broken arm than an 80-year-old.

Because the genotype resists environmental hazards during growth and development, the variation in phenotype among individuals is at its lowest. This makes measuring an individual child's biological age easy and precise. Pediatricians have a wealth of data that provide accurate biomarkers for proper growth and development. Even the simple measurements of body weight and height for individuals between the ages of 2 and 20 are excellent biomarkers for development **(Figure 2.4)**.

Figure 2.3 Risk of developing adult-onset diseases as a function of birth weight. Low birth weight is associated with a greater risk of developing glucose intolerance, a prelude to type 2 diabetes, during adulthood. The relative risk ratio is a statistical estimate for the strength of association between a risk factor (birth weight) and the outcome of interest (glucose intolerance). The higher the number the greater the risk for developing the disease. (Adapted from C.N. Hales et al., *BMJ* 303:1019–1022, 1991. With permission from British Medical Journal.)

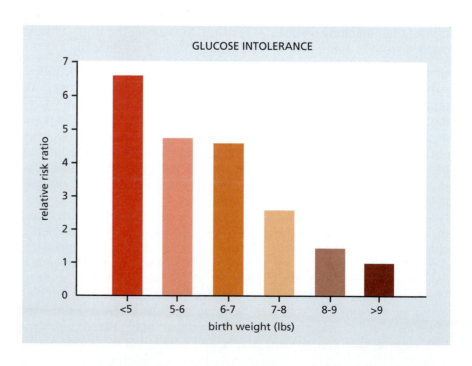

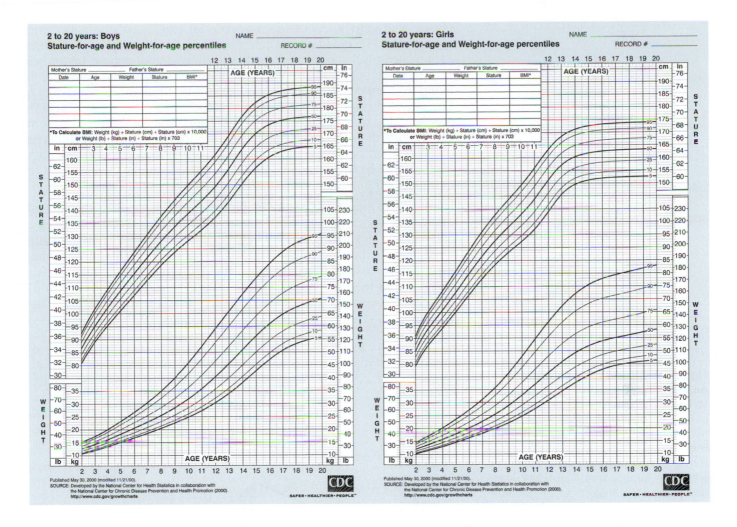

Figure 2.4 Clinical growth charts.
These charts of height and weight provide a mechanism for measuring age-related growth and development. For example, if a 2-year-old child has a body weight that falls along the 75th percentile curve, there is a 97.5% chance that he or she will remain in that percentile at age 20. Any significant deviation can alert health professionals to a possible problem in growth and development. (From the National Center for Health Statistics, in collaboration with the National Center for Chronic Disease Prevention and Health Promotion, 2000.)

These measurements done on an individual over several years accurately describe age-related biological mileposts, as well as whether or not the child's growth and development are on track.

As we age beyond the early reproductive years, our resistance to environmental hazards begins to wane. The immune system becomes less effective, the healing process slows, and the functioning of brain centers that help us avoid hazards, such as maintaining balance, declines. An important point here is that the rate of age-related change in a particular cell, tissue, or organ does not follow any predictable age-related pattern, or at least none that biogerontologists have yet identified. Moreover, the physiological system that may have the least resistance to environmental damage differs from person to person. In other words, as we grow older, the phenotype becomes increasingly unique to the individual and makes finding a biomarker challenging. Separating out how different environments affect the rate of aging has, as yet, been the major challenge in identifying biomarkers.

Lifestyle choices significantly affect phenotype

Humans have the unique ability to control their environment and thus have some say in how the age-related phenotype develops. Let's use skin aging to illustrate this point. Skin undergoes many age-related changes that are genetically determined and are generally accepted as a common part of intrinsic aging. Most notable are the loss of subcutaneous fat, the fat layer between the skin and muscle, and the decline in elasticity. Together, loss of subcutaneous fat and loss of elasticity result in wrinkles. The environment can also affect the development of wrinkles. A person living in the desert will, on average, expose his or her skin to more solar radiation than an individual of similar complexion living in a region of comparable elevation that has more cloud cover. The desert-dwelling human will be at considerably greater risk for skin damage, due to the higher level of radiation, and for the development of wrinkles than the individual living in a location with less sun. The question then becomes, How much skin aging is due to intrinsic factors—that is, biological aging—and how much is due to environment or extrinsic aging?

Humans also have the ability to be proactive in slowing the rate of aging and the development of age-related disease. For example, the first studies evaluating age-related decline in muscle strength were done on sedentary individuals and found that muscle strength declined, on average, 30–50% from age 30 to age 80. While subsequent studies showed that the age-related decline in muscle strength has an intrinsic component, individuals participating in weight-lifting exercises can improve strength **(Figure 2.5)**. This suggests that the age-related

Figure 2.5 Effects of weight training on age-related decline in muscle strength. Individuals between the ages of 60 and 80 years who participated in a two-year weight-training program (brown line) showed significant gains in strength compared with sedentary persons of similar age (orange line). These data clearly demonstrate that the normal age-related loss in muscle strength has an extrinsic component. Note that between weeks 42 and 52, the training was suspended. (Adapted from N. McCartney et al., *J. Gerontol. Ser. A Biol. Sci. Med. Sci.* 51: B425–433, 1996. With permission from Oxford University Press.)

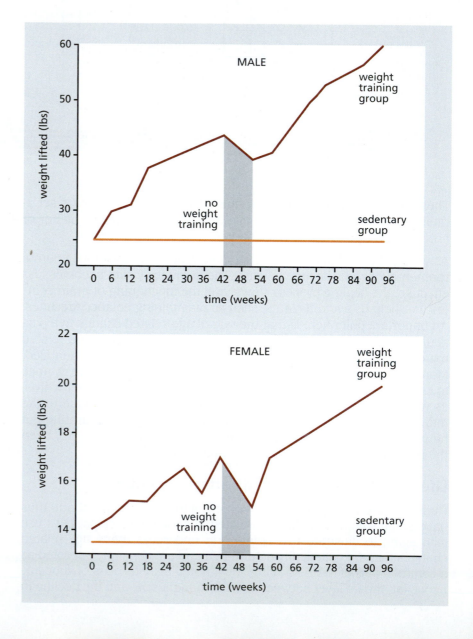

decline in muscle strength also reflects our lifestyle choices. Exercise and proper diet also slow the rate of age-related functional loss in the cardiovascular, neural, and skeletal systems, as well as affect body-weight regulation. The choice to maintain a healthy weight through exercise and diet delays or prevents heart disease, diabetes, osteoporosis, and certain types of cancer. Again, biogerontologists are faced with the problem of determining how much of normal aging results from intrinsic factors and how much from things we do to ourselves.

The epigenome can also affect the rate of aging and longevity

So far, we have introduced the principles of genotype and the age-related phenotype by considering regulation of gene expression at the level of genomic DNA (the sequence of base pairs). A second level of trait development, known as the **epigenome** (*epi-* is Greek for "above" or "in addition to"), also affects development of the phenotype and thus can affect the rate of aging and longevity. The resulting phenotype is called an **epigenetic trait**, a phenotype resulting from changes in a chromosome without alterations in the DNA sequence. The cell uses several methods to regulate gene expression through the epigenome. The two most common are DNA and histone methylation and histone acetylation-deacetylation. DNA and histone methylation processes work by the same mechanism, and in this text both are referred to as DNA methylation. Histones are proteins that help compact the double-stranded DNA to fit it inside the nucleus **(Figure 2.6)**. That is, DNA in its extended strand form is much too big to fit into the nucleus. You will learn in Chapter 5 how the unwinding of DNA from the histone proteins

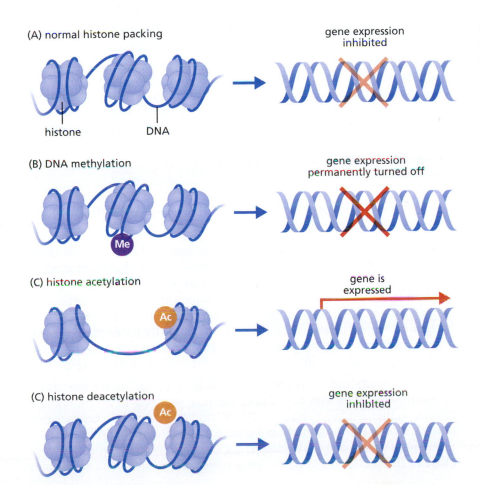

(A) normal histone packing

gene expression inhibited

histone   DNA

(B) DNA methylation

gene expression permanently turned off

Me

(C) histone acetylation

gene is expressed

Ac

(C) histone deacetylation

gene expression inhibited

Ac

Figure 2.6 Mechanisms of epigenetic regulation of gene expression. (A) The DNA in the eukaryotic nucleus is wrapped around histone proteins. Gene expression occurs when the DNA disassociates from the histone. (B) The methylation (Me) of DNA on the histone permanently turns off gene expression for a specific gene by preventing the disassociation of DNA from the histone protein. (C) Histone acetylation-deacetylation is an epigenetic mechanism that assists in the unwinding and winding of the DNA on the histone. Acetylation stimulates gene expression. (D) Deacetylation inhibits gene expression.

regulates gene expression. Simply put, the DNA must unwind from the histone before the gene can be expressed.

About 90% of the epigenome consists of an inherited pattern of methylation (addition of CH_3 groups) of DNA that occurs during fetal development. DNA methylation causes gene expression of that DNA region to be permanently turned off, by preventing the unwinding of the DNA from the histone; this methylation cannot be reversed. DNA methylation is critical to cellular differentiation during fetal development and takes place primarily in non-mitotic cells, cells that do not divide. Methylation of DNA makes sure, for example, that a muscle cell expresses only genes important to muscle function by permanently turning off genes that are used in other types of cells (so that, for example, liver enzymes are not produced in a muscle cell). The mechanism underlying how the methylation pattern leading to epigenetic traits is inherited remains to be fully described.

The other 10% of the epigenome allows for an epigenetic trait to arise sometime during the life span, including during fetal development—that is, these traits are not inherited. This part of the epigenome appears to be influenced heavily by environmental conditions. Histone acetylation probably has its greatest influence at this level of the epigenome. Adding an acetyl group (CH_3CO) to the histone protein unwinds the DNA and encourages gene expression. Removal of the acetyl group, known as histone deacetylation, causes the DNA to rewind onto the histone protein and inhibits gene expression (Figure 2.6C, D). The ability to acetylate and deacetylate the histone means that this epigenetic mechanism for regulating gene expression is, unlike methylation, reversible.

Research into the epigenetic effect on the phenotype has been focused, almost exclusively, on mechanisms associated with DNA methylation during fetal development. Only recently has significant research focused on epigenetic effects that influence the adult phenotype; even less attention has been given to the aging phenotype. There is some evidence that the epigenome affects longevity. As you will learn in Chapter 5, inhibition of gene expression through histone acetylation-deacetylation in simple organisms such as budding yeast and worms can affect life span. Moreover, the finding that these epigenetic effects are influenced directly by population density and food availability indicates that specific environmental conditions are important to the development of the aging phenotype.

Cross-sectional studies compare changes in different age groups at a single point in time

Identifying a reliable marker of biological age in the individual often starts by using cross-sectional investigations. **Cross-sectional studies** compare the average or mean rate of change in a particular physiological system in two or more age groups, known as **cohorts**, at a single point in time. These experimental designs are widely used in biogerontological research. Cross-sectional investigations have several advantages, including simplicity of design and low cost, and they provide good general descriptors of an age-related biological phenomenon.

Cross-sectional studies have added significantly to our knowledge about which biological factors may affect the rate of aging in an individual. However, the results of cross-sectional studies reflect the comparisons of means or averages and are not specific to the individual **(Figure 2.7)**. Some individuals may have values for the study variable that are very close to the mean, while others may have values that fall a significant

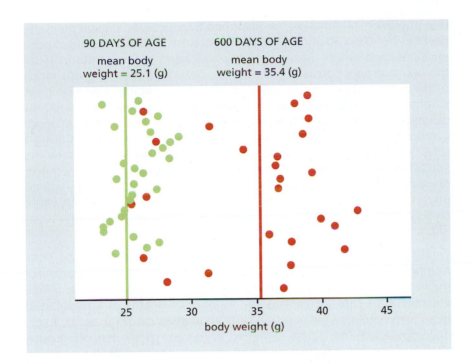

Figure 2.7 Cross-sectional studies reflect comparisons of means or averages and are not specific to the individual. Shown here are two sets of data: the body weights of 30 mice at 90 days of age (green circles) and the body weights of the same mice at 600 days of age (red circles). Note that, compared with the younger mice, the older mice have a much greater spread of values contributing to the mean. (3 mice died before 600 days of age.)

distance from the mean. Moreover, the data spread about the mean becomes larger with age. Even when the data are collected under conditions that largely eliminate variability associated with the environment, significant variability may remain. Because of this intrinsic variability, cross-sectional studies have limited precision in identifying a biomarker for aging.

By definition, a cohort consists of a group of individuals who have, in general, similar life experience. That is, a cohort is more than just people of similar ages. These life experiences, especially those that may affect the rate of aging, may differ significantly between cohorts, and this introduces into the study a variability that cannot be easily controlled. This variability is known as the **cohort effect**. To illustrate the cohort effect, consider the data obtained from a cross-sectional investigation such as that shown in **Figure 2.8**. These data suggest

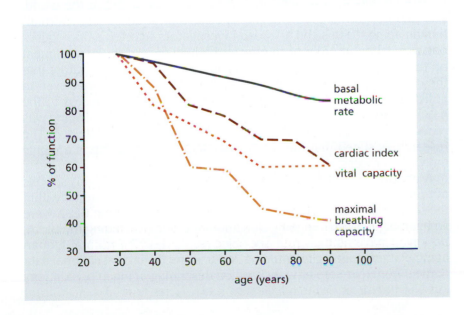

Figure 2.8 A 1959 cross-sectional investigation describing physiological function in different age groups. Basal metabolic rate, cardiac index (a measure of heart function), vital capacity (the largest amount of air expelled after a person's deepest inhalation), and maximal breathing capacity (the volume of air that can be breathed over a few seconds when a person breathes as deeply and quickly as possible) were measured (as a percentage of function at 30 years old) in groups aged between 30 and 90 years. The cohort effect may account for some of the differences in function between the 30- and 90-year-olds. (Adapted from B.L. Strehler, *Q. Rev. Biol.* 34:117–142, 1959. With permission from University of Chicago Press.)

that physiological function, as described by several biological markers, declines between 30 and 90 years of age. Now consider the life experiences of each cohort. At the time of publication of these data in 1959, the oldest group, consisting of 90-year-olds, would have spent much of their life without the benefits of modern medicine. Infections, for example, were treated by removing the tissue affected, and infectious diseases were treated by removing the individual from the population. Conversely, individuals in the group of 30-year-olds were born during a time of rapid advancements in medicine that significantly reduced the spread of disease and prolonged life. There is simply no way to predict or control for the effect of medical knowledge on the age-related phenotypes of these two groups. In other words, the results may more closely reflect the difference in the cohorts' life experiences than any difference in the age-related phenotype.

Finally, conclusions based on cross-sectional analysis may be confounded by the effects of **selective mortality**, that is, the inclusion of individuals who, because of a different genotype, may have a different **mortality rate** (defined as the number of deaths in a population at a given time, in a given group or from a given cause) than the average population. Individuals included in the 60- to 90-year-old groups shown in Figure 2.8 were born when life expectancy at birth was 40–45 years. These individuals had already outlived, at a minimum, 50% of their birth cohort at the time of measurement. Those in the 90-year-old group represented less than 1% of their birth cohort. Thus, these age groups are not a representative sample of the entire age cohort and may have outlived their contemporaries due to some factor(s) not present in those individuals who had died. The results of this cross-sectional investigation may be biased by selecting only "hardy" individuals representing advanced age.

Longitudinal studies observe changes in a single individual over time

The collection of data from the same person over several years is the method used in **longitudinal studies**, a study design that was intended to measure more accurately the rate of aging in an individual. However, longitudinal studies have been no more effective than cross-sectional analysis in identifying a biomarker of aging. This fact has been demonstrated nicely by the Baltimore Longitudinal Study of Aging (BLSA), the longest continually running longitudinal study on aging in the world. The BLSA began in 1958 with the specific aim of describing "normal" human aging. To this end, the lead investigators selected a homogeneous population in order to minimize the influence of the environment on the rate of aging. By 1984, it was apparent to the investigators that the concept of normal aging was a fallacy. They showed that even in individuals who had incomes and educations providing access to the best of health care and information on health care, the rate of aging was random with respect to timing and physiological system affected. The researchers stated that the "BLSA data indicate that aging is a highly individual process. . . . In some variables, individual 80-year-old subjects may perform as well as the average 50-year-old. Aging is highly specific not only for each individual but also for different organ systems within the same individuals" (Shock et al. 1984). The results of the BLSA were so unambiguous that ongoing longitudinal studies, including the BLSA, no longer have as their primary objective to describe normal human aging. Rather, ongoing longitudinal studies and the cross-sectional data derived from them are now focused on describing aging in populations of individuals that share specific attributes **(TABLE 2.1)**. Thus, large-scale longitudinal studies that started as a mechanism for defining aging in the individual have instead been more effective at describing long-term changes in populations of individuals with similar backgrounds.

Even though longitudinal studies are no more effective at identifying a biomarker of aging than are cross-sectional designs, they continue to contribute significantly to the study of aging. An important use of many longitudinal studies has been to demonstrate more accurately than cross-sectional studies the pattern of individual aging in a population. For example, the cross-sectional results shown in Figure 2.8 suggest that the rate of aging after 30 years of age conforms to a linear decline. But, as shown in **Figure 2.9A**, longitudinal analysis finds that functional decline occurs at nonlinear rates and appears to be closely linked to the end of the reproductive life span rather than to calendar time. You will learn in Chapter 3 that the reproductive schedule of a species has a major influence on the rate of aging and longevity.

Another significant contribution of longitudinal studies to the overall progress in aging research is their ability to generate and maintain data from humans with well-documented life histories. Knowing the life history of a subject allows for the identification of environmental factors, such as disease, physical activity, diet, and so on, that can mislead a researcher into concluding that there is an aging effect when, in fact, the change was driven by environmental factors. Such detailed life-history information cannot often be collected in one-time, cross-sectional studies in humans, and this is normally only possible when using laboratory animals as models for human aging **(BOX 2.1)**. As a result of the detailed knowledge about the study subjects, the cohort effect and selective mortality are eliminated in cross-sectional designs that use participants in longitudinal studies. That is, the validity and reliability of results from cross-sectional experiments are significantly enhanced when they are conducted using subjects involved in longitudinal studies (Figure 2.9B).

TABLE 2.1
SELECTED ONGOING LONGITUDINAL STUDIES OF AGING

Name of study	Year started	Age of participants (years)	Number of participants	Gender	Population characteristics
Baltimore Longitudinal Study on Aging	1958	20–100	3000	1958–1978: men only Currently: men and women	Original cohort: white upper-middle-class Currently: inclusion of nonwhite
Honolulu–Asia Aging Study	1991	71–93	3734	Men	Japanese ancestry, with focus on dementia and Parkinson's disease
Longitudinal Study of Ageing in Africa	2004	50+	3500	Men and women	Aging in sub-Saharan Africa; impact of HIV/AIDS on aging
New Mexico Aging Process Study	1978	65–98	780	Men and women	Currently: aging in Hispanics
Normative Aging Study	1963	21–81	2280	Men	95% of population are veterans
Nun Study	1991	75+	678	Women	Focus on Alzheimer's disease; participants are members of the School Sisters of Notre Dame

Figure 2.9 Studies of grip strength using two methods of analyzing longitudinal data. (A) Longitudinal data showing that the rate of aging is nonlinear. The spread of individual data points was fitted to a nonlinear equation by using the entire population. In addition, the graph clearly shows that the start time and amount of decline in grip strength are highly variable. (B) A cross-sectional analysis using means computed from multiple measures made on individuals within each age cohort. The means are represented by the midpoint of the line segments. The direction of each line segment (left to right) indicates a gain or loss in grip strength over the decade of measurement for that age cohort. The length of the line segment represents the length of time the longitudinal data were collected. The graph also shows that the rate of physiological decline is nonlinear with respect to time, in agreement with the longitudinal data in (A). This demonstrates that cross-sectional studies performed within the framework of the longitudinal design eliminate many of the problems associated with earlier cross-sectional analyses. (From D.A. Kallman et al., *J. Gerontol.* 45:M82–88, 1990. With permission from Oxford University Press.)

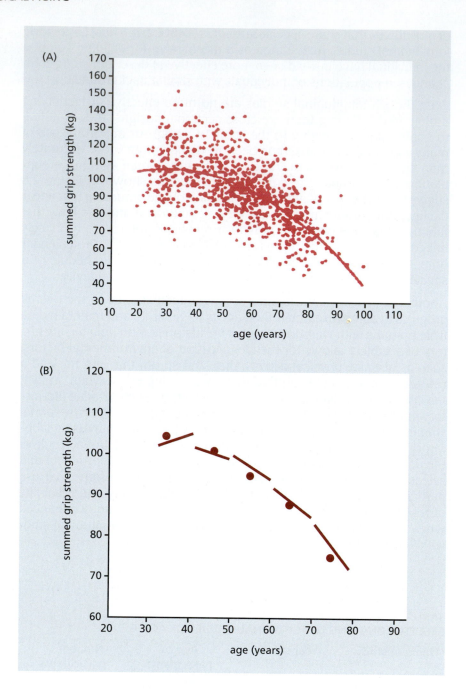

Personal genomics will probably be the key to determining and applying biomarkers for aging

Throughout this section we have emphasized that the rate of aging may be unique to each individual and that this uniqueness poses significant obstacles to establishing a biomarker for aging. Even the best-designed longitudinal studies have been unable to identify a "normal" marker of aging. We noted that the rate of aging may be determined early in the life span, during fetal development. And we saw how epigenetic effects occurring during adulthood influence the phenotype. The current evidence leads to the general conclusion that the aging phenotype may most closely reflect a gene or groups of genes that are highly malleable to environmental interactions throughout the life span, including fetal development. You will see in Chapter 5 that, using animal models,

BOX 2.1   EVALUATING BIOGERONTOLOGICAL RESEARCH IN ANIMALS

As a biological science, biogerontology frequently uses animals as models for humans. Animal models provide a system to measure aging under controlled conditions and to establish the basic mechanisms of a biological process. These types of studies are critical for constructing hypotheses and research methods for aging studies in humans. Thus, performing experiments in laboratory animals, within accepted good practices, is essential to biogerontological research. Knowing some of the good practices used in biogerontological animal research will help you better interpret the validity of research results. Summarized here are four practices to keep in mind when you are reading about aging research that uses laboratory animals, particularly those used as models for human aging.

General practice 1: Researchers should know the life history of the animal model being used

Biogerontologists should have a thorough knowledge of the normal biological events that take place in their animal model throughout its life span, often referred to as its life history. At a minimum, there are three important life-history events that should be known precisely: (1) the approximate age at which the species reaches the end of its developmental growth; (2) the mean life span; and (3) the maximum life span. For the investigator conducting cross-sectional investigations, this information prevents the inclusion of groups of animals that are either too young or too old. Including animals that are too young can lead to conclusions about an aging effect, when, in fact, the observed difference is the result of factors unique to the animal's developmental stage. Using animals that are too old can result in conclusions about aging, when the findings actually apply to only a subset of animals that have reached old age because of a genetic factor not found in the majority of the population—that is, selective mortality.

General practice 2: Researchers should know the reproductive life span of the animal model

Throughout this textbook, and especially in Chapter 3, we emphasize that genes selected for survival to reproductive age influence both the rate of aging and the species' maximum life span. In choosing an animal model for biogerontological research, the researcher should give careful consideration to factors that affect development up to reproductive age, as these factors will significantly affect the rate of aging and longevity. Conversely, knowing the approximate age when the reproductive stage of the life span begins to wane can help an investigator select an appropriate age group to "look" for greatest aging effects.

General practice 3: Researchers should carefully maintain and clearly report the environment in which animals are housed

As you have seen, the environment can affect the age-related phenotype and the extrinsic rate of aging. One of the primary reasons that biogerontologists use animals as models for human aging is so that they can control, to a large degree, the environment to which the animals are exposed. This limits the impact of extrinsic factors on the rate of aging.

Good practices in biogerontological research require that the research animals be maintained behind barriers that provide filtered air, flowing at a positive pressure with respect to the external environment **(Figure 2.10)**. This helps prevent airborne contaminants entering the animals' environment. Bedding and food should be sterilized, and the water slightly acidified and chlorinated. All substances entering the animals' housing should be tested periodically for contaminants. Finally, each colony should contain animals, known as sentinels, that are used to monitor populations for infectious disease pathogens. The sentinel animals are housed with the experimental animals and exposed to the same environmental conditions, and are routinely tested for exposure to infectious agents. Any contamination must be noted when reporting the findings, so the reader can determine whether these issues may have affected the results.

General practice 4: Researchers should assess pathology and cause of death in research animals

No organism dies of old age. Rather, each organism dies of a specific cause or a combination of lethal pathologies. The investigator should have a clear understanding of the biological factors that led to the death of each animal, for two reasons. First, the cause of death or age-related pathology provides researchers with information on how the environment affected their models. The cause of death or age-related pathology should be inherent to the animal, not a consequence of environmental contamination. Second, performing a thorough **necropsy** and histochemical analysis establishes the normal age-related pathologies specific to that species and whether or not those pathologies are also seen in humans. Such knowledge helps other investigators choose the correct models for their experiments.

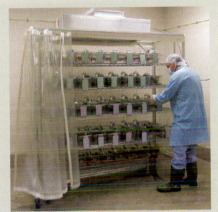

Figure 2.10 A laboratory barrier unit. Barrier units help prevent airborne contaminants from entering the animals' environment. (Courtesy of Jennifer Ruhe, Department of Nutrition, University of California.)

biogerontologists are beginning to identify genes and/or epigenetic effects that respond to environmental changes during fetal and early life development and that also affect the length of life. It is simply a matter of time until **gene homologs** (DNA sequences of one gene that match those of a second gene, even a gene of a different species) are identified and tested in humans. Once genes have been identified that may affect the rate of aging, we can then use **personal genomics**, the analysis of individual genotypes or epigenetic mechanisms, to identify whether "rate of aging" genes or some polymorphisms of those genes exist and are being expressed or inhibited. Having such knowledge will allow health professionals to make recommendations at the individual level about ways to slow the rate of aging.

Personal genomics is currently being used to determine the probability of developing lethal genetic diseases. The limited use of personal genomics reflects the $10,000–$20,000 price tag for having a genome sequenced. Experts agree, however, that within 10–15 years the price will drop to a level at which widespread use of personal genomics will become possible. We now stand at a time in history when our knowledge of the factors that determine the rate of aging is sufficiently advanced that by using current technology—personal genomics—identification of a reliable marker to measure aging in the individual will soon be accomplished.

MEASURING BIOLOGICAL AGING IN A POPULATION

Measurement of biological aging in an individual, although the most desirable and potentially the most accurate form of measurement, is highly variable and remains difficult to accomplish. Estimating the rate of aging in a population, by contrast, can be accomplished rather easily and has significantly less variability. However, population aging has less accuracy because, usually, the descriptor for the rate of aging is the change in *mortality rate* in the population with increasing age. Recall that aging and the risk of dying, although related, may not be the same, especially for species with a postreproductive life span. Moreover, an organism can die as a result of nonbiological aging events, such as trauma or disease, but its death will still be recorded in the dataset used in statistical calculations of the species' life span.

Do the problems associated with the use of mortality as a measure of aging in a population invalidate its use in the study of biological aging? Not by a long shot. Population mortality rates have been used for more than two hundred years in describing aging at the organismal level. Indeed, an analysis of mortality rates in the oldest cohort of a population has changed the way in which biogerontologists approach the study of aging and longevity (see the discussion at the end of this section). Many biogerontologists use the results from mortality analysis in a population to develop and test hypotheses that may predict individual aging or aging at the molecular level. Therefore, you need to have a basic understanding of the derivation, use, and limitations of mortality rate in describing aging within a population.

Mortality rates estimate the number of deaths in populations

Determining the **crude mortality**, the death rate in a total population without regard to age, is fairly simple, provided one has an accurate count of the population size and the number of deaths (Equation 2.1).

Mortality, $M = \dfrac{D}{P}$ (2.1)

where

D = number of deaths in the population

P = population size

This mortality rate, M, is normally converted to a standard measure—for example, deaths per 1000 or 100,000 individuals per year—for ease of interpretation and comparison.

Measuring the number of deaths in a human population in most countries poses few problems, because law requires formal recordings of deaths. However, the size of the population at any given time can only be estimated, using data derived from the most recent census. In the United States, a population census is conducted every 10 years. To reduce error associated with estimating the actual size of a population, human mortality rates are calculated as statistical probabilities by using actual death rates applied to a hypothetical population size, such as 100,000. Although human mortality rates are considered estimates, they have been shown to be extremely precise and should be considered accurate descriptions of the death rate in a population.

Mortality rates can be determined for nonhuman populations as well. Measurements of population size and number of deaths in short-lived species monitored under laboratory conditions can be extremely precise, and, thus, so are the mortality rates. However, population size and deaths are difficult to measure in long-lived wild populations. Imagine trying to count the population size and deaths of a particular bat species (bats can live up to 20 years) living in the United States—an impossible task. Therefore, wildlife demographers typically confine their work to small areas, say 2–3 acres, housing a high-density population, then extrapolate the mortality rate to the general population.

Life tables contain information on mortality, life expectancy, and the probability of dying

A **life table** describes the death characteristics of a population at specific ages or age intervals. There are two types of life tables. A **cohort life table** follows the death characteristics of a population of a single birth cohort that can be observed throughout the life span. The cohort life table is useful for species having short life spans or for historical populations having accurate birth and death records. Following a live birth cohort in long-lived species, including humans, throughout the actual life span is not practical. In these cases, the current or period life table is used. A **current life table** applies the current death characteristics of a living population to a hypothetical birth cohort, usually 100,000 **(TABLE 2.2)**. These values are then used to forecast mortality statistics over the life span of an actual population, based on data collected at a single time point. In either case, cohort or current life table, the mechanics of constructing the values given in the life table are the same. Life tables can be either complete or abridged. A **complete life table** is defined as one in which the age interval is one year. Use of all other age groupings defines an abridged life table—such as the life table presented in Table 2.2.

Life tables, cohort or current, consist of seven columns. Because a life table is used throughout the world as the primary mechanism to estimate the rate of death in a population, the column names and variable designations are standardized by convention. Column 1 contains the age (in the case of a complete life table) or age interval (in the case of an abridged life table). With the exception of the first age interval in

TABLE 2.2
ABRIDGED LIFE TABLE FOR THE TOTAL UNITED STATES POPULATION, 2006

1	2	3	4	5	6	7
Age interval $(x, x + n)$[1]	Probability of dying during age interval (q_x)	Number of people surviving to age x (l_x)	Number of people dying during age interval (d_x)	Person-years lived during age interval (L_x)	Total person-years lived after the beginning of age interval (T_x)	Expectation of life (e_x)
0–1	0.006713	100,000	671	99,409	7,770,850	77.7
1–4	0.001138	99,329	113	397,045	7,671,441	77.2
5–9	0.000694	99,216	69	495,891	7,274,396	73.3
10–14	0.000822	99,147	81	495,587	6,778,505	68.4
15–19	0.003214	99,065	318	494,627	6,282,918	63.4
20–24	0.004998	98,747	494	492,532	5,788,291	58.6
25–29	0.005033	98,253	495	490,029	5,295,759	53.9
30–34	0.005583	97,759	546	487,470	4,805,730	49.2
35–39	0.007389	97,213	718	484,380	4,318,260	44.4
40–44	0.011381	96,495	1098	479,916	3,833,880	39.7
45–49	0.017264	95,397	1647	473,118	3,353,964	35.2
50–54	0.025576	93,750	2398	463,087	2,880,846	30.7
55–59	0.036064	91,352	3295	448,955	2,417,759	26.5
60–64	0.054578	88,057	4806	428,979	1,968,804	22.4
65–69	0.079166	83,251	6591	400,600	1,539,825	18.5
70–74	0.121699	76,661	9330	361,363	1,139,225	14.9
75–79	0.195009	67,331	13,130	305,372	777,862	11.6
80–84	0.302509	54,201	16,396	230,960	472,489	8.7
85–89	0.447212	37,805	16,907	146,101	241,530	6.4
90–94	0.617641	20,898	12,907	69,775	95,429	4.6
95–99	0.782678	7991	6254	21,745	25,654	3.2
100 and over	1.000000	1737	1737	3909	3909	2.3

[1] n = a single unit of time.

a human life table, in which the interval 0–1 years must be used, the selection of the age or age interval is at the discretion of the statistician constructing the table; with human data, it is customary to use 5-year intervals for an abridged table. Column 2 contains the probability of death (q_x), also known as age-specific mortality. Column 3 shows the number of persons from the original 100,000 of the hypothetical cohort who are alive at the beginning of the age interval (l_x). Column 4 shows the number dying (d_x) for each interval. Column 5, person-years lived (L_x), represents the total time (in years) lived between two birthdays for the population at each age interval. Column 6 contains the total number of person-years that would be lived (T_x) after the beginning of the age interval x to $x + n$. Finally, Column 7 gives the life expectancy (e_x). The **life expectancy** is the average number of years remaining to be lived by those surviving to that age.

The information contained in the life table is the primary tool of the demographer and has many uses for the trained specialists, uses that are well beyond the scope of this text. Here, we are concerned primarily with how the rates of mortality, primarily q_x, and the shape of the curves derived from the calculation of mortality rate help to describe aging in a population. The Appendix contains a more detailed description of how a life table is constructed, including more precise definitions and the method of calculating each variable.

Age-specific mortality rate rises exponentially

The **age-specific mortality rate** (q_x) is a useful measurement in aging research (Equation 2.2). It is calculated as follows:

Age-specific mortality rate, $q_x = \dfrac{d_x}{l_x}$ (2.2)

where

d_x = number of deaths in the population during a specified time period

l_x = number of people surviving to age x

x = age or age interval

Age-specific mortality rate simply describes the chance (probability) of dying within a specific age range. For example, Table 2.2 shows a q_x value of 0.011 for the age interval 40–44 years. This means that you have a 1.1% chance of dying during this age interval. Graphing the age-specific mortality allows us to visualize the rate over a lifetime. As you can see in **Figure 2.11**, the probability of death between ages 1 and 70 increases slowly. After age 70, age-specific mortality increases and continues to increase at an apparently constant rate until the entire population has died.

The shape of the age-specific mortality graph suggests an exponential function. Benjamin Gompertz noted this in 1825 and suggested that age-specific mortality rate accelerates in populations that have a significant number of individuals that survive to the age of reproduction. Thus, the function that describes the rate of mortality in a population bears his name: the **Gompertz mortality function (Equation 2.3).**

Gompertz mortality function, $m(t) = q_x e^{G(t)}$ (2.3)

where

$m(t)$: = mortality rate as a function of age at time t

q_x = age-specific mortality rate

e = mathematical constant

$G(t)$ = Gompertz mortality constant at time t

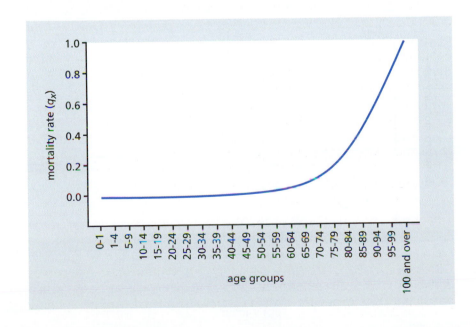

Figure 2.11 Age-specific mortality rate, q_x, for the total United States population, 2006. (Data from E. Arias, United States life tables, 2006, *Natl Vital Stat. Rep.* 58:1-40, 2010. With permission from the National Center for Health Statistics.)

This function can be rewritten algebraically:

$$\ln m(t) = \ln q_x + G(t) \tag{2.4}$$

As shown in **Figure 2.12**, the Gompertz mortality rate describes a constant increase in mortality with age, thus providing a marker of biological aging in the postdevelopment population. Note that the mortality rate of the age interval 0–1 years does not conform to the exponential increase characterizing the other age groups. In a human population, the death rate of individuals from birth to 1 year of age defines the **infant mortality rate** and will always be significantly higher than that of other childhood intervals, due to the inherent risk of birth and infancy. The impact of the infant mortality rate on aging is discussed in later chapters. Also note the slight rise in mortality that occurs around the ages of 10–15 and 15–20. Gerontologists refer to this rise as the "I-think-I-am-immortal" effect, because of the propensity of males to engage in risky behaviors that cause death. It is also known as the "stupidity or testosterone bump."

Age-independent mortality can affect the mortality rate

Gompertz assumed that all members of the population, even the very young, died of natural causes due to **age-dependent mortality**, or biological aging. However, death can occur as a result of accidents or other environmentally induced trauma, such as infections, that do not reflect biological aging. Moreover, current understanding of age-dependent mortality does not, for the most part, include deaths before puberty. Death that is not the result of biological aging is known as **age-independent mortality** and can be included as a variable in the Gompertz mortality function. Generally, age-independent mortality is included in the Gompertz mortality function equation only when comparing the mortality rates of different populations of a species that have significantly different environmental influences on death—war, infectious disease, and so on **(Figure 2.13)**. In addition, most research has shown that age-independent mortality varies considerably before puberty and then becomes constant in stable populations that are not exposed to excessive environmental insults **(Figure 2.14)**. Therefore, calculation of the Gompertz mortality function (Equation 2.3) will not generally include data for age-specific mortality before puberty. In other words,

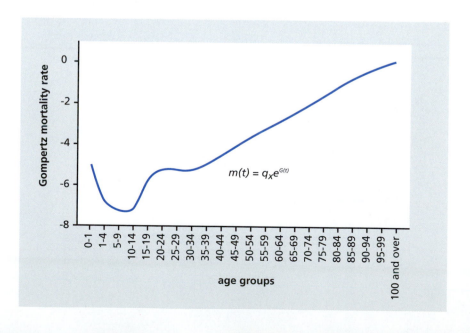

Figure 2.12 Gompertz mortality rate, $m(t)$, for the total United States population, 2006. (Data from E. Arias, United States life tables, 2006, *Natl Vital Stat. Rep.* 58:1–40, 2010. With permission from the National Center for Health Statistics.)

$$m(t) = q_x e^{G(t)}$$

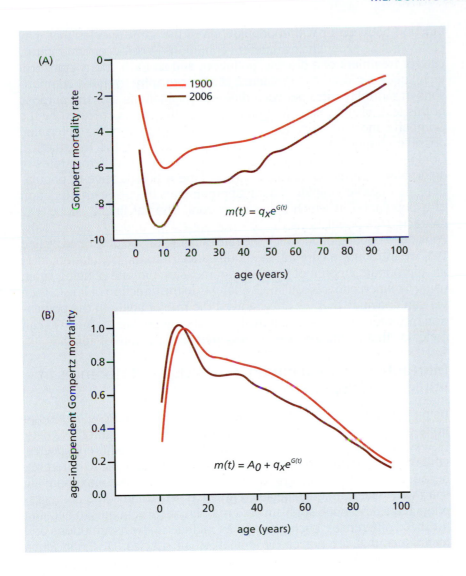

Figure 2.13 Comparison of the Gompertz mortality function and the age-independent Gompertz mortality function. These graphs show the mortality rates in the United States in 1900 and 2006 by graphing (A) the standard Gompertz mortality function and (B) the standard Gompertz mortality with a constant, A_0, added to account for age-independent mortality (the lowest mortality rate for each population was used as age-independent mortality). Note that the differences in early life mortality shown in (A) are partially eliminated by correcting for age-independent mortality. (Data from F.C. Bell and M.L. Miller, Life Tables for the United States Social Security Area 1900–2010, Pub. No. 11-11536, Washington, DC: Social Security Administration, 2005; data from E. Arias, United States life tables, 2006, *Natl Vital Stat. Rep.* 58:1–40, 2010. With permission from the National Center for Health Statistics.)

calculation of the function begins with data at the lowest mortality rate, usually around the time of puberty.

Because age-dependent and age-independent mortality have been defined and can be dealt with mathematically, we now have a method for evaluating the rate of aging by using the Gompertz mortality

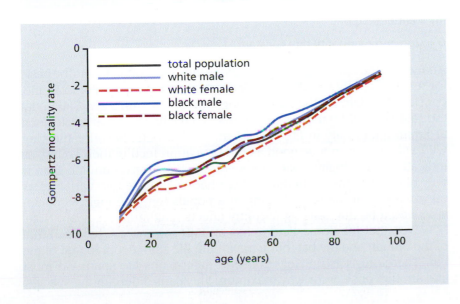

Figure 2.14 Gompertz mortality plots for different United States populations, 2006. The lowest mortality rate in all groups occurs around the age of 10 years. Although mortality rates for the four populations differ during young adulthood, the mortality rates after age 40 are similar. (Data from E. Arias, United States life tables, 2006, *Natl Vital Stat. Rep.* 58:1–40, 2010. With permission from the National Center for Health Statistics.)

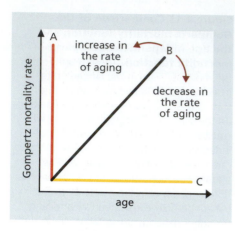

Figure 2.15 A change in slope of the Gompertz mortality rate provides a measure of change in the rate of aging in a population. As the slope moves toward line A, the rate of aging increases. As the slope moves toward line C, the rate of aging decreases.

function between or within populations of the same species that may or may not be experiencing different conditions—socioeconomic status, cure or treatment of a disease, pollution, and so forth. Many statistical techniques are used in the science of **demography** to make comparisons in mortality rates between different populations. For the purpose of this text, visually examining the slope of the lines generated by the Gompertz mortality function provides an easy method to determine possible changes in the rate of aging. For example, the hypothetical Gompertz plots shown in **Figure 2.15** illustrate three different rates of aging based on mortality rate analysis. Line A is indicative of populations that age very rapidly. Line B approximates the Gompertz function of a population in which aging is gradual. The flat line C, slope = 0, characterizes populations that do not appear to undergo aging. Now consider a population in which the rate of aging can be described by line B. If we introduce a cure for, say, lung cancer into the population, after an appropriate amount of time we can measure the rate of aging again using Gompertz analysis. If line B changes in the direction of line C, then we can conclude that we have slowed the rate of aging in the population by treating lung cancer. If the line shifts toward line A, then we can conclude that curing lung cancer has increased the rate of aging.

Mortality-rate doubling time corrects for differences in initial mortality rates

Using the Gompertz analysis for measuring the rate of aging in populations of a single species has proved very useful to biogerontologists. However, comparisons between the rates of aging of different species, while possible with Gompertz analysis, can be difficult to interpret, because population size, age-independent mortality, and maximum life span vary widely among species **(Figure 2.16)**. While some demographers have developed sophisticated mathematical schemes to account for these differences, such procedures include many assumptions and techniques that are not easily applied to all species. Dr Calab Finch suggested in 1990 that the difficulties associated with using Gompertz analysis for comparisons between species could be partially reduced by the simple calculation of the **mortality-rate doubling time** (MRDT), or the time required for the mortality rate in a population to double, given by **Equation 2.5**.

Mortality-rate doubling time, $MRDT = \dfrac{\ln 2}{G} = \dfrac{0.693}{G}$ (2.5)

where

$\ln 2$ = natural log of 2

G = Gompertz mortality rate constant

Analysis of the rate of aging that includes the MRDT often forms the basis for testing hypotheses about the individual rate of aging. For example, consider the data for the human, dog, and bat in **TABLE 2.3**. The initial mortality rate (IMR) tells us that humans have become extremely good at limiting mortality in the early years of life. Bats, because they are wild animals, are not as good as humans at protecting their young, as indicated by the higher IMR. Moreover, the bats' maximum life span is only about 20% that of humans. If we had only these two variables as our markers for the rate of aging in a population, one logical conclusion would be that IMR significantly affects maximum life span. Note, however, that the domestic dog has an IMR somewhere between that of the human and the bat, but a maximum life span equal to that of the bat. Therefore, it is unlikely that IMR determines life span. We must find another answer for why maximum life spans differ among species.

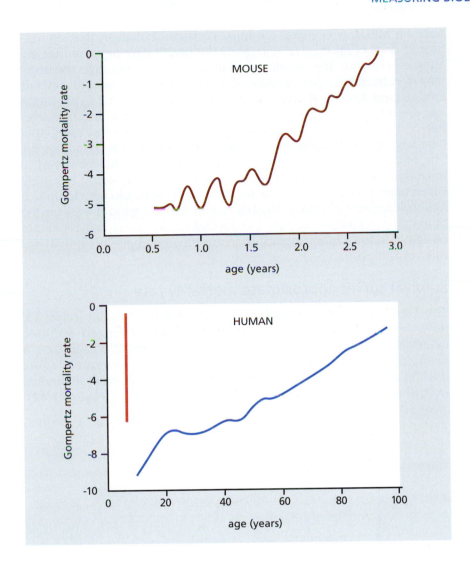

Figure 2.16 Comparison of Gompertz mortality rates of two different species. These graphs show the Gompertz mortality plots for a population of mice and for the total United States population in 2006. Although the small size of the mouse population causes significant variation in the data, note that the pattern of mortality after a specific age (1.5 years in the mice, 25–30 years in the humans) is similar and describes a gradual rate of aging. However, differences in age-independent mortality and maximum life span can be difficult to compare between species. If we were to plot the mouse data on the same x-axis scale as the human data, the result would be the red line on the human plot. (Bottom data from E. Arias, United States life tables, 2006, *Natl Vital Stat. Rep.* 58:1–40, 2010. With permission from the National Center for Health Statistics.)

Addition of the MRDT to the analysis gives some insight on what might be happening.

Although bats have a high IMR, their average MRDT of 5 years suggests a slower aging rate than that observed in domesticated dogs, which

TABLE 2.3
INITIAL MORTALITY RATE (IMR), MORTALITY-RATE DOUBLING TIME (MRDT), AND MAXIMUM LIFE SPAN OF VARIOUS SPECIES

Species	IMR	MRDT (years)	Maximum life span (years)
Human	0.0002	8	120
Dog	0.02	3	10–20
Laboratory mouse	0.01	0.3	4–5
Laboratory rat	0.02	0.3	5–6
Bat	0.36	3–8	11–25
Herring gull	0.2	6	49
Turkey	0.05	3.3	12.5
Quail	0.07	1.2	5
Fruit fly	1	0.02	0.3
Nematode	2	0.02	0.15
Rotifer	6	0.005	0.10

have an MRDT of 3 years. We must consider, however, that domesticated dogs often receive several of the health care benefits that are afforded humans. This would, in all likelihood, contribute to the dogs' long lives by reducing age-independent mortality. Bats are wild and thus tend to have a rather short postreproductive life span, due to predation and diseases. In other words, compared with dogs, bats have high age-independent mortality. Nonetheless, the MRDT of 5 years suggests that bats have a rate of aging much closer to that of humans than dogs do. The MRDT of bats is also significantly greater than that observed in mice and rats, the two most common species used in aging research. Given appropriate environmental conditions that would help to reduce age-independent mortality, the bat would make a good model for the human for evaluating possible factors that influence the individual rate of aging. Indeed, many laboratories are breeding bats in captivity for this very purpose.

Survival curves approximate mortality rate

The construction of a life table and analysis of mortality rates are time-consuming and generally complicated. Moreover, an accurate determination of mortality rates requires large population samples, a luxury that most biologists do not have. Therefore, biogerontologists use the simpler **survival curve**, a graphical representation of survival over time, to estimate the rate of aging in a population **(Figure 2.17)**.

Figure 2.17 Comparison of survival curves for two different species. Survival curves are generated from a life table (see Table 2.2). Column 1 (age) is plotted against the conversion of column 3 (l_x) to percentage of life span. The patterns of survival for the mouse population and for the human population in 2006 are similar, suggesting that survival curves in a small population are a good approximation of true mortality rates. (Bottom data from E. Arias, United States life tables, 2006, *Natl Vital Stat. Rep.* 58:1–40, 2010. With permission from the National Center for Health Statistics.)

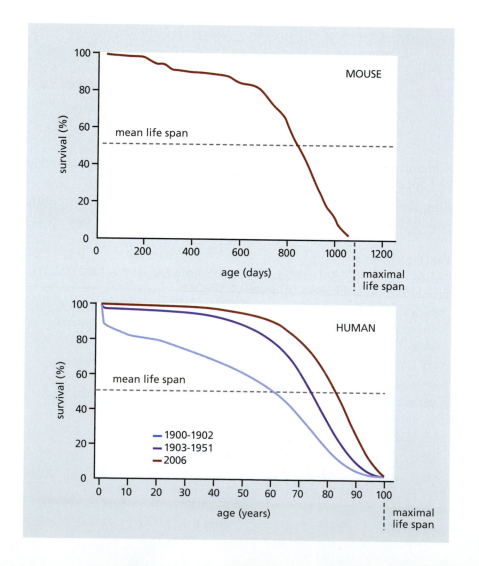

The same general assumptions concerning the rate of aging that are applied to the life table can be used for the survival curve. Indeed, comparing the shape of the survival curve with the log of mortality rate demonstrates that as the nature of mortality changes, so too does the shape of the survival curve **(Figure 2.18)**.

The data used to construct a survival curve also provide two important variables—mean life span and maximum life span—that are used extensively in biogerontology. **Mean life span** corresponds to the arithmetic average of the life span in the population. **Maximum life span** reflects the age of death of the longest-lived individual in the population. Biogerontologists often use the upper 10% of survival as their marker of maximum life span because survival curves are normally generated from numerically small populations and few individuals reach very old age. Both mean life span and maximum life span have specific meanings with regard to factors affecting the shape of the survival curve.

Mean life span is a measure of extrinsic aging in populations. Biogerontologists compare mean life spans of populations of a single species that are exposed to different environments or experimental treatments to determine whether external factors affect the rate of population senescence. Note that deaths in the first 50% of the populations described in Figure 2.17 occur over a long period of time. On the other hand, the second 50% of the populations die within a short percentage of the life span. Thus, mean life span is a relative measure of the

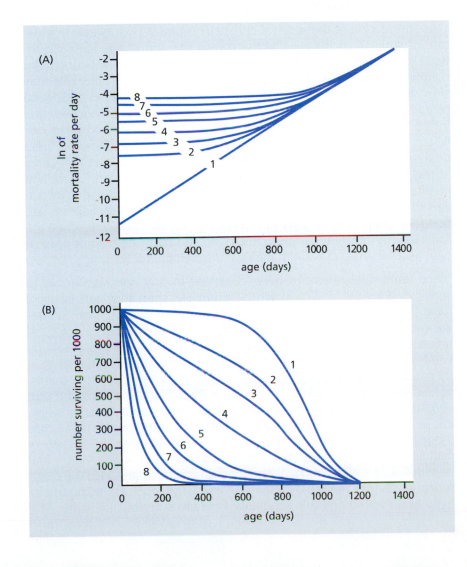

Figure 2.18 Comparison of Gompertz mortality rate plots and survival curves. (A) Gompertz plot. (B) Survival curve. Numbers on the lines indicate the same populations in both graphs. For example, line 1 in (A), the mortality rate expected for a gradually aging population, corresponds to line 1 in (B), the expected survival curve for the same population. As the pattern of the Gompertz mortality rate changes, so does the pattern for survival, indicating that survival curves are a good approximation of mortality rate. (From G.A. Sacher, in Handbook of the Biology of Aging [C.E. Finch and L. Hayflick, eds.], New York: Van Nostrand Reinhold Company, 1977. With permission from John Wiley and Sons)

survival rate during development and maturity phases and the impact of these two life phases on senescence. Note how the change in early life survival of the human population shown in Figure 2.17 changes the mean life span. A positive change in mean life span would suggest that some external intervention slowed the extrinsic rate of aging and allowed a greater proportion of the population to enter senescence. Conversely, a negative change in mean life span denotes an increase in the extrinsic rate of aging and fewer individuals that will senesce. The fewer the individuals reaching senescence, the greater the likelihood that any conclusion made concerning maximum life span may be influenced by selective mortality.

Maximum life span provides the biogerontologist with some information about the genetic or intrinsic rate of aging. You will learn in the coming chapters that life span potential has evolved from genes selected for successful reproduction, although those genes have not yet been identified. Thus, any change in maximum life span following some treatment or experimental procedure can be taken to mean that some fundamental property of intrinsic aging, most likely genetic, has been altered. Notice that the maximum life spans of the three human populations in Figure 2.17 are identical, even though the shapes of the survival curves are different. That is, maximum life span was not affected by the extrinsic rate of aging.

Deceleration of mortality rate at the end of life suggests the possibility of longevity genes

Close inspection of the survival curves in Figure 2.17 suggests that the rate of survival increases slightly toward the end of the life span for these populations. The change in rate at the tail of survival curves is a common observation and suggests that the late-life mortality rate may slow rather than remain constant as predicted by Gompertz. Several researchers have investigated this possibility and found a deceleration in mortality rate in late life, when the populations studied have sufficient numbers of individuals left at the end of life to allow accurate analysis. The first to recognize this phenomenon was Dr. James Carey at the University of California, Davis. He used more than 1.2 million Mediterranean fruit flies to demonstrate that mortality rate declines in flies in old age **(Figure 2.19)**. The implications of these findings are that the Gompertz mortality function does not accurately describe the trajectory of mortality at all stages of life and that one cannot predict, with accuracy, the maximum life span for a population.

Since immortality has not been observed in any species known to senesce, does the finding that maximum life span may not be fixed have practical applications? Most likely not. But deceleration of late-life mortality has huge theoretical implications for biogerontological investigators who study the evolution and genetics of aging. As you will learn in the next chapter, evolutionary theory predicts that longevity has arisen from genes selected for successful reproduction; that is, genes determine longevity. The data in fruit flies are consistent with evolutionary theory and show that a subset of the population has different longevity characteristics than the majority of the population. Since longevity is determined by the genome, it follows that this subset of fruit flies must have different genes that allow them to live longer and to senesce at a different rate. The work in fruit flies provides the theoretical basis that molecular biologists and geneticists are using to test hypotheses about possible candidate genes that modify life span.

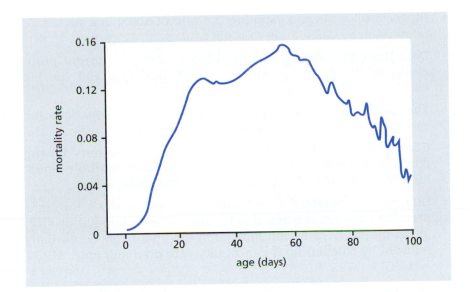

Figure 2.19 The mortality rate for 1.2 million fruit flies. From 0 to 20–30 days of age, the rate shows the typical exponential increase expected for a population kept under laboratory conditions—that is, little age-independent mortality. However, the analysis of this large population of fruit flies shows that mortality slows and decelerates in late life, unlike the constant mortality rate predicted by Gompertz. (From J.R. Carey, P. Liedo and J.W. Vaupel, *Exp. Gerontol.* 30:605–629, 1995. With permission from Elsevier.)

ESSENTIAL CONCEPTS

- Biogerontologists have traditionally used the passage of time as the standard unit of measurement for expressing age-related change. However, trying to determine whether possible time-related differences seen at the chosen time points have any meaning for biological aging can be difficult and often leaves many unanswered questions.

- Due to differences in our interactions with the environment, the rate of biological aging is highly individualized.

- The intrinsic rate of aging reflects an individual's genotype and can be measured with a fair amount of accuracy. The extrinsic rate of aging, a result of the environment's interaction with the genotype—that is, the phenotype—has infinite possibilities and can be challenging to measure.

- Regulation of gene expression that results in the phenotype can also be influenced by epigenetic effects.

- Differences between the phenotypes of individuals increase with age. As we grow older, the phenotype becomes more and more unique to the individual.

- Recent research has suggested that the trajectory in the rate of aging may be established during fetal and early life development.

- Cross-sectional investigations compare the means of different age groups at a single point in time. Cross-sectional studies have limited precision, due to factors such as intrinsic variability, the cohort effect, and selective mortality.

- Longitudinal studies have shown that the rate and timing of age-related physiological functional decline are highly variable. Longitudinal studies have also shown that age-related physiological decline is nonlinear with respect to time.

- Personal genomics may soon be used to identify whether "rate of aging" genes exist and how these genes are expressed.

- Biogerontologists can estimate the rate of aging in a particular population by calculating mortality rates.

- Data from life tables can approximate both the current probability of dying, or age-specific death rate (q_x), and the life expectancy of an individual (e_x).

- The Gompertz mortality function, $m(t) = q_x e^{G(t)}$, describes the rate of mortality in a population.

- Comparison of mortality rates between populations having different environmental conditions that affect mortality must include a measure of age-independent mortality.

- Difficulties associated with using Gompertz analysis for comparisons between species, due to differences in population size, initial mortality rate, and maximum life span, can be partially reduced by calculating the mortality-rate doubling time, $MRDT = \ln_2/G$.

- Survival curves are graphical representations of survival over time and can be used to estimate the rate of aging in a population when construction of life tables and analysis of mortality rates are impractical.

- Current data indicate that the rate of survival increases slightly toward the end of the life span in many populations, suggesting that the late-life mortality rate may slow.

DISCUSSION QUESTIONS

Q2.1   Define genotype and phenotype. Which one has the most influence on the intrinsic rate of aging? Which has the most influence on the extrinsic rate of aging?

Q2.2   What are some of the factors that make identification of a biomarker for human aging challenging?

Q2.3   Consider the following experimental design for a study on how aging might affect hearing. Subjects in three age groups, 20–49, 50–69, and 70–90 years, will be exposed to different levels of sound, and their ability to hear the sounds will be compared. Is this an example of a cross-sectional or longitudinal design? Explain how the cohort effect and selective mortality will affect the results and conclusions.

Q.2.4   Has longitudinal analysis of human aging been more effective than cross-sectional analysis at identifying a biomarker of human aging? Why or why not? Why are the impacts of selective mortality and the cohort effect reduced in cross-sectional studies conducted with subjects participating in longitudinal studies?

Q2.5   Explain why a biomarker for human aging may require the use of personal genomics. How will personal genomics be used to improve the health of the aging population?

Q2.6   Why should a biogerontologist who uses laboratory animals to model human aging be concerned with the reproductive life span of the animal?

Q2.7 What are the three factors that make interspecies comparisons of mortality rates difficult when using the Gompertz model?

Q2.8 Consider the graphs in **Figure 2.20**, showing survival curves for two groups of mice. Which group of mice would you suspect to have the slower rate of aging? Why? What are the approximate mean and maximum life spans for each group?

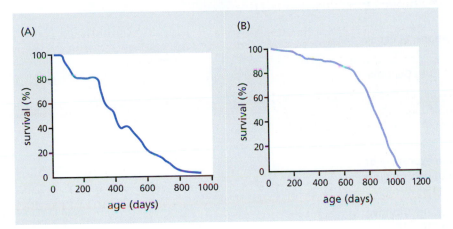

Figure 2.20

Q2.9 Which is considered a good measure of the extrinsic rate of human aging: mean or maximum life span? Why?

Q2.10 How does the observation that the mortality rate of a population decelerates in late life support the concept of selective mortality?

FURTHER READING

MEASURING BIOLOGICAL AGING IN THE INDIVIDUAL

American Federation of Aging Research. www.afar.org

Barker DJ, Godfrey KM, Fall C, et al. (1991) Relation of birth weight and childhood respiratory infection to adult lung function and death from chronic obstructive airways disease. *BMJ* 303:671–675.

Barker DJ, Winter PD, Osmond C, et al. (1989) Weight in infancy and death from ischaemic heart disease. *Lancet* 2:577–580.

Ferrucci L (2008) The Baltimore Longitudinal Study of Aging (BLSA): a 50-year-long journey and plans for the future. *J Gerontol A Biol Sci Med Sci* 63:1416–1419.

Johnson TE (2006) Recent results: biomarkers of aging. *Exp Gerontol* 41:1243–1246.

Khoury MJ, McBride CM, Schully SD, et al. (2009) The scientific foundation for personal genomics: recommendations from a National Institutes of Health–Centers for Disease Control and Prevention multidisciplinary workshop. *Genet Med* 11:559–567.

McCartney N, Hicks AL, Martin J & Webber CE (1996) A longitudinal trial of weight training in the elderly: continued improvements in year 2. *J Gerontol A Biol Sci Med Sci* 51:B425–433.

Miller RA (2006) Principles of animal use for gerontological research. In Handbook of Models for Human Aging (Conn PM ed), pp 21–31. New York: Elsevier.

Newman AB (2010) An overview of the design, implementation, and analyses of longitudinal studies on aging. *J Am Geriatr Soc* 58:S287–289.

Ruggiero C, Metter EJ, Melenovsky V, et al. (2008) High basal metabolic rate is a risk factor for mortality: the Baltimore Longitudinal Study of Aging. *J Gerontol A Biol Sci Med Sci* 63:698–706.

Shock NW, Greulich RC, Andres R, et al. (1984) Normal human aging: the Baltimore Longitudinal Study of Aging. www.grc.nia.nih.gov/blsahistory/blsa_1984/index.htm

Strehler BL (1959) Origin and comparison of the effects of time and high-energy radiations on living systems. *Q Rev Biol* 34:117–142.

MEASURING BIOLOGICAL AGING IN POPULATIONS

Arias E (2010) United States life tables, 2006. *Natl Vital Stat Rep* 58:1-40.

Carey JR (2003) Longevity: The Biology and Demography of Lifespan. Princeton: Princeton University Press.

Carey JR, Liedo P, Orozco D & Vaupel JW (1992) Slowing of mortality rates at older ages in large medfly cohorts. *Science* 25:457–461.

Chiang CL (1984) The Life Table and Its Applications. Malabar, FL: Robert E. Krieger Publishing Company.

Finch CE (1990) Longevity, Senescence, and the Genome. Chicago: University of Chicago Press.

Gavrilov LA & Gavrilov NS (1991) The Biology of Lifespan: A Quantitative Approach. Newark, NJ: Harwood Academic Publisher.

Gompertz B (1825) On the nature of the function expressive of the law of human mortality, and on the mode of determining the value of life contingencies. *Phil Trans R Soc* 115:513–585.

Motulsky HJ (2010) Comparing survival curves. In Intuitive Biostatistics. Oxford: Oxford University Press, pp 210–218.

EVOLUTIONARY THEORIES OF LONGEVITY AND AGING

3

"NOTHING IN BIOLOGY MAKES SENSE EXCEPT IN THE LIGHT OF EVOLUTION."
-THEODOSIUS DOBZHANSKY, BIOLOGIST
(1900-1975)

Charles Darwin gave humanity a new way of thinking that forever changed our view of life. Darwin suggested that nature alone is responsible for the immense variation in life forms found on Earth. His theory of evolution by natural selection remains largely unchanged and is at the intellectual core of all discoveries in the biological sciences, even 150 years after the theory was first introduced at a meeting of the Linnean Society of England, on July 1, 1858.

Darwin expanded his thoughts on evolution by natural selection in *On the Origin of Species*. In this text, he describes how variations in the traits of individuals within a species provide the material responsible for the great diversity of life on Earth. Thus, it should come as no surprise that biogerontologists have looked to theories of evolution to form the biological basis for the existence of an extended post-reproductive life, aging, senescence, and longevity.

In this chapter, we examine how evolutionary theory and evolutionary biology are being used to explain why and how we age. We begin with a brief look at some historical evolutionary theories related to longevity and aging, then explore in more detail some current evolutionary models.

FOUNDATIONS OF EVOLUTIONARY THEORIES OF LONGEVITY AND AGING

The evolutionary theories of longevity and aging have their roots firmly planted in basic concepts of evolutionary biology. In this section, we briefly explore topics in general evolution that significantly influenced the development of evolutionary models related to longevity and aging.

IN THIS CHAPTER . . .

FOUNDATIONS OF EVOLUTIONARY THEORIES OF LONGEVITY AND AGING

EVOLUTION AND LONGEVITY

TESTING EVOLUTIONARY MODELS OF LONGEVITY

EVOLUTION AND AGING

Weismann established the separation between soma and germ cells

At the time of Darwin's death in 1882, evolution by natural selection had not yet been fully accepted by the biological sciences community. Although most biologists of the time agreed that Darwin was undoubtedly correct, the theory still had many holes that prevented its complete acceptance. One such problem was the mechanism by which traits were passed between generations. Most scientists of Darwin's time believed that the soma (body) cells transmitted the properties of heredity directly to the germ (sex) cells, or **gametes**. It was not until the great German theorist August Weismann (1834–1914) proved otherwise that our current concept of a division between somatic cells and germ cells was established. This set the basis for a theory on aging. In one experiment, Weismann mated mice that had had their tails cut off, prior to reproductive age, for 22 generations. Each mated pair produced mice with tails **(Figure 3.1)**. This experiment showed that body cells were not communicating with the sex cells and that gametes alone transfer the chemicals of inheritance to the next generation.

The separation between soma and germ cells suggested to Weismann a division of labor within the organism: somatic cells existed solely to support the germ cells and their function of passing on the "stuff" of inheritance. Based on this evidence, Weismann theorized that the soma's job was to ensure that the individual lived long enough to reproduce. Once this job was done, there was no further need for the soma, and aging and death of the organism would follow. As you will see later in this chapter, Weismann's thoughts on aging form the basis of what would later be called the **trade-off hypothesis**, which suggests that cost of successful reproduction is mortality.

Figure 3.1 Weismann's experiment establishing the separation between germ line (sex cells) and soma (body cells). Weismann discovered that removing the tails of mice before the start of reproduction had no effect on the appearance or length of tails in the offspring, generation after generation.

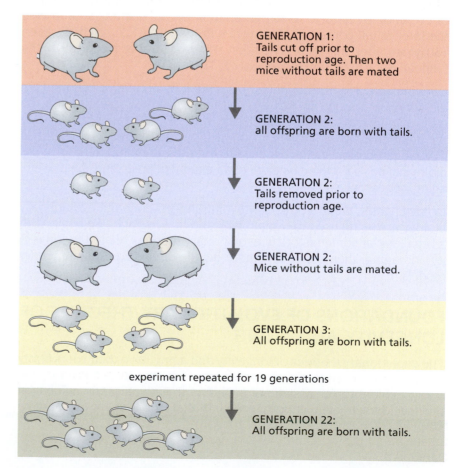

GENERATION 1:
Tails cut off prior to reproduction age. Then two mice without tails are mated

GENERATION 2:
all offspring are born with tails.

GENERATION 2:
Tails removed prior to reproduction age.

GENERATION 2:
Mice without tails are mated.

GENERATION 3:
All offspring are born with tails.

experiment repeated for 19 generations

GENERATION 22:
All offspring are born with tails.

Weismann proposed that aging is a nonadaptive trait

In Weismann's first writing on the subject of evolution and aging, he states that

> . . . in regulating duration of life, the advantage to the species, and not to the individual, is alone of any importance. This must be obvious to anyone who has once thoroughly thought out the process of natural selection. It is of no importance to the species whether or not the individual lives longer or shorter, but it is of importance that the individual should be enabled to do its work towards the maintenance of the species. This work is reproduction or the formation of a sufficient number of new individuals to compensate the species for those which die. As soon as the individual has performed its share in this work of compensation, it ceases to be of any value to the species, it has fulfilled its duty and it may die. (Weismann 1891)

This early writing by Weismann suggested that aging arose through natural selection as a positive adaptation that worked for the good of the species, a general concept known as **group selection**. That is, selection occurs at the level of the group rather than the individual. Weismann originally believed that aging is selected to rid the group of old and useless individuals who are no longer capable of reproduction but continue to use limited and valuable resources, such as food and water. Ridding the group of nonreproducing adult members who do not contribute to continuation of the species allows resources to be preferentially allocated to the reproductively active members of the group, and the **fitness** of the remaining group increases.

Weismann's logic of the group benefit in ridding the species of older, nonreproducing adult members has obvious appeal and remains a common misconception. Evidence from studies in molecular evolution has conclusively established that the individual—or, more precisely, the gene—is the focus of adaptation, precisely what Darwin originally believed. Only reproductively active individuals influence the makeup of the genome, and this makeup is driven exclusively by the need of the individual to achieve reproductive age and pass on the gene. In other words, the gene does not "know" that in later years, the group would need to rid itself of old people. In fact, some evidence now exists to suggest that menopause (reproductive aging) and the extended post-reproductive life span may be an adaptation that increases fitness (**BOX 3.1**).

In time, Weismann became aware that his original view on group selection as the basis for his theory of aging conflicted with Darwin's proposal that natural selection works on the variation among individuals within a species. Weismann had to either reject Darwin's natural selection theory or revise his own view on the evolutionary basis of aging. He chose the latter. Weismann formulated his new theory based on the thought that as soon as a trait becomes useless to an individual, natural selection no longer acts to either remove or maintain the trait. Recall that Weismann saw the post-reproductive period as having no value to the organism. Since most of the physical problems of aging occur after reproduction, the traits of aging neither increase nor decrease fitness, so aging and/or senescence are neutral to the forces of natural selection. Weismann referred to such traits as **nonadaptive traits**. The neutrality of aging allowed Weismann to retain his suggestion that a postreproductive period in multicellular organisms is useless, while at the same time maintaining his strict Darwinist view.

BOX 3.1   THE GRANDMOTHER HYPOTHESIS AND FEMALE LONGEVITY

Throughout history, human grand-mothers have been portrayed as kind and loving members of an extended family who provide care to their grandchildren. Many cultural anthropologists believe that the grandmother–grandchild relationship separates our species from all others, including nonhuman primates. A few anthropologists now suggest that the relationship between the grandchild and the grandmother transcends a simple emotional connection and may have had a role in the evolutionary development of longevity. The **grandmother hypothesis** posits that care given by grandmothers to children in early societies allowed their daughters to have more children and thus increase overall fitness for the species. The longer the grandmother lived, the more children her daughter could have, and the greater the fitness. That is, longevity—or, at least, the long post-reproductive life span observed in humans—was selected because of its benefits to reproductive success.

The grandmother hypothesis is an extension of the theory of G.C. Williams that menopause might be an adaptation that increases overall fitness for the species. Williams suggested that in evolutionary history, older mothers ran a greater risk of dying during childbirth and thus would be unable to provide for their surviving children. Death of these offspring would result in decreased fitness for the species. If, however, older women were physiologically unable to reproduce because of menopause, older mothers could focus on providing resources to the offspring that already existed. Similarly, the grandmother hypothesis suggests that a long postreproductive life span evolved as a result of the grandmother's help in rearing their daughters' older offspring. This would allow the mother to focus greater resources on the younger offspring that were still totally dependent on her. The grandmother was essential to the survival of her grandchildren. Thus, a long postreproductive life span increased fitness.

Studies evaluating demographic data collected in the eighteenth and nineteenth centuries have generally supported the grandmother hypothesis. The researchers selected several measures of fitness, including (1) mothers having more offspring; (2) mothers reproducing earlier and more often; and (3) mothers having shorter interbirth intervals. The researchers found that the number of grandchildren correlates directly with the life span of the grandmother **(Figure 3.2)**: the greater the age of the grandmother, the more grandchildren. Investigators also found that their measures of fitness were enhanced when the grandmother lived in the same house as her daughter and provided care for the grandchildren, compared

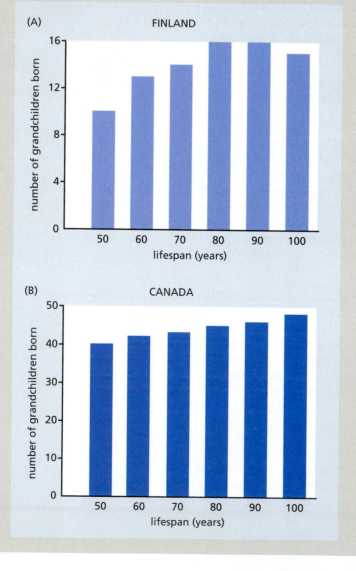

Figure 3.2 Relationship between female lifespan and total number of grandchildren contributed to the following generation. Data are shown (A) for Finland (1702–1823) and (B) for Canada (1850–1879). The rise in slope for each population is the equivalent of postreproductive women gaining two extra grandchildren for every 10 years the women survived beyond age 50. (From M. Lahdenpera et al., *Nature* 428:178–181, 2004. With permission from Nature Publishing Group.)

BOX 3.1   THE GRANDMOTHER HYPOTHESIS AND FEMALE LONGEVITY

with families that did not include the grandmother. Based on these findings, the authors concluded that "our results lend strong support for the hypothesis that prolonged female postreproductive lifespan is adaptive, to our knowledge revealing for the first time the substantial fitness benefits that females accrue by living beyond reproductive age."

Other studies have not found a "grandmother benefit" for species fitness, but have demonstrated that inclusion of the grandmother as a caregiver for her daughter's children shortened the interbirth interval. If a grandmother was able to care for recently born children, the mother could reproduce quickly and have more children than a woman without grandmother care. More children means greater fitness, and the short interbirth interval would sooner or later be dominant in the population.

The grandmother hypothesis can be very appealing and does have scientific credibility. Nonetheless, there are some questions that need to be answered before this hypothesis receives widespread acceptance. First, were there enough women having long postreproductive life spans throughout our evolutionary history to increase fitness? Recall from Chapter 2 that the average life span throughout most of history has hovered around 35–40 years of age

and that less than 2% of the population lived beyond 60 years of age. Children born into families without grandmothers would have vastly outnumbered children in families having grandmothers. Thus, **selection pressure**—the events altering the genetic composition of individuals (in this case, a grandmother providing care)—for a gene giving rise to a long postreproductive life span or enhanced longevity may not have been strong enough to influence gene selection.

Second, is it possible that older male and female siblings contributed to the care of dependent children? If, for example, the care was provided primarily by an older sibling, the selection pressure for a longevity gene would not just be weak, it would be nonexistent. Given the scarcity of old people in prehistoric populations, it seems more likely that care of dependent children would fall to an older sibling rather than a grandmother. At the very least, there would have been considerable variation in who cared for the dependent child.

Finally, what were the family structure and responsibilities? The overwhelming majority of early societies were patriarchal. It was the male, not the female, who provided the most important care—the gathering of food—for the dependent

child. The grandmother may have provided emotional, nursing, and palliative care for dependent children, an important job to be sure, but insignificant compared with the biological necessity of food. Without the male providing food for the family, there wouldn't have been a grandmother on which to base a hypothesis.

There is little doubt that grandmothers have played an important role in the care of grandchildren throughout history. It is even possible that grandmothers contributed more to child rearing during evolutionary history, allowing their daughters to have a greater number of children. Whether the greater number of children increased species fitness and resulted in the adaptive trait of a long postreproductive life span remains to be seen. The data supporting the grandmother hypothesis are neither extensive nor compelling and rely almost exclusively on correlative rather than causal analysis. However, we should not dismiss out of hand any hypothesis for which at least some supportive research exists. With supportive mathematical models and highly controlled laboratory research, we may just find that grandma has been more important to the family than just being able to prepare a great dinner.

The appearance of neutral or nonadaptive traits has often caused confusion when seen in the light of Darwin's ideas of variation, fitness, and adaptation. This confusion undoubtedly arises as a result of the exclusive focus on natural selection as the only force of evolution. But even Darwin recognized that traits could become "fixed" without the force of natural selection:

> Variations neither useful nor injurious [to reproduction] would not be affected by natural selection, and would be left either a fluctuation element, as perhaps we see in certain polymorphic species, or would ultimately become fixed... (Darwin, 1859)

Thus, aging could have arisen without an influence on reproduction. As you'll see, Weismann's thoughts on nonadaptive aging set the basis for theories that predict that the neutral trait of aging is either a by-product of genes fixed for a reproductive advantage or a random expression of genes in older age groups.

Population biologists developed logistic equations to calculate population growth

Mendel's principles of inheritance were not introduced formally into evolutionary theory until the early 1900s. Therefore, neither Darwin nor Weismann was able to incorporate Mendel's findings into their theories. Mendel provided the answer to the question of how variation, underlying Darwin's fundamental principle of natural selection, arose in a species. According to Mendel, variation occurred because the **alleles** for (that is, different versions of) a specific gene transmitted by each parent to the offspring can have slightly different forms. The question then became, "At what rate and by what mechanisms do these alleles become dominant in a population?" Research into such questions gave rise to a new form of evolutionary analysis, **population genetics**, a science that focuses on the origin of allele variation in a population.

Evolutionary theories of aging and longevity rely on some basic principles of population genetics to better explain how life span in a population is affected by the frequency at which an allele that enhances reproduction appears in a species' genome. Determining the rate at which an allele appears in a population requires a basic understanding of **reproductive potential**—a species' relative capacity to reproduce itself under optimal conditions—and the population's growth.

Unicellular (single-cell) species, such as *Saccharomyces cerevisiae*; simple multicellular organisms, such as *Caenorhabditis elegans*; and cells from multicellular species grown in cultures—all are being used to understand the basic mechanisms underlying aging and longevity. Although *S. cerevisiae*, *C. elegans*, and cells growing in culture propagate by different mechanisms, they have similar patterns of growth. These growth patterns have a significant effect on the population's life span.

The rate of growth in populations of simple organisms reflects the birth rate minus the death rate and is called the **intrinsic rate of natural increase**, *r*. However, reproduction and growth in all species, simple and complex, are constrained by environmental factors such as food, space, and temperature that significantly influence the population's growth and reproduction potential. These and other constraints are collectively known as the **carrying capacity of a population**, *K*, or the constraints placed on population size due to environmental factors. These concepts are used in the **Verhulst-Pearl logistic equation** (Equation 3.1) to describe population growth for any population, particularly populations that are constrained by lack of mobility and/or are maintained under highly controlled conditions.

Verhulst-Pearl logistic equation:

$$\frac{dN}{dt} = rN \frac{(K - N)}{K} \tag{3.1}$$

where

N = population size

r = intrinsic rate of natural increase

K = carrying capacity of the population

Figure 3.3 illustrates how constraints on simple organisms can affect growth and reproduction and influence gene selection. Growth in the cell population during the initial stages of culture is slow, due to a small population of "parent cells." As the population of cells increases, so does the growth rate, since there are more "parent cells" to produce offspring. Growth of the population during this period is driven almost exclusively by the intrinsic rate of natural increase, since food is abundant and

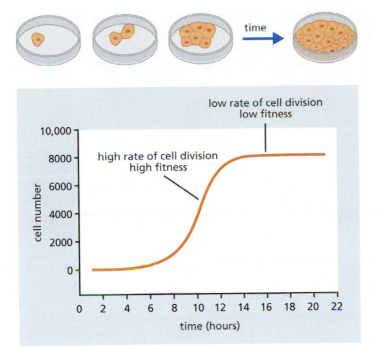

Figure 3.3 Population growth in a hypothetical cell population, as described by the Verhulst-Pearl logistic equation. In this cell population, the constraint on population growth is represented by space (size of the container) and food. Note that the period of high fitness occurs early in the life span of the culture, when cell division is greatest. Low fitness occurs late in the life span, when population growth slows and plateaus.

space is plentiful. As food supply and space begin to shrink, the rate of reproduction slows, as does the population growth. Finally, the space and food necessary to support the size of the population run out. If more food and space are not provided, the population of cells will senesce and die. The logistic equation of population growth becomes important when we discuss replicative senescence in Chapter 4.

This simple example demonstrates a fundamental principle of population genetics that has great importance to the evolution of aging and longevity: The age at which high rates of reproduction occur have the highest level of Darwinian fitness, due simply to the number of individuals that are reproducing. As reproduction slows, fitness declines. Thus, alleles that convey traits important to survival and reproduction are selected over alleles that impart longevity. This important concept helped population geneticists formulate theories of how genes selected for survival also determine the length of life.

A population's age structure describes Darwinian fitness in complex eukaryotes

More advanced eukaryotes are not constrained by environmental conditions to the same degree as described in our previous example. The mobility of complex eukaryotes allows animals to search for food and water, find shade on hot days, and move to insulated locations on cold days. Therefore, for complex eukaryotes, mobility means that environmental constraints, or K factors, are variable. Determining the rate at which alleles are fixed in complex eukaryotes that have variable K factors requires a more complex set of equations than the logistic equations used for simple organisms. These equations describe a population's reproductive contribution to future generations, or species fitness, at any given time during the life span and constitute a method known as **age-structure analysis**.

Age-structure analysis can be used to determine whether a particular age group within a population contributes to the action of natural selection on that population. It can also be used to answer the question, "At what age is an individual most likely to pass on traits beneficial to the species' survival?" Knowing when specific alleles are most likely to be selected provides information on whether aging, senescence, and/or longevity arose through natural selection.

In Chapter 2 you were introduced to the concept of age-structure analysis in the calculation of survival (l_x), using variables in the life table. As a component of fitness, survival predicts how many individuals are available for reproduction in a specific age group. Because survival declines with each successive age group after the onset of reproduction, it would seem that fitness might also decline. However, survival provides only a marker of potential fitness. To calculate actual fitness in a population, we need to include a measure of the reproduction rates. This measure is known as **fecundity**, m_x.

For age–structure analysis, the population is stratified into age groups. The process of age grouping has no set standard but primarily reflects the reproductive characteristics of the population. Periodic-breeding populations, such as animals that breed only once per season (most birds and reptiles), are typically grouped by the number of breeding seasons. Populations in which reproduction occurs continuously (most mammals) are generally grouped by a convenient time interval, in days, months, or years.

Reproduction rate describes age-specific fitness in breeding populations

Age-structure analysis in periodic-breeding populations is simplified by the fact that offspring are produced at regular intervals—in the breeding season. Algebraic equations can be used to calculate both the total population and the breeding season–specific reproduction rate. The numerical results of these equations are referred to as the **net reproduction rate (Equation 3.2)**. The net reproduction rate for a population (R_0) is the sum of all breeding season–specific survival rates (l_x) multiplied by breeding-specific fecundity (m_x).

$$\text{Net reproduction rate, } R_0 = \sum_{x=0}^{x=\infty} l_x m_x \qquad (3.2)$$

where

R_0 = potential number of offspring produced by a newborn over its life span

l_x = survivorship, calculated from a life table

m_x = fecundity

The net reproduction rate provides a measure of reproductive power and, thus, fitness. It is most often used in analysis of periodic-breeding populations. When the net reproduction rate is applied to animals with breeding seasons, it is referred to as the **breeding season–specific reproduction rate** and indicates which groups have the most influence on growth and fitness. For example, consider the hypothetical seasonal-breeding population shown in **Figure 3.4**. Predation and environmental hardships cause survivorship to decline in each successive breeding season. However, different rates of sexual development of individuals within the population cause fecundity to increase during the first few breeding seasons. The increase in fecundity early in the population's life history will outpace the decrease in survivorship. The inverse relationship between survivorship and fecundity during the first few breeding seasons results in an increase in the breeding season–specific

survivorship (*l*)	fecundity (*m*)	breeding season	breeding season-specific reproduction rate
16%	0	6	0
33%	30%	5	0.1
50%	40%	4	0.2
67%	75%	3	0.5
83%	24%	2	0.2
100%	0	1	0

$$R_0 = \sum_{x=0}^{x=\infty} l_x m_x = 1$$

Figure 3.4 Age–structure analysis using the net reproduction rate in a hypothetical, seasonal-breeding population at equilibrium ($R_0 = 1$). The shaded blocks represent population size. The highest breeding season–specific reproduction rate is 0.5, in breeding season 3, reflecting an increase in fecundity. The lowest breeding season–specific reproduction rate is 0, in the youngest (breeding season 1) and longest-lived (breeding season 6) animals. That is, fitness is greatest in the younger population and lowest in the older population.

reproduction rate. As survivorship, fecundity, and reproduction rate decline during each successive breeding season, the population is eventually left with a small group of long-lived individuals that do not reproduce. Once again, natural selection favors alleles that enhance survival to reproductive age rather than traits that add to the length of life of the species.

Equations describing populations of species that continuously reproduce require integration to determine the rate of growth. The equation describing population growth in continuously breeding populations was proposed by the statistician Alfred Lotka, who built on the work of the eighteenth-century Swiss mathematician Leonhard Euler. Thus, the equation bears both their names **(Equation 3.3)**. Note that Equation 3.3 is simply the integration of Equation 3.2 and thus imparts the same basic information on reproductive potential and fitness.

General form of Euler-Lotka equation of population growth:

$$1 = \int_0^\infty e^{-rt} l_{(x)} m_{(x)} \tag{3.3}$$

where

e = the mathematical constant

r = intrinsic rate of population growth at time$_t$

$l_{(x)}$ = survivorship at time$_x$

$m_{(x)}$ = fecundity at time$_x$

Fisher described the relationship between reproductive potential and Darwinian fitness in populations

The **Euler-Lotka equation** set the basis for R. A. Fisher's development of a mathematical model that could be applied to theories of the evolution of aging and longevity. Fisher (1890–1962) was the first to suggest that solving for r, the intrinsic rate of natural increase, in Equation 3.3 provides a measure of fitness for an individual in a population. Since Fisher's primary interest was predictive statistical analysis, he was more concerned with how the current population growth could predict future individual fecundity and fitness. To this end, Fisher derived a measure, the **reproductive value**, v_x, which he suggested predicted an individual's future reproductive contributions relative to the reproductive output of the total population **(Equation 3.4)**.

Reproductive value, $v_x = \frac{e^{rx}}{l_x} \int\limits_{x}^{\infty} e^{-rt} \, l(t)m(t)$ (3.4)

where

v_x = reproductive value of an individual at time x

e = the mathematical constant

r = intrinsic rate of population growth

l_x = survivorship

$l(t)$ = remaining total survivorship of the population at time t

$m(t)$ = remaining fecundity of the population at time t

For Fisher, v_x provided insight into answering the question, "At what age will an individual in a continuously breeding population have the greatest fitness?" As shown in **Figure 3.5**, the reproductive value provides a measure of potential fitness by estimating the future reproductive contribution of a specific age group.

Although Fisher helped to establish the foundations for evolutionary models of longevity and aging, his only allusion to a relationship between reproductive value and aging came from his observation that v_x declines at about the same time that mortality begins to increase **(Figure 3.6)**. He suggested, for the first time, that longevity might be a by-product of alleles fixed for survival to reproductive age. A more formal application of Fisher's age–structure analysis to the evolution of longevity would have to wait until the verbal postulation of Sir Peter Medawar and the mathematics of W.D. Hamilton, as described below.

EVOLUTION AND LONGEVITY

The extrinsic rate of aging leads to a decline in the force of natural selection

The first serious application of Weismann's and Fisher's hypotheses on longevity came from Sir Peter Medawar (1915–1987). Medawar used a simplified version of age–structure analysis to establish that the life span of all matter, inorganic and organic, results from extrinsic factors. Moreover, demonstration of the **extrinsic rate of aging**—the rate of aging in a population due to environmental hazard—established what

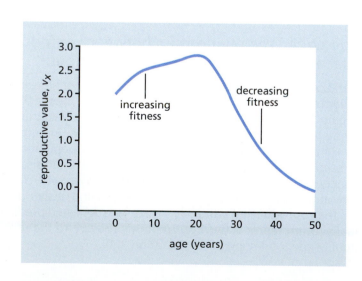

Figure 3.5 Reproductive value (v_x) of females in the Commonwealth of Australia in 1911. This graph was created by inserting recorded birth and death rates in Equation 3.4. As expected, the reproductive value is highest at young ages and decreases as the population ages, approaching the end of the reproductive life span. Note that an increase in v_x indicates an increase in fitness, and a decrease in v_x corresponds to a decrease in fitness. (Adapted from R.A. Fisher, The Genetical Theory of Natural Selection, Oxford: Clarendon Press, 1930.)

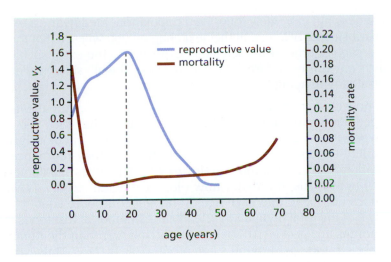

Figure 3.6 Reproductive value and mortality rate of Taiwanese females in 1906. Note that the mortality rate begins to increase at roughly the same time that reproductive value, v_x, begins to decrease, at the point indicated by the dashed line. (Adapted from R.A. Fisher, The Genetical Theory of Natural Selection, Oxford: Clarendon Press, 1930.)

many believe to be the fundamental principle underlying the evolutionary basis of longevity: the force of **natural selection** declines with age.

In 1951, Medawar proposed a thought experiment that demonstrates this principle. Imagine a start-up laboratory that has a "population" of 1000 test tubes. Although the test tubes do not age, each month, exactly 10% of the test tubes are broken in random accidents—extrinsic, environmental factors. Personnel in the laboratory replace the broken tubes so that the total population is always 1000 at the beginning of each month. That is, the total population of test tubes is at equilibrium. Fortunately for our demonstration, the director of the laboratory requires that all test tubes be marked with the date they entered the population. Thus, we can follow the life history of the original population of 1000, as shown in **Figure 3.7**.

The age-structure distribution illustrated in Figure 3.7 describes a population in which the survivors of any age group always outnumber the survivors in the subsequent age group. In this population, greater survivorship of the test tubes of one group compared with the next group cannot be due to increased fragility, because there is no aging—that is, no intrinsic factor that increases the probability of death. Death occurs simply because, with increasing age, the test tubes are exposed for a longer time to extrinsic hazards of the environment and thus have an increased risk of being broken. Through this description of a

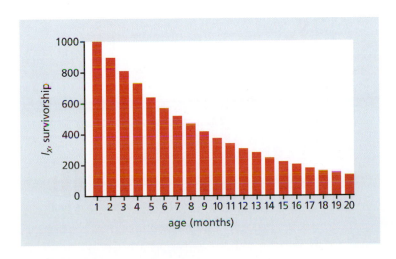

Figure 3.7 Life history of a test tube "population" in Medawar's thought experiment. This graph shows the number of surviving test tubes (survivorship, l_x) from an original immortal population of 1000, assuming a fixed probability of death of 10% per month. (Adapted from P.B. Medawar, An Unsolved Problem of Biology, London: H.K. Lewis & Company, 1952.)

non-senescing, extremely long-lived population, Medawar established the extrinsic rate of aging. In the next chapter, you will see that the simple passage of time, with increasing exposure to the extrinsic hazards of one's environment, forms the very basis for the physiological decline associated with aging.

We'll now change the parameters a little. Instead of having the laboratory personnel replace the test tubes, we assume that each age group in the population described in Figure 3.7 has been, miraculously, given the ability to reproduce. As luck would have it, they reproduce at a constant rate of 10% per month **(Figure 3.8)**. The tubes still do not age, and all individuals, no matter how old, have equal fertility. However, that does not mean that each age group will contribute equally to the renewal of the total population. Clearly, the number of test tubes in the 1–2 months age group (10% of 900) is much larger than the number of test tubes in the 13–14 months age group (10% of 273). Thus, younger populations make a greater contribution to the replacement of the population, not because they are more virile, but simply because they have not been exposed to environmental hazards as long as the older populations.

In the next step of this demonstration, the test tubes are no longer intrinsically immune to aging. At some age they become feeble and lose the ability for self-reproduction. Does the age at which reproduction begins to wane matter to the survival of our test tube population? It most certainly does. Imagine that the loss of reproduction occurs early in the life span, say, at 2–3 months of age. The consequences will be catastrophic to the continuation of the species and the test tubes will soon be gone from our laboratory. If, however, senescence and loss of reproduction occur at 15–16 months, the results will clearly decrease the total number of offspring but will not have a significant impact on continuation of the species. In other words, whether or not the older age group contributes to the gene pool of test tubes becomes irrelevant. The force of natural selection declines as fitness approaches zero.

Medawar theorized that aging arose as a result of genetic drift

Medawar's explanation for the evolutionary process by which aging and/or longevity became fixed in the organism had the same paradox that troubled Weismann. That is, if the force of natural section declines

Figure 3.8 Relationship between number of offspring and age of Medawar's test tube "population." Assuming that the test tubes reproduce at a constant rate of 10% per month and that the probability of death remains fixed at 10%, younger populations will produce more offspring than the older groups, because there are more younger test tubes than older test tubes. This simple observation by Medawar describes the reason for the decline in the force of natural selection with age. (Adapted from P.B. Medawar, An Unsolved Problem of Biology, London: H.K. Lewis & Company, 1952.)

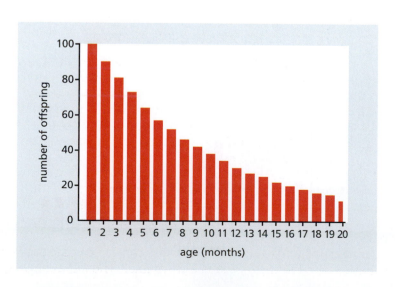

with age, and aging occurs after the start of reproduction, then how did the physiological declines associated with aging become part of the organism's genetic makeup? Recall that Weismann solved the paradox by suggesting that aging was neutral with respect to reproduction. Because Weismann's thoughts were developed before Mendel's principles of inheritance were known, he could not explain his reasoning in terms of classical genetics. Medawar, on the other hand, lived at a time when the principles of inheritance had been firmly established and most scientists believed in **genetic determinism**, the process in which genes were selected only by natural section and only for a specific biological process. (We now know that there are other mechanisms for gene selection and that one gene can be involved in several related but separate processes.) Thus, Medawar needed a theory that could explain how aging, a process contrary to survival and reproduction, could arise and be regulated by genes in the absence of natural selection. He based his explanation for the appearance of aging on the principle of **genetic drift**, a process in which genes can be fixed in a small population as a result of the random sorting of alleles at the time of **meiosis**. Medawar's theory on the evolution of aging has come to be known as the **mutation accumulation theory of senescence**.

Let's look at an example. Say that a small, older population has two alleles of a gene that codes for a post-reproductive life span: a dominant allele causing short post-reproductive life span (*SL*) and a recessive allele causing a longer post-reproductive lifespan (*ll*). As shown in **Figure 3.9**, this small, senescent population separates genetically into a classical Mendelian distribution: ¼ recessive and ¾ dominant (¼ *llll*, ½ *SLll*, ¼ *SLSL*). Because of the high extrinsic rate of aging and low fitness of the older population, only two mating pairs (four individuals) are capable of reproduction. These mating pairs arise purely by chance. In a path leading to fixing of the *ll* allele (left path in Figure 3.9), both individuals of one mating pair contain two copies of the recessive allele; both individuals of the other pair contain one copy each of the *ll* and *SL* alleles. For the completely recessive mating pair (*llll*), the only possible offspring are also completely recessive. It is also possible that the mating pair with mixed alleles produces an offspring that is completely recessive, through random sorting of alleles at the time of meiosis. Thus, purely by chance, but nonetheless a real possibility, the recessive *ll* allele becomes the dominant gene *LLLL*. A similar statistical strategy can be demonstrated for elimination of the *ll* allele (shown in the right path in Figure 3.9).

Medawar proposed that aging and longevity arise separately in post-reproductive populations

Recall that Medawar lived in a time when genetic determinism was completely accepted. If a biological event was observed—and aging was viewed as a biological event—then there was a gene regulating that process. Therefore, Medawar needed to explain in genetic terms how senescence, the slow decline in function during the post-reproductive period, was regulated. He reasoned that a single gene could not cause senescence, as a single gene with massive deleterious effects would eventually be eliminated from the genome. Rather, senescence more closely reflected the fixation of hundreds if not thousands of genes that had small, non-lethal, but negative effects. Medawar's thoughts on the evolutionary basis of aging arose from the example of the disease Huntington's chorea. Huntington's chorea is a neurological disorder resulting from a recessive mutation that is expressed at midlife (late thirties to mid-forties) and is always fatal. Thus, if a single gene

Figure 3.9 Gene fixing due to genetic drift and mutation accumulation. In this population, *SL* is a dominant allele causing short postreproductive life span and *ll* is a recessive allele causing long postreproductive life span. The left path illustrates how chance mating and random sorting of alleles can cause the recessive *ll* allele to become the dominant gene *LLLL*. The right path illustrates how the recessive *ll* allele can be eliminated from the population.

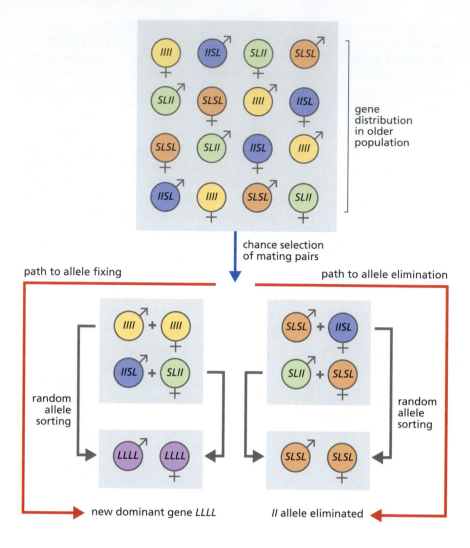

were responsible for aging, those carrying the Huntington's chorea gene would have been eliminated from the population before genetic drift could fix the gene in the genome. On the other hand, non-lethal, deleterious genes could become fixed (1) because selection pressure against these genes would be low, due to the small population involved in passing on the gene, and (2) because of genetic drift.

Medawar's writing does not make clear whether he understood that extrinsic factors leading to the breakdown of molecular structures could result in the aging of organic life. Most likely, and in keeping with the science of his era, Medawar would have thought that organic matter would not be subject to the same physical forces that caused non-organic matter to senesce. We now know that thermodynamic forces, including entropy, also apply to organic matter (see Chapter 4). However, Medawar established the foundations for the theory that aging arises as a stochastic process resulting from thermodynamic forces, whereas longevity evolved through natural selection as a by-product of genes important to survival to reproductive age.

Hamilton's force of natural selection on mortality refined Medawar's theory

The foundation of mutation accumulation lies in the observation that a trade-off exists between fecundity and mortality; highest fecundity, and therefore fitness, occurs at a time of lowest mortality. In suggesting the possibility of mutation accumulation, Medawar relied on Fisher's parameter, r, and its derivative, the reproductive value, v_x, as a measure of fitness. Medawar did not, however, explain his theory in mathematically explicit terms, a critical step in theoretical evolutionary biology. This next step in the scientific explanation of the evolution of longevity would be taken by W.D. Hamilton (1936–2000).

Hamilton recognized that integrating reproductive potential across the entire life span was an imprecise method for measuring fitness. He hypothesized that any population, even a stable population, can fluctuate in size at different ages, depending on age-specific mortality. Fitness was more likely to reflect reproductive potential at specific ages than the total integrative value over the life span, as suggested by Fisher's reproductive value.

To quantify the force of natural selection, Hamilton derived expressions for the change in fitness with respect to age-specific mortality. Like Fisher's reproductive value, v_x, Hamilton's force of natural selection on mortality, s_x, declines after the start of reproduction **(Figure 3.10A)**. However, unlike v_x, s_x is stable and highest before reproduction. Because extrinsic factors make the force of natural selection on

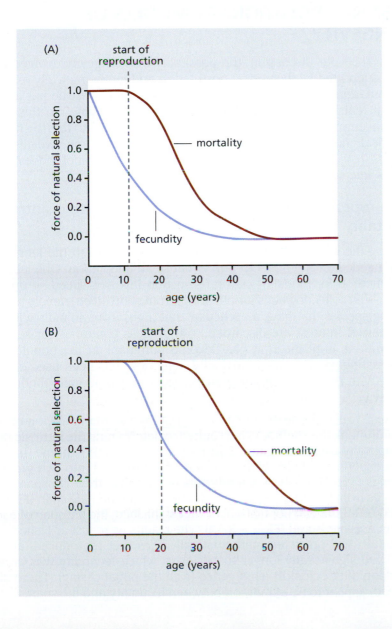

Figure 3.10 Hamiltion's force of natural selection on mortality and fecundity. (A) The force of natural selection on mortality, s_x, acts on a hypothetical human population. Note that mortality (brown line) is highest before the start of reproduction (indicated by the dashed line). (B) A small subset of the same human population has genes coding for late-life reproduction. Since reproduction in this subset population starts later than that observed in the total population, the force of natural selection on mortality is delayed. (Adapted from W.D. Hamilton, *J. Theoret. Biol.* 12:12–45, 1966. With permission from Elsevier.)

mortality highest prior to reproduction, evolution would have selected genes that were necessary for surviving to reproductive age. Hamilton's method for assessing aging and longevity agreed with previous work indicating that selection pressure would be extremely weak for genes that regulated aging or longevity. Therefore, if genes existed for longevity and/or aging, they must be related to genes selected for survival to reproductive age.

Hamilton's mathematical theory on longevity was a monumental breakthrough in understanding how life span, not aging, evolved. Imagine, for instance, that some individuals within a small population had genes coding for late-life reproduction (Figure 3.10B). The force of natural selection on mortality would start later in the life span. From a theoretical standpoint, if there is no reproduction, there cannot be a trade-off between fecundity and mortality. For any given age throughout reproduction, this small group would have greater fecundity at older ages. Over evolutionary time, and because of genetic drift, genes that impart a long life span could be fixed within the genome. Hamilton provided the mathematical theory to suggest that longevity could have evolved.

TESTING EVOLUTIONARY MODELS OF LONGEVITY

Thus far in our discussion, the possibility that longevity evolved has been supported by theoretical postulation and mathematically explicit computations. All theories, no matter how sound, must be put to the test in real life before complete acceptance is possible. To this end, several empirical investigations have been conducted to test the mathematical models of life-span evolution. In this section, we discuss results from a few investigations that link the evolution of longevity to genes selected for reproduction.

Late-reproducing organisms have a lower rate of intrinsic mortality

Testing theories of evolution can be challenging, given the length of time needed for natural selection to cause an adaptation. One laboratory method for modeling natural selection involves speeding up the process by collecting eggs (or offspring) from short-lived, rapidly reproducing species that have a particular trait. Eggs from animals without the trait of interest are discarded. The process continues for several generations until the trait has become dominant in the population. This testing process, called **artificial selection**, has been used extensively with *Drosophila melanogaster* to test the evolutionary models of longevity.

Because artificial selection often does not involve manipulation of the environment, this method is best suited to determining the intrinsic rate of mortality caused by genes currently present in the genome. In one such experiment, eggs were collected and separated into two groups: (1) eggs from flies that began reproduction early in their life span; and (2) eggs from flies that began reproduction late in their life span **(Figure 3. 11)**. After the two populations were established, the life span of each group was measured **(Figure 3.12)**. The results were just as Fisher and Hamilton had predicted mathematically: flies that reproduced late in life lived significantly longer than flies in which fecundity was greatest soon after **eclosion** (emergence from the pupal casing). There did, indeed, appear to be a trade-off between fecundity and life span.

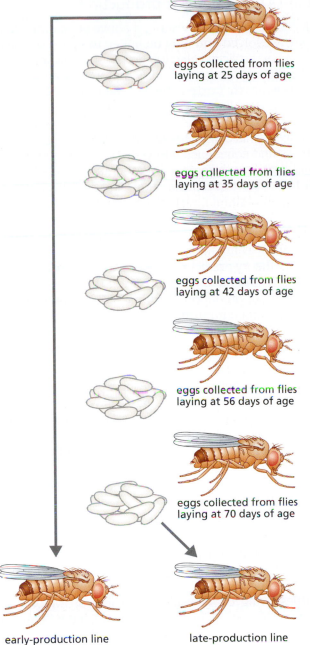

eggs collected from flies
laying at 25 days of age

eggs collected from flies
laying at 35 days of age

eggs collected from flies
laying at 42 days of age

eggs collected from flies
laying at 56 days of age

eggs collected from flies
laying at 70 days of age

GENERATIONS

early-production line

late-production line

Figure 3.11 The use of artificial selection to create early- and late-reproduction lines in *D. melanogaster.* Eggs were collected and separated into two groups: eggs from flies that began reproduction early in their life span and eggs from flies that began reproduction late in their life span. Eggs from both groups were allowed to hatch, pass through their larval and pupal stages to produce adult flies, and begin reproduction. This process continued until two distinct populations of flies were formed, which took almost three years.

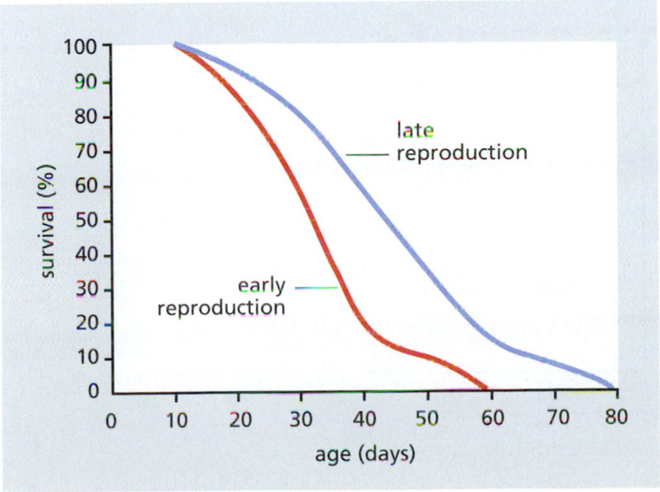

100
90
80
70
60
50
40
30
20
10
0

survival (%)

late
reproduction

early
reproduction

0   10   20   30   40   50   60   70   80

age (days)

Figure 3.12 Survival curves for female *D. melanogaster* artificially selected for early and late reproduction. The early-reproduction line (red) has an average age at highest fecundity of 4–5 days, and the late-reproduction line (blue) has an average age at highest fecundity of 8–12 days. These results demonstrate that genes selected for their timing at the start of reproduction also affect the length of the life span. (Adapted from M.R. Rose, *Evolution* 35:1004–1010, 1984. With permission from Wiley-Blackwell.)

Genetic drift links life span to reproduction

The artificial selection experiment described above eloquently demonstrates that longevity evolved and may be linked to genes selected for reproduction. However, artificial selection does not test whether or not environmental conditions, the primary driving force of natural selection, can work to fix genes that extend the life span. To do this, we need to see whether there is a similar trade-off between reproduction and life span when extrinsic factors, such as differences in predation and starvation, are introduced. This can be tested under laboratory conditions. For example, Sterns and colleagues allowed eggs from two groups of flies to hatch and progress through the larval and pupal stages into adulthood. Prior to the start of reproduction, 90% of the adults in one group are killed, a method simulating high predation. In the other group, the density of flies in the housing chamber is reduced to a level at which all flies have sufficient food and no predation throughout reproduction, a method simulating environmental conditions leading to a stable population. Evolutionary theory predicts that flies subject to high predation would reproduce early and have more offspring than flies reared in a stable population without predation. And, indeed, this prediction was supported by experimental data from this type of experiment. In other words, the methods of this experiment simulated natural selection; flies exposed to high or low predation adapted to their environment.

Of course, we are interested in whether or not genes selected for a particular reproductive scheme can determine the length of life. Evolutionary models of longevity predict that flies with early reproduction (the high-predation group) should have a faster rate of mortality than late-reproduction flies (the low-predation, stable group). Again, this prediction was supported by the data **(Figure 3.13)**. Thus, both the intrinsic and extrinsic rates of mortality are linked to the timing of reproduction, which, in turn, determines the length of the life span. The mathematical and verbal predictions of Fisher, Medawar, and Hamilton are supported.

It is important to demonstrate the trade-off between reproduction and life span in wild populations, not just in laboratory experiments. Investigations in wild populations compare the reproduction schedule and life span of a single species living in two different environments. For example, turtles living on an island with low predation can be compared with the same species living on the mainland with high predation. In most cases, populations with high predation (high extrinsic

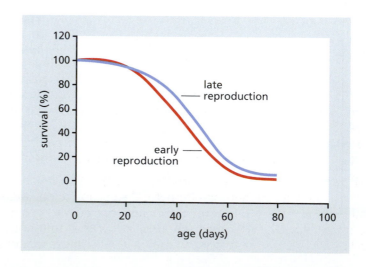

Figure 3.13 Relationship between extrinsic mortality rate and duration of life span. As described in the text, flies are subject to controlled environmental conditions. Simulation of high predation leads to the development of early-reproducing flies, while simulation of low to no predation results in a line of flies with late reproduction. Data collected from such experiments show that late-reproducing flies have slightly longer mean and maximum life spans.

mortality) have significantly shorter life spans than populations of the same species living in areas with low predation (low extrinsic mortality). This finding has held for several different land-based phyla, including birds, mammals, and reptiles. However, similar studies in fish have not supported the trade-off hypothesis. The different result in fish than in land-based phyla demonstrates that theories, while supported by laboratory tests, may not always follow the prediction in the wild. Nonetheless, there is enough evidence from the study of wild populations to generally support the evolutionary basis of longevity.

Results from testing the evolutionary theory of longevity have changed research in biogerontology

Before laboratory testing of the evolutionary theory of longevity, biogerontology was an observational science lacking clear and precise definitions of aging and longevity (see the discussion in Chapter 1). Without these critical definitions, identification of biological mechanisms underlying aging and longevity was difficult, due in part to researchers defining aging and/or longevity to fit their hypotheses, rather than the other way around. Moreover, biogerontology lacked consensus as to the direction of research. The results from laboratory experiments on the evolution of longevity described here separated the biological processes of aging and longevity; more precise definitions were now possible. In addition, the understanding that longevity evolved through genes selected for reproductive success and that aging reflects unregulated, random, and stochastic mechanisms led to a much clearer research direction for biogerontology.

Separation of the evolutionary origins of aging and longevity revolutionized research in biogerontology. Biogerontologists were now able to formulate hypotheses based on an accurate description of how a species' longevity became part of its overall life history. To this end, geneticists looking for genes affecting longevity could narrow their search to genes that connect longevity to the reproduction schedule for the species. In Chapter 5, you'll learn that many genes have been identified that have dual effects—that is, they were selected for reproductive success but also affect longevity. And in Chapter 7, you'll learn that results from the laboratory testing of the evolutionary theory of longevity established a solid foundation for, and helped to give rise to, the study of **gerontological biodemography**, a science that integrates biological knowledge with demographic research on human longevity and survival.

EVOLUTION AND AGING

As we have seen, the longevity of a species is closely linked to genes selected for survival to reproductive age. Longevity has evolved. Conversely, the work of Medawar, Fisher, and Hamilton demonstrated that the slow decline in physiological function that is aging could not have arisen through natural selection. Aging has not evolved. This does not mean that genes have no role in the aging process. It simply means that genes related to the aging of a cell, tissue, organ, or organism were not subjected to the forces of evolution for that specific purpose.

Recall that Medawar's theory of mutation accumulation relies on evolution by genetic drift rather than on natural selection to explain aging in evolutionary terms. His theory predicted that certain genes exist for the specific purpose of aging. We now know that aging does not have a basis in genetic determinism and, instead, reflects random and stochastic mechanisms (see Chapter 4). In this section we explore two theories of aging—antagonistic pleiotropic and disposable soma—that evolutionarily account for age-related decline in physiological and biological function.

Antagonistic pleiotropy is a special case of general pleiotropy

Pleiotropy is a genetic mechanism in which a single gene produces more than one trait. For example, wild-type *D. melanogaster* has smoothed, curved bristles on its thorax, whereas the *singed* mutation, a single-point mutation, results in Drosophila with short, twisted bristles. In addition, the female *singed* mutant is sterile and produces ill-formed eggs (compared with the wild type) that never hatch. Therefore, in *D. melanogaster*, the lack of expression in one gene produces two phenotypes.

G.C. Williams combined the general concepts of pleiotropy with Medawar's declining force of natural selection to suggest a mechanism by which senescence can arise in a population in which genes are selected for reproductive success. This theory, **antagonistic pleiotropy**, predicts that genes conveying a benefit for fitness early in life will be selected even though they may be disadvantageous in later life. Williams summed up antagonistic pleiotropy as follows:

> The selective value of a gene depends on how it affects the total reproductive probability [that is, over the entire life span]. Selection of a gene that confers an advantage at one age and a disadvantage at another will depend not only on the magnitudes of the effects themselves, but also on the times of the effects. An advantage during the period of maximum reproductive probability would increase the total reproductive probability more than a proportionately similar disadvantage later on would decrease it. So natural selection will frequently maximize vigor in youth at the expense of vigor later on and thereby produce a declining vigor. (Williams 1957)

Williams reaffirmed the importance of the declining force of natural selection as the basis for aging, and he placed prime importance on the timing of the effect. As we have seen, effects occurring early in the reproductive life span are strongly selected for, because of the large number of individuals involved in reproduction. If that same gene conveys a negative trait late in the life span, the lack of selection pressure (due to the limited number of viable reproductive members of the species) will allow the gene to be expressed and is, for all intents and purposes, selectively neutral. The gene will be selected because the benefit occurs during a time of high fitness, even if the detriment of the gene for the individual is significantly greater than the benefit provided early in the life span. That is, the trade-off between benefit and detriment of a pleiotropic gene will always favor the benefit, if it occurs at the right time in the reproductive life span.

At least two examples suggest the possibility of antagonistic pleiotropy in nature. First, the calcification of bone during fetal and childhood development imparts a reproductive advantage (protection of internal organs, body stability, etc.) and thus high fitness. Several genetic mechanisms that include hundreds of genes have been selected to ensure proper bone calcification. However, these same genes can be detrimental later in life, causing calcification of arteries during the post-reproductive period, which can lead to coronary artery disease and myocardial infarction. Second, the Drosophila experiments described earlier in the chapter also provide evidence of antagonistic pleiotropy. In both experiments, the long-lived flies had fewer eggs, and the eggs where deformed (compared with those of the short-lived flies). Thus, the genetic mechanism that regulates normal egg production and development in short-lived flies must also have produced the problems associated with the eggs in the long-lived flies.

The disposable soma theory is based on the allocation of finite resources

Recall that August Weismann theorized that over evolutionary time, the soma had become mortal in order to support the immortal germ line. It was not until the early 1980s that scientists would pick up on Weismann's theory and develop a hypothesis to explain the underlying mechanism for the trade-off between the soma and the germ line. This hypothesis, first developed by Thomas Kirkwood, is known as the **disposable soma theory**.

The disposable soma theory is based on the evolutionary principle that all environments have finite resources, and organisms compete for those resources. The organisms that are the most efficient at using the available resources will survive, while inefficient organisms will die off. For example, imagine a time when **protozoa** (unicellular organisms) were beginning to evolve into **metazoa** (multicellular organisms). This would have been a time of great variation in the types of metazoa that were beginning to appear. The selection pressure would become more intense as the limited resources, such as food, were used up. The metazoan species that used the resources most efficiently would be the species to survive and pass on their genetic material to the next generation.

So what is the most efficient use of resources? The disposable soma theory suggests that the best use of resources is to give highest priority to the cells responsible for the continuation of the species, namely, the cells of reproduction, or the germ line. Supporting cells, those of the soma, would need only enough resources to ensure their primary job: supporting the survival of the germ line to the point of reproduction. That is, the soma could be disposed of once reproduction had occurred.

But, where and how are those resources being spent? Clearly, the production of a gamete would cost, in terms of energy, no more than the production of, say, a liver cell. Rather, the disposable soma theory predicts that early metazoans used available resources preferentially to maintain repair mechanisms of the DNA in the germ line. This suggestion arises from the current observation that it takes a lot of energy to ensure that the DNA sequence is correct. So, if we assume that an organism had to make an "evolutionary choice" between accuracy in the DNA of the germ cell or some function of a somatic cell, maintaining the DNA of the germ cell would be the best chance for the survival of the species.

Unlike antagonistic pleiotropy, the disposable soma theory has not been tested experimentally. Rather, the disposable soma theory has been shown to be theoretically possible by describing how well it fits with established general evolutionary theory. In particular, this theory can be viewed as a special case of John Maynard Smith's optimality theory. **Optimality theory** predicts that an individual will optimize a behavior so that the cost associated with that behavior is minimized in accordance with the local environment.

Optimality theory can be illustrated by species that lay eggs exposed to predation. The problem for these species is how to maximize the number of eggs that survive to hatching without investing more energy for the reproductive process than the environment will allow. Too few eggs, and they may all be lost to a predator; too many eggs, and insufficient nutrition to support egg production may result in "bad" eggs. The individual must arrive at a compromise to optimize survival of its offspring.

With regard to aging, at some point in evolutionary time, organisms had to "make a decision" about how much energy they should give to reproduction and how much to maintaining the soma. Too much energy diverted to accurate transmission of the genome to the next generation would result in not enough energy for maintaining the soma. The organism might not live long enough to successfully reproduce. Too much energy given to somatic maintenance, and the individual might live forever, but the accuracy of the genome would be compromised, eventually causing the extinction of the species. The disposable soma theory predicts that an organism will optimize resources so that there is high fidelity in the DNA of the gamete, with the leftover resources directed to maintenance of the soma. At some point during the post-reproductive period, extrinsic aging will cause an accident to occur in the soma. The soma will not have the resources necessary to repair the function, and aging will ensue.

ESSENTIAL CONCEPTS

- August Weismann suggested that somatic cells need only live long enough to ensure survival to reproductive age. Once that job was done, there was no further need for the soma—and aging ensued.

- Weismann believed that because the detrimental effect of aging occurred after the start of reproduction, it was neutral with respect to selection pressure; that is, he though that aging neither increased nor decreased fitness.

- Population genetics defines two basic principles that affect the growth of a population: (1) the intrinsic rate of natural increase, r; and (2) the carrying capacity of a population, K.

- The values of r and K are used in the Verhulst-Pearl logistic equation, $\frac{dN}{dt} = rN\frac{K-N}{K}$, to describe population growth.

- Many species have K factors that are variable. For populations with variable K factors, age-structure analysis is used to determine population growth and its impact on fitness.

- The logistic equation and age-structure analysis show that species fitness is greatest at the time of the greatest rate in population growth (reproduction). This means that alleles that convey traits important to survival to reproductive age will be selected over alleles that impart longevity or aging.

- Sir Peter Medawar demonstrated that the force of natural selection declines with age.

- Genetic drift predicts that genes neutral to the force of natural selection can be fixed in a population as a result of the random sorting of alleles at the time of meiosis. Medawar suggested that genetic drift could work to select genes for aging.

- W.D. Hamilton established the mathematical foundation for the evolutionary theory of longevity by using a value he called the force of natural selection on mortality, s_x. He suggested that genes that impart overall longevity for the individual must be related to genes selected for survival to reproductive age.

- Results of laboratory experiments support the mathematical and verbal predictions of Fisher, Medawar, and Hamilton and lead to the conclusion that both the intrinsic and extrinsic rates of aging are linked to the timing of reproduction, which, in turn, determines the length of the life span.

- G.C. Williams's theory of antagonistic pleiotropy predicts that genes that convey a benefit for fitness early in life will be selected even though they may be disadvantageous later in life.

- T.B. Kirkwood's disposable soma theory is based on the evolutionary principle that all environments have finite resources, and organisms compete for those resources. Organisms that are the most efficient at using the available resources will survive.

DISCUSSION QUESTIONS

Q3.1 August Weismann proposed the principle of the separation between somatic and germ cells. He also suggested that aging is a nonadaptive trait. Discuss why these two principles might establish the theoretical basis of current thought on the evolution of aging and longevity.

Q3.2 Consider the two graphs of population distribution shown in **Figure 3.14**. Figure 3.14A represents the logistic growth of cells, and Figure 3.14B represents the population distribution of a population of ducks in which egg laying begins in breeding season 3. In terms of fitness, discuss the fundamental principles important to the evolution of longevity and aging that these data demonstrate.

(A)

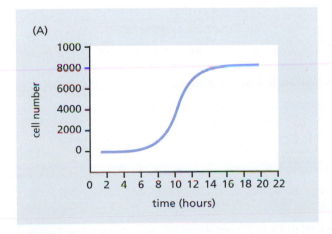

(B)

Figure 3.14

Q3.3   Discuss the importance of reproductive value, as proposed by Fisher, to evolutionary predictions made by Medawar and Hamilton.

Q3.4   Explain why Medawar's test tube example establishes the important principle that natural selection declines with age.

Q3.5   Discuss the basic process of genetic drift and how genetic drift could influence aging and longevity.

Q3.6   Explain why the force of natural selection on mortality is always highest before the start of reproduction. Also, explain why the force of natural selection on mortality can be used to demonstrate the relationship between genes selected for survival to reproductive age and the genetic basis of longevity.

Q3.7   Discuss how artificial selection was used to empirically test whether the intrinsic rate of aging is connected to the timing of reproduction.

Q3.8   G. C. Williams speculated that "natural selection will frequently maximize vigor in youth at the expense of vigor later on and thereby produce a declining vigor." Describe the evolutionary theory of aging that Williams was speculating about.

Q3.9   In terms of the disposable soma theory of aging, discuss the follow statement: "The cost of immortality of the germ line is somatic mortality."

Q3.10 Our current understanding of the evolution of longevity and/or aging demonstrates how scientists rely on their predecessors when searching for the truth about biological phenomena. Beginning with Charles Darwin, draw a concise timeline of the major events and people related to the important concepts that led to our current understanding of the evolution of longevity and aging.

FURTHER READING

FOUNDATIONS OF EVOLUTIONARY THEORIES OF LONGEVITY AND AGING

Darwin C (1958 [1859]) On the Origin of Species, p 495. New York: Signet Classic.

Dawkins R (1989) The Selfish Gene, p 352. Oxford: Oxford University Press.

Fisher RA (1930) The Genetical Theory of Natural Selection, p 272. Oxford: Clarendon Press.

Haldane JBS (1942) New Paths in Genetics, p 206. New York: Harper and Brothers.

Hawkes K (2004) Human longevity: the grandmother effect. Nature 428:128–129.

Lahdenpera M, Lummaa V, Helle S, et al. (2004) Fitness benefits of prolonged post-reproductive lifespan in women. Nature 428:178–181.

Lotka AJ (1956) Elements of Mathematical Biology, p 465. Mineola, NY: Dover Publications.

Weismann A (1891) Essays upon Heredity and Kindred Biological Problems, p 471. Oxford: Clarendon Press.

EVOLUTION AND LONGEVITY

Burke MK & Rose MR (2009) Experimental evolution with Drosophila. Am J Physiol Regul Integr Comp Physiol 296: R1847–1854.

Charlesworth B (1970) Selection in populations with overlapping generations: I. The use of Malthusian parameters in population genetics. Theor Popul Biol 1:352–370.

Hamilton WD (1966) The moulding of senescence by natural selection. J Theor Biol 12:12–45.

Medawar PB (1952) An Unsolved Problem of Biology, p 24. London: H.K. Lewis and Company.

Reznick DN, Bryant MJ, Roff D, et al. (2004) Effect of extrinsic mortality on the evolution of senescence in guppies. Nature 413:1095–1099.

TESTING EVOLUTIONARY MODELS OF LONGEVITY

Stearns SC, Ackermann M, Doebeli M & Kaiser M (2000) Experimental evolution of aging, growth, and reproduction in fruitflies. Proc Natl Acad Sci USA 97:3309–3313.

EVOLUTION AND AGING

Kirkwood TB (1977) Evolution of ageing. Nature 270: 301–304.

Kirkwood TB (2002) Evolution of ageing. Mech Ageing Dev 123:737–745.

Kirkwood TB & Holliday R (1979) The evolution of ageing and longevity. Proc R Soc Lond B Biol Sci 205:531–546.

Partridge L & Barton NH (1993) Optimality, mutation and the evolution of ageing. Nature 362:305–311.

Rose MR (1984) Laboratory evolution of postponed senescence in Drosophila melanogaster. Evolution 38:1004–1010.

Rose MR (1991) Evolutionary Biology of Aging. Oxford: Oxford University Press.

Rose MR, Burke MK, Shahrestani P & Mueller LD (2008) Evolution of ageing since Darwin. J Genet 87: 363–371.

Rose M & Charlesworth B (1980) A test of evolutionary theories of senescence. Nature 287:141–142.

Williams GC (1957) Pleiotropy, natural selection and the evolution of senescence. Evolution 11:398–411.

CELLULAR AGING

4

"INSIDE EVERY OLDER PERSON IS A YOUNGER PERSON WONDERING WHAT HAPPENED."

-JENNIFER YANE, ARTIST (1854-1900)

Cells are the fundamental units of life. Every living organism descends from nothing more than a single cell. Even you, one of the most complex organisms on the planet, began as a single cell, when an egg produced by your mother was fertilized by another single cell, a sperm. All the other cells in your body arose from this single fertilized egg. The study of cell function—cell biology—provides the fundamental concepts that describe the origins of life and the mechanisms that sustain it. Our ultimate understanding of how and why we age will come from the study of cells.

In this chapter, we discuss aging at the cellular level. We explore possible mechanisms underlying cellular aging that lead to a decline in function and determine the life span of the entire organism. We begin with a brief review of essential concepts in general cell biology that pertain to the aging of cells.

THE CELL CYCLE AND CELL DIVISION

Eukaryotic cells duplicate and divide through an orderly sequence of events known as the **cell cycle**. In this section, we review the basics of the cell cycle; in the next section, we'll look at some of the mechanisms involved in cell cycle control.

The cell cycle consists of four phases plus one

The eukaryotic cell cycle has four distinct phases, as shown in **Figure 4.1**: G_1 **phase** (gap 1), **S phase** (synthesis phase), G_2 **phase** (gap 2), and **M phase**. An additional phase, G_0, indicates that the cell has exited the cycle. The G_1, S, and G_2 phases are collectively known as **interphase**. Most of the life of a normal mitotic cell is spent in the G_1 phase.

During G_1, the cell prepares for DNA replication by increasing the concentration of enzymes and other proteins necessary for chromosomal duplication. G_1 also serves as a checkpoint that gives the cell time to determine whether the conditions for replication are optimal. The S phase is the period of time during which DNA replication takes place. G_2 is similar to G_1 in that it prepares the proteins necessary for the next phase, in this case **mitosis**, and determines whether the cell should

Figure 4.1 The cell cycle. The cell cycle consists of four phases: G_1, S phase, G_2 (collectively known as interphase), and M phase. The G_0 phase indicates that the cell has exited the cycle.

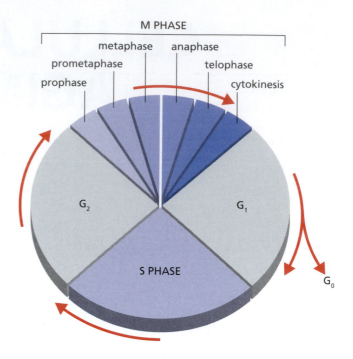

proceed to this next phase. The M phase is the final phase in which one cell becomes two. Cells in the G_0 phase are metabolically active but have left the cell division cycle.

DNA replication occurs during the S phase

DNA must be replicated before the cell divides. This replication process occurs during the cell's S phase. The base pairs of DNA in eukaryotes are in long, linear, double-helical strands contained in the chromosomes. The double helixes are held together by hydrogen bonds that are individually weak but collectively strong. Thus, significant amounts of energy are required to break apart the single strands of the helixes for DNA replication, energy amounts that are greater than could be associated with a single enzymatic reaction. The eukaryotic cell has solved the energy problem by dividing the process of strand separation into thousands of individual reactions, starting at sites along the DNA called **replication origins (Figure 4.2)**. The replication origins tend to be areas with several A–T base-pair sequences (A–T bonds are weaker than G–C bonds).

Replication origins on the double-stranded DNA recruit specialized proteins called **origin recognition complexes (ORCs)**. The binding of the ORCs to the double-stranded DNA marks the end of G_1 and the beginning of S phase. The phosphorylation of proteins within the ORCs breaks apart the complex and causes separation of the DNA strands at the replication origins. The DNA is now ready for duplication.

Cell division occurs during the M phase

Once the cell-division regulatory proteins of G_2 have received the appropriate intracellular signals, the cell moves into the M phase and completes the process of making two new cells. The M phase is divided into six stages **(Figure 4.3)**. The first five stages are traditionally known as **mitosis**. The final stage, **cytokinesis**, is the period in which one cell becomes two. During the first stage of mitosis, **prophase**, **centrosomes** separate and form the **spindle poles** and begin moving apart to opposite sides of the nuclear envelope. As the two centrosomes

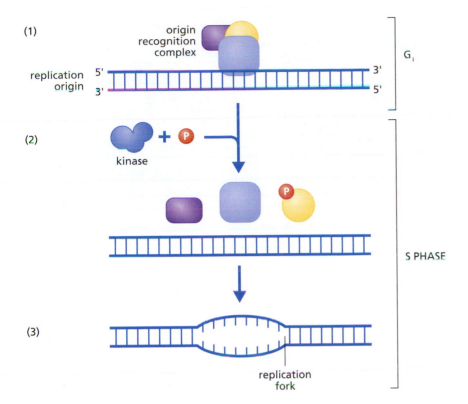

(1)

origin recognition complex

replication origin

G_1

(2)

kinase

S PHASE

(3)

replication fork

Figure 4.2 Preparation of DNA for replication during S phase. (1) Origin recognition complexes (ORCs) bind to replication origins on the double-stranded DNA during G_1. (2) Kinases phosphorylate the ORC proteins, causing degradation of the complex and separation of the DNA strands. (3) Separation of the DNA strands forms replication forks at the replication origin.

move apart, the **mitotic spindle** begins to grow in various directions. **Condensin**, a protein complex, begins to compact the chromosomes during this phase. Next, the nuclear envelope begins to break down, and the spindle attaches to the **sister chromatids** during the second stage of mitosis, **prometaphase**.

Prophase and prometaphase prepare the sister chromatids for separation. Final preparation for chromosome separation begins during **metaphase**, when the chromatids align at the equator, halfway between the two spindle poles. At the same time, the spindle poles begin a sort of tug-of-war, pulling the sister chromatids toward themselves. Actual separation of the chromosomes occurs in **anaphase**, when the **anaphase-promoting complex (APC)** releases proteolytic enzymes targeted at breaking the cohesin linkage that holds the sister chromatids together. Breakdown of the cohesin linkage results in separation of the chromatids and migration of the two new sets of chromosomes toward the spindle poles. During the next phase, **telophase**, the nuclear envelopes develop around the new sets of chromosomes and enlarge. The chromosomes decondense and begin gene transcription of nuclear and cytosolic proteins.

The final stage, cytokinesis, begins when a contractile ring composed of **actin** and **myosin** (the contractile proteins also found in skeletal muscle) forms at the center of the dividing cell. The contractile ring causes a furrow to form perpendicular to the spindle. Actin and myosin can generate significant force; this causes a "pinching" of the furrow and separation of the two new cells. Each new cell contains a full complement of chromosomes and organelles.

REGULATION OF THE CELL CYCLE

Nothing is more important to the living organism than the accurate duplication of its DNA and proper cell division. Inaccuracy in either can lead to altered cellular function and, in some cases, to diseases

Figure 4.3 The six stages of M phase.
(1) Prophase: the two centrosomes separate, and the mitotic spindle begins to expand. (2) Prometaphase: the nuclear envelope begins to break down, and the mitotic spindle attaches to the kinetochores. (3) Metaphase: sister chromatids align at the equator of the spindle, and the spindle begins to contract. (4) Anaphase: the sister chromatids separate to form two new daughter chromosomes, and the mitotic poles move apart. (5) Telophase: a contractile ring forms in the middle of the cell, and a nuclear envelope forms around each of the two new sets of daughter chromosomes. (6) Cytokinesis: the contractile ring "pinches" off the two new daughter cells.

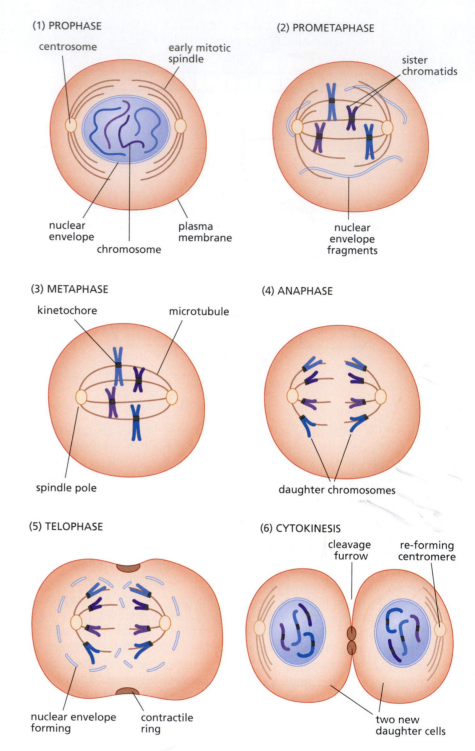

(1) PROPHASE
centrosome
early mitotic spindle
nuclear envelope
chromosome
plasma membrane

(2) PROMETAPHASE
sister chromatids
nuclear envelope fragments

(3) METAPHASE
kinetochore
microtubule
spindle pole

(4) ANAPHASE
daughter chromosomes

(5) TELOPHASE
nuclear envelope forming
contractile ring

(6) CYTOKINESIS
cleavage furrow
re-forming centromere
two new daughter cells

such as cancer. Eukaryotic cells have developed a complex system of regulatory steps to ensure accurate duplication of DNA. Included in this regulation are various "checkpoints" that are designed to evaluate whether the process should continue to the next step. If the regulatory system functions correctly and detects errors in the DNA, chromosome formation and cell division are halted until the errors in the DNA are repaired. Pauses in DNA replication resulting from this checkpoint regulation are often referred to as **molecular brakes**.

In this section, we discuss briefly the molecular mechanisms that underlie the regulation of the cell cycle, a process often called the cell cycle control system. We'll discuss this regulation in the same order as the cell cycle, from the extracellular signals that initiate the cell cycle to the separation of daughter cells during cytokinesis.

S-cyclins and cyclin-dependent kinases initiate DNA replication

DNA replication is initiated by the expression of nuclear proteins known as **cyclins**. Extracellular mitotic signals, through agents such as growth hormones, initiate the cell cycle by inducing the synthesis of nuclear mitotic signal proteins. The type of nuclear mitotic signal proteins will vary, depending on the type of extracellular mitotic signal. That is, many different types of S-cyclin proteins can be expressed. The nuclear mitotic signal protein binds to a **promoter region**, the region in the DNA that "turns on" the gene, inducing expression of the S-cyclin protein **(Figure 4.4)**. The S-cyclin binds to a **cyclin-dependent kinase (Cdk)**, a phosphorylating enzyme always present in the nucleus. The cyclin-Cdk complex catalyzes the phosphorylation of other proteins that signal the nucleus to proceed into the S phase and begin DNA replication. Phosphorylation of proteins as a method for cell signaling is common in eukaryotic organisms. One such signaling pathway, found in the cells of a worm, was the first biochemical mechanism discovered that links reproduction to the length of the life span, as is thoroughly discussed in the next chapter.

Once the cell has entered the S phase, replication of DNA stimulates the expression of cellular enzymes called **proteases**, which separate and degrade the cyclins bound to the Cdk. This final step ensures that DNA replication occurs only once during the cell cycle.

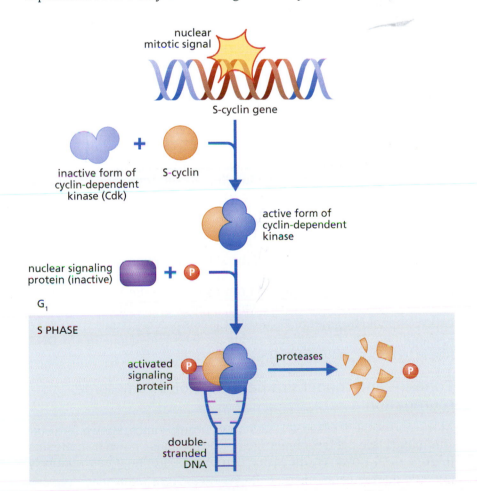

Figure 4.4 Cyclins initiate DNA replication. Extracellular mitotic signals bind to the cell, which, in turn, generates a nuclear mitotic signal that induces the expression of S-cyclin. The binding of S-cyclin to cyclin-dependent kinase (Cdk) activates this enzyme and causes the phosphorylation of other proteins, which signal the cell to enter S phase and begin DNA replication. Once DNA replication begins, the Cdk proteins are inactivated by the proteolytic degradation of S-cyclins.

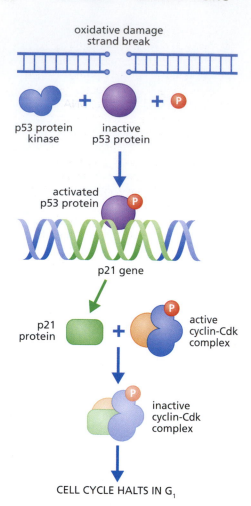

Figure 4.5 The p53 checkpoint of G₁.
When DNA damage, such as damage by the superoxide radical (•O₂⁻), is detected, repair mechanisms activate the p53 protein kinase. The phosphorylated p53 protein binds to the regulatory region of the p21 gene, stimulating expression of the p21 protein. In turn, the p21 protein binds to the cyclin-Cdk complex, deactivating the enzyme and temporarily halting the cell cycle in G₁. Once the DNA has been repaired, proteases degrade the p21 protein and the cell moves into S phase.

The p53 pathway can prevent DNA replication at the G₁-to-S phase transition

As in the other phases of the cell cycle, there are checkpoints to ensure that the DNA is accurately copied during synthesis. One such checkpoint, the **p53 pathway**, helps prevent damaged DNA from being copied. For example, suppose that damage has occurred in a section of DNA due to an attack by a free radical (free radicals are explained later in this chapter). The DNA repair mechanisms detect the error, begin repair of the section, and activate a protein kinase that phosphorylates the p53 nuclear protein. Activation of p53 causes it to bind to the promoter region of the gene for another protein, p21. The p21 protein deactivates the cyclin-Cdk complex through negative inhibition and temporarily halts the cell cycle **(Figure 4.5)**. When the repair is complete, proteases are activated that separate the p21 from the cyclin-Cdk complex, and the cell cycle proceeds to the S phase. The p53 pathway has been implicated as having a role in cell senescence through an interaction with telomere shortening (see below).

Many proteins are involved in the replication of DNA

Once the checks on DNA accuracy are complete and the cyclin-Cdk complex has been activated, replication of DNA can begin. The DNA replication machinery contains several proteins that work in a specific order to replicate the chromosome **(Figure 4.6A)**. One such protein, **helicase**, uses energy released from ATP hydrolysis to separate the DNA into the single strands used as templates for synthesis. Single-strand binding proteins keep the two strands apart. Another protein, **primase**, creates the short sequence of RNA needed to prime the replication process. The helicase and primase make up a replication complex called the **primosome**. When the DNA strands have been separated and primed with RNA, DNA **polymerase** begins synthesis of the new DNA by adding **nucleotides** to the 3′ end of the new strands. Synthesis of a new strand always occurs in a 5′→3′ direction, the only way DNA polymerase can move.

DNA replication moves in both directions from the replication origins. Thus, the primosome creates DNA single-strand templates that are oriented in both the 5′ to 3′ direction and 3′ to 5′ direction (Figure 4.6). The 5′ to 3′ orientation is called the **leading strand template** and the 3′ to 5′ orientation is known as the **lagging strand template**. Since DNA polymerase adds nucleotides only to the 3 end of the growing DNA strand (moving in the 5′ to 3′ direction on the parent strand) synthesis of the new **leading strand** is relatively straight forward. However, how does DNA polymerase replicate a new strand from the parent lagging strand that is oriented in the 3′ to 5′ direction? The replication machinery solves this problem by moving backward on the 5′→3′ lagging strand template (Figure 4.6B) to produce the **lagging strand**. As DNA polymerase moves backward, it produces small DNA segments called **Okazaki fragments** (Figure 4.6C). This process requires repeated RNA priming. The Okazaki fragments are later joined together by **DNA ligase** to form a continuous strand.

Cohesins and condensins help control chromosome segregation

At the end of the S phase, the two new chromosomes, known as sister chromatids, are tightly bound together by proteins known as **cohesins** **(Figure 4.7)**. However, the sister chromatids are large and relatively

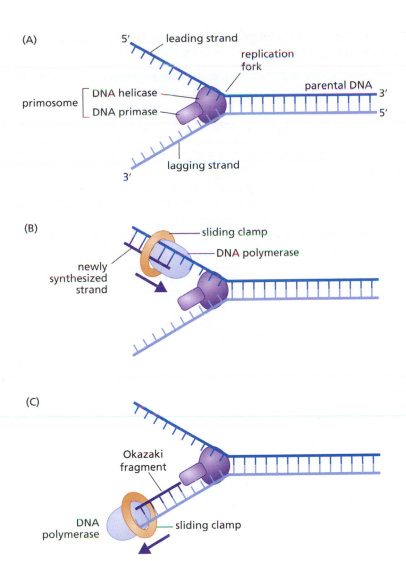

(A)

5′   leading strand

replication
fork

DNA helicase

primosome

DNA primase

parental DNA
3′

5′

lagging strand

3′

(B)

sliding clamp

DNA polymerase

newly
synthesized
strand

(C)

Okazaki
fragment

DNA
polymerase

sliding clamp

Figure 4.6 DNA replication.
(A) Helicase separates the double helix of the DNA into single strands. DNA primase prepares the RNA primer needed to begin the replication process. (B) On the leading strand template, DNA polymerase synthesizes the new strand in a 5′→3′ direction and needs only one primer to start the process. (C) On the lagging strand, DNA polymerase moves backward, producing small DNA segments called Okazaki fragments. The Okazaki fragments are "backstitched" together to complete the new DNA strand. The sliding clamp holds the DNA polymerase on the parent strand while allowing for its movement along the template.

unorganized, a physical condition that inhibits efficient separation. Proteins known as **condensins** are assembled in the G_2 phase and used in the M phase to reduce the size of the chromatid and enhance the strand-separation process. Another complex that is important for the proper separation of the chromosomes is the **centromere**. The centromere holds the two chromatids together and is the site where the **kinetochore** forms, the structure that attaches the centromere to the mitotic spindle.

The metaphase-to-anaphase transition marks the final checkpoint in the cell cycle

Regulation of the transition between the G_2 and M phases occurs through the cyclin-Cdk complex. This process is similar to that described in Figure 4.4; however, M-cyclin rather than S-cyclin is expressed. Once this cell cycle control point is passed, the M phase begins, and the final regulatory step in the cell cycle occurs at the metaphase-to-anaphase transition. Whereas the cyclin-Cdk complex uses a phosphorylation-dephosphorylation regulatory system, the metaphase-to-anaphase transition involves a **proteolytic process**. The primary protein of this process is the **anaphase-promoting complex (APC)**, a member of the **ubiquitin ligase** family of enzymes. Many enzymes in this family participate in reactions that mark improperly folded proteins for destruction.

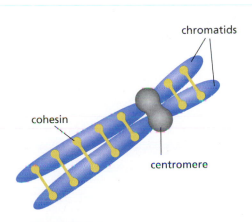

chromatids

cohesin

centromere

Figure 4.7 Sister chromatids as they appear at the end of S phase. The centromere holds the two sister chromatids together. Cohesins also help to hold the chromatids together.

As you will learn in Chapter 9, ubiquitin ligase may have a significant role in the development of Alzheimer's and Parkinson's diseases.

The APC marks for destruction the protein **securin**, which is bound to the cohesins holding the sister chromatids together. Securin prevents proteases from breaking apart the sister chromatids before metaphase. Removal of securin from the cohesins allows proteases present in the cell to separate the sister chromatids. In turn, the mitotic spindle begins to contract toward the centrosomes, and the independent chromatids begin to move apart. The APC also marks for destruction any remaining cyclins, once again ensuring that DNA replication occurs only once. The cell cycle now progresses unimpeded through telophase and cytokinesis, producing two new, identical cells.

Fully functional cells can exit the cycle at the G_0 phase

Initiation of the cell cycle in multicellular organisms appears to occur only when the cell receives an extracellular signal from substances called **mitogens**. Mitogens come in a variety of forms, including hormones, neuropeptides, and steroids, and they arise from locations anatomically distant from the cell or as factors secreted by neighboring cells. If a cell does not receive a mitogenic signal at the right time, it will dismantle the cell cycle control mechanisms and enter a modified version of the G_1 phase known as G_0. The G_0 phase allows the cell to carry out its physiological role without having to divert energy to cell division. Some adult cells can re-enter G_1 when given the appropriate mitogenic signal. The transition between G_0 and G_1 can occur quickly and continually, as is observed in human intestinal cells (3- to 4-day turnover rate), whereas other cell types remain mostly in G_0 and divide infrequently (liver cells may divide only once per year). Although fully mitotic cell types outnumber post-mitotic cell types, the absolute number of cells in multicellular organisms that remain permanently in G_0 (that is, are post-mitotic) is much greater.

TABLE 4.1 provides examples of cell types with various lengths of stay in G_0. Fully mitotic cells can replicate throughout the adult life span; semi-mitotic cells replicate infrequently; and post-mitotic cells do not replicate in the adult. The mechanisms underlying the transition between G_1 and G_0 are not well understood but are thought to play a key role in replicative senescence.

TABLE 4.1 EXAMPLES OF FULLY MITOTIC, SEMI-MITOTIC, AND POST-MITOTIC MAMMALIAN CELLS

Fully mitotic cells	Semi-mitotic cells	Post-mitotic cells
Fibroblasts	Liver cells	Heart cells
Glial cells	Visual cells	Skeletal muscle cells
Keratinocytes	Hair follicles	Brain cells
Vascular smooth muscle cells		
Lens cells		
Endothelial cells		
Lymphocytes		
Intestinal cells		
Germ cells		
Stem cells		

REPLICATIVE SENESCENCE

Theories of cellular aging predict that altered cellular function regulates or "drives" an organism's aging. Because organismal aging reflects a gradual decline in function after the reproductive period, the decline and eventual loss of cell division has generally been viewed as characterizing cellular aging, a period known **replicative senescence**. In this section, we consider whether cells have a finite life span and, if so, how this affects whole-organism senescence or longevity.

A mistake delayed the discovery of cell senescence for 50 years

In 1912, Alexis Carrel took a piece of chicken heart and placed it in a container on top of a concoction of chicken plasma and liquefied chicken embryonic tissue. The plasma contained the proteins that formed the matrix on which cells could grow. The liquefied embryonic tissue provided the nutrients necessary for growth, although the exact nature of the components in the fluid was not known. After a couple of days, new cells appeared on the piece of chicken heart. With the help of Montrose Burrows, Carrel became the first person to successfully grow normal cells outside an animal's body.

Over the next few days, the growth of the culture slowed and then stopped, when the liquefied chicken embryonic tissue was exhausted and the cells ran out of space. Carrel removed some of the cells, placed them in a new container, and added fresh chicken embryonic tissue and plasma. As before, the cells in the new culture divided until they reached the carrying capacity of the dish. The cell culture that was started in 1912 was cultured continuously until Carrel's death in 1946, leading to the dogma that cells were immortal. However, scientists learned that Carrel's method for liquefying chicken embryonic tissue did not remove all the live cells from the serum. That is, the cell culture was not immortal; it was simply a culture to which new cells were being added with each feeding. Despite this, the dogma that cells grown *in vitro* are immortal was widely accepted and rarely challenged for almost 50 years. Indeed, so strong was the belief that cells grown in culture were immortal that evidence suggesting otherwise was viewed as being the result of technical error and universally attributed to a human mistake. Thus, it was understandable why, in 1961, Leonard Hayflick and Paul Moorhead, who worked at the Wistar Institute in Philadelphia, were cautious about reporting their findings that *in vitro* cultures of human embryonic fibroblasts were dying after several population doublings.

Hayflick's and Moorhead's research findings created the field of cytogerontology

Hayflick and Moorhead were interested in the underlying biochemical mechanism that transformed normal cells into malignant, cancerous tumors. They chose to work with human embryonic fibroblasts (which was legal in 1961) because, as they reasoned, these cells would not have had significant environmental exposure that might confound the intrinsic biological mechanism(s) responsible for the development of cancer. They no longer grew cells directly on chopped-up pieces of tissue, as did Carrel. Instead, cells from embryonic tissue were separated from each other by the connective tissue–digesting enzyme **trypsin** and placed on a growth medium in the culture flask **(Figure 4.8)**. The cells were then suspended in a medium containing the necessary nutrition and cellular growth factors.

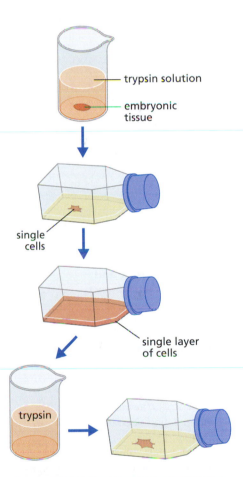

trypsin solution

embryonic tissue

single cells

single layer of cells

trypsin

Figure 4.8 Method used to determine the life span of cells in culture. Fibroblasts are separated into individual cells by trypsin digestion. The cells are then placed into a culture flask containing cell-free fetal bovine serum. When the cells completely cover the medium in the flask and there is no more room to grow (confluence), they are again digested with trypsin, and a few of the detached cells are re-plated as before. Each passage of cells from trypsin digestion to confluence is referred to as one population doubling.

After a few months of working on the experiment, and after several population doublings, Hayflick and Moorhead found that some of the cells were no longer dividing. Meticulous observation showed that subcultivations would cease to divide after 40–60 population doublings. In subsequent years, the finite number of cell population doublings would be known as the **Hayflick limit**. However, the dogma that cells were immortal was so ingrained in the thinking of biologists that the scientific community did not accept this first experimental evidence of finite cell population doubling. Hayflick and Moorhead received criticism from experts in the field, suggesting that their cell cultures might have chromosomal damage causing the death of the cell population and that this damage might be greater in one sex than the other. In answer to this criticism Hayflick and Moorhead performed one more experiment, in which chromosomes from both sexes were evaluated for structural integrity and whether gender had an effect on survival. In Hayflick's own words, here is what they did **(Figure 4.9)**:

> . . . we mixed equal numbers of human female cells [determined by sex chromosome analysis] at the tenth subcultivation level (young cells) with human male cells at the fortieth subcultivation level (old cells). We also continued to subcultivate the unmixed cell cultures of each sex. . . . After about thirty more subcultivations we looked at all three cultures. In the mixed culture we found only female cells present. The male cells in the mixture had died several weeks before, when they had reached their maximum of fifty subcultivations. The pure male cell culture had died several weeks before we decided to look at the mixture. The pure female cell culture, like the female cells in the mixture, were still luxuriating. (Hayflick & Moorhead 1961)

Little doubt was left that cells grown *in vitro* did indeed have a finite life span. The prediction of an evolutionary theory of senescence with a finite life span for cells was, at least for cells in culture, accurate. So important were these results that a new field of study was born—**cytogerontology**, the study of cell senescence (now known as replicative senescence).

Cells in culture have three phases of growth

Beyond proving that cells were mortal, Hayflick and Moorhead's experiments established, for the first time, the life-span characteristics of cells in culture. Cell cultures derived from a small population of

Figure 4.9 Hayflick and Moorhead's experiment to determine whether cells grown in culture have a finite life span. Hayflick and Moorhead tested three different cultures. (A) A mixture of equal numbers of human male cells (*blue*) in their 40th population doubling and human female cells (*red*) in their 10th population doubling. After an additional 30 population doublings, only the female cells continued to divide. (B) Female cells started at the 10th population doubling and kept for 30 additional population doublings. Female cells were still dividing after 40 population doublings. (C) Male cells started at the 40th population doubling. No viable male cells were left after 30 additional population doublings.

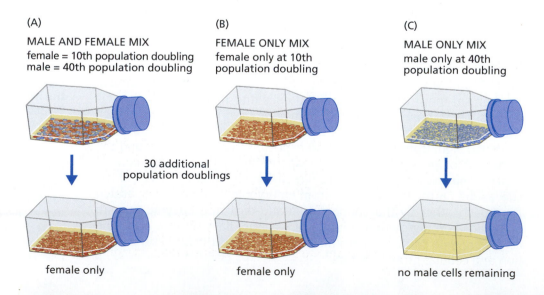

(A)
MALE AND FEMALE MIX
female = 10th population doubling
male = 40th population doubling

(B)
FEMALE ONLY MIX
female only at 10th
population doubling

(C)
MALE ONLY MIX
male only at 40th
population doubling

30 additional
population doublings

female only

female only

no male cells remaining

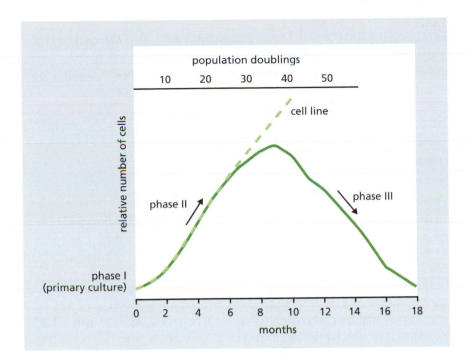

Figure 4.10 Life history of mitotic cells in culture, as originally described by Hayflick and Moorhead. Phase I begins with the primary culture and is characterized by a slow rate of proliferation and population doubling. Cells in phase II grow rapidly to a peak at about 75% of their maximal population doubling. Cells in phase II can immortalize spontaneously or through an intervention and become a cell line. An increase in population doubling time and the death of the culture occur in phase III. (From L. Hayflick and P.S. Moorhead, *Exp. Cell Res.* 25:585–621, 1961. With permission from Elsevier.)

homogeneous embryonic cells grow in three distinct phases, as shown in **Figure 4.10**. Phase I is characterized by a rather slow growth period occurring immediately after the cells are removed from the donor and plated on the first growth medium. This slow growth may last for 2–3 months, during which the first 10–12 population doublings may occur. Phase I most likely reflects a period of adaptation of the cells to the new *in vitro* environment. The slow growth of phase I gives way to the rapid and constant proliferation rate of phase II, which may last for an additional 8–9 months of cell division (30–40 population doublings). During phase II, mutations can be introduced that change the culture from one with a finite life span to one with an infinite life span. Finally, the cell culture enters phase III, a period of declining cell proliferation rate that ends in the loss of cell division. Hayflick and Moorhead originally proposed that phase III ends with the culture's loss of proliferation at about 12 months. Subsequent studies found that a subpopulation of cells will cease replication but remain physiologically viable for extended periods of time.

The criterion for the end of the replicative life span of a cell population in culture has been widely accepted as the failure of the population to double within four weeks. This definition does not mean that all cells in the population have lost the capacity to divide, just that the majority have. Cells, like whole organisms, age at non-uniform rates. Indeed, in old, non-doubling populations, many individual cells retain the capacity for division. (The opposite is also true: young populations that have doubled only a few times have some cells that are no longer capable of division.) If clones are removed from the flask in which the population has not doubled within four weeks, new populations can be cultivated that have the replication characteristics of phase II. That is, the cell population defined as having reached the end of its proliferative life span may contain individual cells that are still capable of division. In fact, some of the clones will start populations that have the same proliferative life span as the population from which they were derived. Much like the life-span characteristics of a human population, cell populations have significant variation in individual rates of aging and individual life span.

TABLE 4.2
FETAL FIBROBLASTS' POPULATION DOUBLING RATES AND MAXIMUM LIFE SPAN FOR SEVERAL SPECIES

Species	Population doublings	Species' maximum life span (years)
Bat	16–29	3–10
Chicken	35–40	6–10
Horse	30–40	35–40
Human	45–60	115–120
Mouse	12–15	4–5
Rabbit	21–27	10–15

Interspecies comparisons suggest that population doubling rates vary with the species **(TABLE 4.2)**. Moreover, population doubling rates do not seem to be strongly correlated with length of life. The common laboratory mouse lives an average of 3–4 years, with fibroblasts doubling about 15 times. Human beings have a maximum life span of 115–120 years and have fetal fibroblasts that double 45–60 times **(Figure 4.11)**. However, the chicken, with a maximum life span of 6–10 years, has a population doubling time closer to that of the long-lived human than the short-lived mouse.

Not all cell types have a finite population doubling time. Fibroblasts isolated from some mouse tissues display infinite population doubling while in phase II growth, a process known as spontaneous immortalization. Primate and avian cells rarely immortalize spontaneously in culture, but they can become immortal through manipulation of their genomes. For example, transfection (the introduction of DNA from the cells of one species or virus into a cell of a different species) of human or other primate fibroblasts by the Simian virus 40 (SV40) T-antigen gene leads to immortalization of the population. As discussed in **BOX 4.1**, cancer cells also undergo an infinite number of population doublings. Cells that are immortal are referred to as **cell lines** and have found wide use in biology because of their ease of replication and uniform genetic makeup.

Senescent cells have several common features

Replicative cessation is the major characteristic of late-passage cells—cells that are part of a population approaching its Hayflick limit—but several other features are also commonly observed in senescent cell populations, including morphological alterations, cell function alterations, arrest in cell division, and immunity-associated functions **(TABLE 4.3)**. Senescent cells typically display cellular enlargement, including

TABLE 4.3
PHENOTYPE OF A SENESCENT CELL POPULATION

Arrest in cell division	Lengthening of cell cycle 　　Molecular brakes at G_1-S phase 　　Cells remain responsive to extracellular mitogenic signals
Cell function alterations	Decrease in proteins associated with DNA replication Decrease in RNA synthesis and associated proteins Decrease in overall rate of protein synthesis
Immunity-associated function	Increase in cellular "junk"—remnants of nonfunctional proteins 　　Decrease in the general function of the cellular catabolic machinery 　　Increase in secretion of pro-inflammatory cytokines
Morphological alterations	Enlarged cells 　　Multinucleated cells 　　Breakdown of extracellular matrix

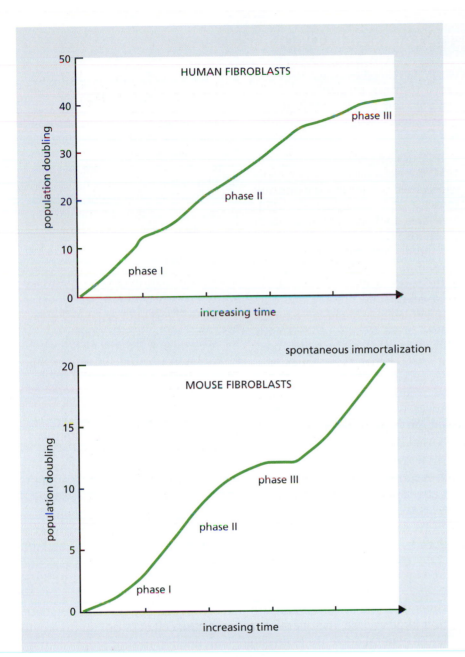

Figure 4.11 Fetal fibroblast doubling times in hypothetical populations of human and mouse cells. Human fibroblast cultures have a maximum of 45–50 doublings. Conversely, the pace of cell division in mouse fibroblasts slows after 15–20 doublings, and then the cells often spontaneously immortalize.

an increase in size of the nucleus and multinucleated cells. Moreover, the distance between cells becomes greater, due to an increase in the cellular secretion of extracellular matrix proteases and collagenase. The decrease in viable extracellular matrix proteins may also cause a slowing of replication, as the cell will not have sufficient material to anchor itself.

The slowing of replication seems to be due to a lengthening of the cell cycle that most likely reflects the activation of molecular brakes at the G_1-S phase interface. This inhibition of replication in senescent cell populations during G_1 is supported by the observation that the intracellular mitogenic signal remains responsive to extracellular mitogens. Such observations have led to the suggestion that cell senescence may be caused by errors in the replication mechanism, although consistent data supporting this theory have yet to be obtained.

Cells in a senescent population undergo a general decrease in function that includes decreases in the rate of DNA, RNA, and protein synthesis.

BOX 4.1 THE CELLS OF HENRIETTA LACKS: HERS OR OURS?

Techniques employed by modern molecular cell biologists have provided immortal cell lines for just about every conceivable research question. Cell lines constructed by genetic engineering or started from malignant tumors can be obtained commercially or through cell-line repositories maintained by the national government. There was a time, however, when such cell lines did not exist. That all changed on a winter's day in 1951.

Henrietta Lacks, 34, mother of five, arrived at a Baltimore hospital on February 9, 1951, to begin radiation treatment for a malignant cervical tumor. Before the physician covered the tumor with radium, George Gey, a resident at the hospital, requested a small sample of the tumor. Gey was interested in how cancer cells grow, and he had not been able to culture a long-term colony of cells from other tumor types. Henrietta Lacks's cervical tumor cells were different. Within hours after the cells were placed on a growth medium, they began to grow. Gey soon found that he could culture these cells indefinitely, and he began to offer subcultivations,

which he named HeLa for *Henrietta Lacks*, to other scientists. Indeed, it was HeLa cells that Jonas Salk used to test his polio vaccine prior to human testing.

Henrietta Lacks died on October 4, 1951, but HeLa cells remain the most widely used cell line in the world and continue to provide scientists with a reliable tool for specific research questions. Oddly, however, it was not until 1975 that David Lacks, Henrietta's husband, and the rest of the Lacks family found out about HeLa cells. Quite possibly, the only reason the family found out was that a need arose for the Lacks family's DNA. HeLa cells were so widely used that they began to contaminate other cell lines. Researchers needed DNA from Henrietta Lacks's close relatives to develop an assay that could distinguish HeLa cells from other cell types.

There is no doubt that many people and companies have profited from the use and/or sale of HeLa cells. The Lacks family, however, has not realized a penny from Henrietta's cells, nor is it likely they ever will.

The US Supreme Court ruled that cells removed from an individual are no longer the property of that individual. Rather, the cells and/or tissue belong to the person responsible for the removal of the cells. Particulars as to who will profit from the sale of the tissue must be specified at the time of informed consent. Because there was no informed consent when Henrietta's cells were removed, the Lacks family is not entitled to compensation.

The cells of Henrietta Lacks must be given credit for much of our understanding of cell biology. In addition, her cells are directly responsible for spawning a debate about who owns the tissue after the donor has died. This debate, while still ongoing, has led to regulation concerning the rights of donors. Patient's and donor's rights are now integral parts of the scientific landscape, and we have Henrietta Lacks's cancer cells to thank for that. The discussion about ethical treatment of tissues and cells will become even more important as genetic engineering techniques progress to the ultimate point of human cloning.

Given that the synthesis rates of these vital cell components regulate, to a large degree, the overall function of the cell, it is not surprising to find a general decrease in physiological function in senescent cell populations. Moreover, senescent cell populations are characterized by an increase in intracellular "junk." This so-called junk reflects a decrease in the ability of senescent cells to break down and metabolize nonfunctional proteins. Finally, senescent cells seem to have an increased secretion of proteins, such as inflammatory cytokines, normally found in cells that have been injured.

Replicative senescence can be used to describe biological aging

From the first description of cellular senescence provided by Hayflick and Moorhead, many scientists have questioned the value of an *in vitro* model as an appropriate descriptor of aging in whole organisms. While there can be little doubt that an organism's aging must reflect cellular dysfunction, the fact remains that most cells in adult eukaryotic organisms are post-mitotic. Thus, it is not clear how a model of aging based on the cessation of cell division adds valuable information about a process that occurs mainly in the absence of cell division. Although supporters of replicative senescence often point to the fact that population doubling rates correlate with a species' life span, these correlations are weak at best and do not occur in all species. In addition, mitotic somatic cells taken from old individuals can often show robust population doubling rates.

Most biogerontologists would agree that cellular replicative senescence, per se, does not directly cause aging in organisms that are primarily post-mitotic. Evaluating the relationship between cellular replicative senescence and organismal aging is not, however, the only value of using cell cultures as a model of aging. Rather, cell cultures can provide researchers with a highly controllable and predictable environment in which to describe a variety of aging events at the cellular level that are not directly related to the population's replicative history. We are just beginning to see results from investigations using the cellular replicative senescence model to evaluate alterations in membrane function, mitochondrial damage, and other phenomena related to rate of aging rather than longevity. One exciting area of research where replicative senescence has the potential for offering significant value is the interface between aging and disease. These investigations, only recently begun, are asking the important question, "What are the molecular mechanisms that underlie a cell's transition between normal function and physiological states that increase the risk of disease?"

THE CAUSE OF CELLULAR AGING: ACCUMULATION OF DAMAGED BIOMOLECULES

The mechanisms underlying replicative senescence are not yet known and remain at the theoretical level. Theories providing an explanation for the process of replicative senescence fall into two categories, depending on the process being described and the evolutionary evidence underlying that mechanism. If mechanisms leading to loss of function with age are the main concern, replicative senescence will be described by random or stochastic events. If cell longevity is the primary concern, there should be genetic or programmed mechanisms leading to the death of the cell. In later sections, we explore random-event mechanisms ("Oxidative Stress and Cellular Aging") and a programmed theory of replicative senescence ("Telomere Shortening and Replicative Senescence").

Most biogerontologists agree on one point: cellular aging is the result of an intracellular accumulation of damaged proteins. Therefore, in this section, we explore how and why the accumulation of damage to biomolecules occurs and why this damage may lead to altered rates of aging.

Biomolecules are subject to the laws of thermodynamics

Like all matter in the universe, biomolecules are subject to the **laws of thermodynamics**, physical laws governing the relationship between work, energy, and heat. The **first law of thermodynamics** applies to the conversion of energy from one form to another. When converting energy from one form to another, the total energy of the system before and after the conversion remains the same. That is, energy is neither created nor destroyed. In the conversion of energy from one form to another, we run up against the second law of thermodynamics: the transfer of energy from one form to another is not 100% efficient, and some of the energy becomes unusable. Unless the system receives an input of new usable energy to replace the amount of unusable energy lost as heat, the system will become randomly rearranged—that is, will move in the direction of disorder (entropy). It is this movement toward increasing entropy that causes the accumulation of damaged proteins that leads to aging.

The total energy of a biological process is called **enthalpy (*H*)** and equals usable energy (**free energy, *G***) plus unusable energy (**entropy, *S***). This relationship is described in Equation 4.1.

Total energy of a system, $H = G + TS$ (4.1)

Where

H = enthalpy, the total energy in a system

G = usable energy, or free energy

S = unusable energy, or entropy

T = temperature of the system

Because the energy resides in the bonds of molecules, H, G, and S cannot be measured directly in a living biological system. We can, however, measure the *change*, represented by the Greek letter delta (Δ), in each quantity, as long as we know the temperature at which the reaction occurs. We start by determining whether a reaction releases or consumes usable energy. That is, $\Delta G = G_{products} - G_{reactants}$. If the reaction increases usable energy, the sign of ΔG is positive; if the reaction decreases usable energy, the sign of ΔG is negative. Since the first law tells us that the total energy of a reaction does not change, any change in G must be accompanied by an equal and opposite change in H and S, but H cannot be measured. Therefore, we can measure ΔG and ΔS and solve for ΔH (**Equation 4.2**).

Change in usable energy of a chemical reaction,

$\Delta G = \Delta H - T\Delta S$ (4.2)

Where

ΔG = change in free energy of a reaction ($G_{products} - G_{reactants}$)

ΔH = total amount of energy added to or released by the system

ΔS = change in entropy

T = temperature of the system

Life requires the constant maintenance of order and free energy

Biological processes use molecules with significant usable energy (highly ordered) in reactions that result in an organism's productivity, growth, and repair. In these reactions, and because of the second law, unusable energy is released, which lowers the amount of usable energy needed to carry out additional biochemical reactions that sustain life. As usable energy decreases, entropy and the disorder of the system (or unusable energy) must increase, according to the first law. To restore the order, and thus maintain life, organisms must constantly supply new usable energy to replace that lost to entropy. Let's use the contraction of a muscle to illustrate this principle of thermodynamics in biological processes.

Muscle cells (muscle fibers) convert the chemical energy of **adenosine triphosphate (ATP)** into mechanical energy called contractions **(Figure 4.12)**. The conversion of ATP to ADP (adenosine diphosphate) + P_i (inorganic phosphate) releases usable energy that drives the contraction. Because of the second law of thermodynamics, unusable energy in the form of heat is also released. In fact, 80% of the energy derived

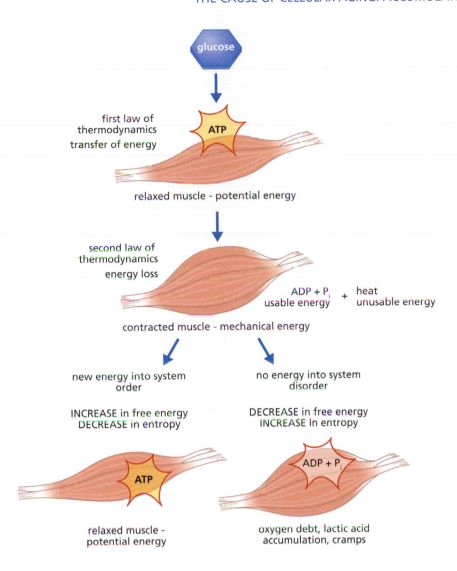

Figure 4.12 The first and second laws of thermodynamics applied to muscle cells. Muscle cells convert glucose energy into ATP energy in preparation for contraction, as dictated by the first law of thermodynamics. The conversion of ATP to ADP + P_i releases usable energy that drives contraction. Because of the second law of thermodynamics, heat is also released. If more glucose enters the system, replacing energy lost to contraction, the muscle relaxes in preparation for the next contraction. If no glucose is available for conversion to ATP, the muscle remains in the contracted state.

from breaking the phosphate bond is released as heat; entropy and disorder increase and usable energy decreases with every contraction of a muscle fiber. If the system provides a mechanism that replaces the usable energy lost to entropy, then the muscle can relax in preparation for the next contraction. That is, the order lost to entropy is restored, as seen in Figure 4.12, with conversion of ADP to ATP (that is, more bonds, more usable energy, more order). Conversely, if that energy lost to entropy cannot be restored, the muscle fiber remains in the contracted state, resulting in a lack of oxygen, and cells will begin to die. If additional free energy never comes—that is, if we are dead—the laws of thermodynamics work to completion and the muscle fiber achieves energy equilibrium.

The mechanism underlying aging is the loss of molecular fidelity

Recall that the atomic structure of a molecule determines its functions. We know this principle as the structure-function relationship. If even one amino acid in a protein does not appear in the proper sequence, then the biological activity of that protein can be reduced. Maintaining the proper structure of a molecule, known as **molecular fidelity**, is critical for survival of the organism and requires a constant input of usable energy. In other words, organisms are constantly battling the increase in entropy, and successful organisms are those that win the battle by maintaining molecular fidelity.

The ability to restore the free energy lost to entropy and maintain order requires an enormous investment of resources by the organism in order to survive to reproductive age. At the very basic level, successfully surviving to reproductive age simply reflects genes that have been selected to maintain the molecular fidelity of proteins. However, as you learned in Chapter 3, there would be no reproductive advantage to maintaining this high level of investment in resources after reproductive age has been achieved. The laws of the universe would eventually rule, and the slow march toward energy equilibrium would lead to a loss in molecular fidelity, disorder of the cell, and a decrease in cell function due to accumulation of damaged proteins. Cellular aging reflects the accumulation of damaged proteins caused by the basic laws of the universe—the disorder of increasing entropy.

Aging reflects the intracellular accumulation of damaged biomolecules

There is now overwhelming evidence to indicate that the passage of time results in changes to molecular structure. While the mechanisms underlying these changes remain unclear, we can say with confidence that the cellular accumulation of damaged biomolecules "speeds up" after sexual maturation. Since structure determines function, the older cell becomes less efficient in carrying out normal operations. We can think of somatic cellular aging in terms of a balancing act between the maintenance of molecular fidelity (order) and entropy (disorder) **(Figure 4.13)**. Due to the force of natural selection, genes have been selected that maintain molecular fidelity during development, a period in which highly functional cells are critical. With the passage of time, the extrinsic rate of aging, in accordance with the disposable soma theory (see Chapter 3), causes damage to biomolecules. Because the proteins that help to maintain molecular fidelity are also subject to the second law, the ability to repair or replace damaged biomolecules declines, resulting in the intracellular accumulation of damaged biomolecules over time. When entropy begins to outpace the maintenance of molecular fidelity, the cell can no longer maintain normal function, and cell death occurs.

We have been purposely vague about the precise type of damage that may cause the molecular infidelity leading to cellular aging. The reason for this is that the exact type of damage occurring to the cell is largely unimportant to the process of aging. Both the type and amount of damage vary considerably among cell types, and no single type of damage process can be considered the primary cause of cellular aging. For example, in this chapter we use damage caused by oxygen-centered free radicals to describe how random events in normal metabolism lead to the accumulation of damage and the appearance of aging. Other chapters describe how the addition of glucose to proteins, known as **glycosylation**, changes the protein's structure and leads to accumulation of damage. In Chapter 9, you will learn that the misfolding of proteins can result in insoluble aggregates that may be the precursors to neurological disease. It would be difficult to demonstrate that age-related damage caused by oxygen-centered free radicals is more important to the aging process than are glycosylated proteins or insoluble protein aggregates.

The important message to take away from this discussion of the cause of the accumulation of damage is not what type of damage accumulates

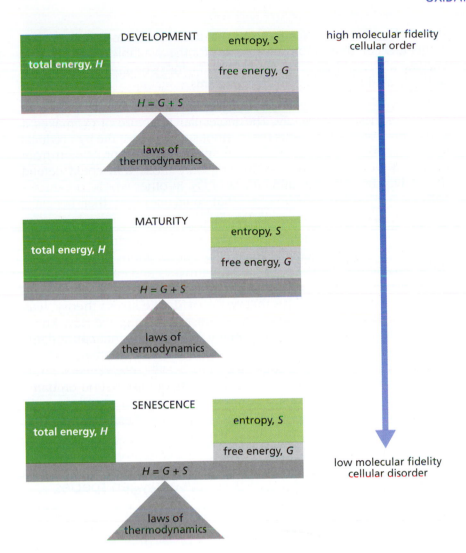

Figure 4.13 Molecular fidelity and energy balance during the life span. Molecular fidelity and cellular order are maintained during the run-up to reproductive age (development). Replacement of usable energy outpaces the amount of energy lost to entropy. As we age beyond the start of reproduction (maturity), increasing entropy leads to a loss in molecular fidelity and the accumulation of damaged proteins—cellular disorder. Senescence is characterized by low molecular fidelity and increased cellular disorder, as entropy increases faster than the introduction of new usable energy.

but that cellular damage occurs in accordance with the basic laws of the universe. As such, we have no more control over the deteriorating effects of aging within the human body than we do with any other object in the physical world. We may be able to stop one type of damage, only to find that another type has started. This will continue until the cell succumbs to the second law. Death of the organism occurs when enough cells of a particular organ or organs have stopped functioning as a result of molecular infidelity. You will learn in Chapter 10 that there are ways in which we can slow the rate of aging and live longer, healthier lives, but aging will eventually prevail and the universe will claim us.

OXIDATIVE STRESS AND CELLULAR AGING

Somewhere between 2.5 and 3 billion years ago, cells of blue-green algae (cyanobacteria) began to convert atmospheric carbon dioxide (CO_2) into glucose, using the energy of sunlight. During this process, known as photosynthesis, oxygen (O_2) was released into the atmosphere. As the O_2 began to accumulate, two critical events occurred that led to a greater diversity of life. First, the radiant energy of the sun converted O_2 into ozone (O_3), which accumulated in the upper atmosphere and trapped harmful ultraviolet rays. Life was now able to leave the protection of the oceans and take up residence on land. Second, because oxygen-based metabolism, known as **aerobic metabolism**, is much more efficient than metabolism without oxygen (**anaerobic** metabolism), single-cell organisms that used aerobic metabolism grew bigger and competed more successfully for the available resources. The success of these single-cell aerobic organisms over anaerobic cells led to the evolution of multicellular organisms.

Unfortunately for many organisms, the accumulation of oxygen in the atmosphere was deadly. The molecular structure of O_2 makes it extremely toxic to organisms that cannot safely reduce the by-products of aerobic metabolism to water. Organisms without protection from oxidation would soon die off, leaving only those that could defend themselves from the harmful effects of O_2. In other words, organisms with the highest fitness were those that could use O_2 for their benefit.

In 1956, Denham Harman theorized that as cells age, some by-products of aerobic metabolism known as **oxygen-centered free radicals**, products of oxygen metabolism having one or more unpaired electrons, escape the normal degradation pathways and cause damage to biomolecules. In turn, these damaged molecules accumulate in the cell and lead to cellular aging. His theory, the **oxidative stress theory**, was once thought to be the primary mechanism of aging. We now know that oxidative stress is but one of many mechanisms that cause damage to a cell that leads to age-related dysfunction, but damage caused by oxidative stress remains one of the most heavily researched areas in biogerontology. Because of the vast amounts of information on damage caused by oxygen stress, we are using this process as one example of a random event that leads to the accumulation of cell damage. This process should not be seen as the primary or the only cause of cellular damage leading to aging.

Oxidative metabolism creates reactive oxygen species

Before explaining the nature of oxygen-centered radicals, it might be helpful to briefly review oxidation-reduction reactions that convert organic fuels into cellular energy. The flow of electrons from one compound to the next forms the basis of how we create usable cellular energy from the nutrients we ingest—fats, **carbohydrates**, and proteins—and determines the reactivity of a compound. A substance is oxidized when it loses one or more electrons; it is reduced when it gains one or more electrons. In any reaction involving electron flow between two substances, oxidation and reduction have to work together. When one substance loses an electron (oxidation), the other substance gains an electron (reduction). Oxidized compounds tend to be highly reactive, because they are seeking electrons from other compounds to stabilize their electron configuration. Reduced compounds tend to be more stable than oxidized compounds, because their electron configuration is closer to the ground state.

Oxidation-reduction in biological processes can be characterized by the loss or gain of oxygen and hydrogen. If a compound gains oxygen or loses hydrogen, it has been **oxidized**. If it loses oxygen or gains hydrogen, it has been **reduced**. In biochemistry, oxidation-reduction of compounds is usually associated with a third molecule in processes known as coupled oxidation-reduction reactions. **Figure 4.14** illustrates the coupled oxidation-reduction involving pyruvate and lactate. Lactate is oxidized (loss of hydrogen) to form pyruvate through the reduction (gain of hydrogen) of **nicotinamide adenine dinucleotide (NAD$^+$)** to NADH + H$^+$. Conversely, pyruvate is reduced to lactate when NADH + H$^+$ is oxidized to NAD$^+$.

This brief description of oxidation-reduction reactions might also help you to understand how antioxidants work, an important concept in

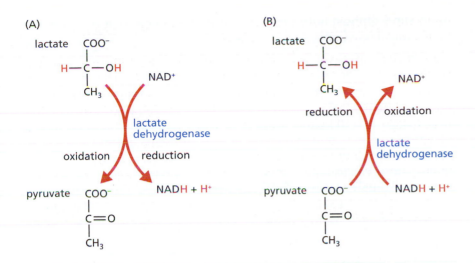

(A)   (B)

Figure 4.14 Coupled oxidation-reduction reactions involving pyruvate and lactate. (A) In the oxidation of lactate to pyruvate, two hydrogens (*red*) of lactate are stripped by the oxidized form of nicotinamide adenine dinucleotide (NAD^+), creating the reduced form, $NADH + H^+$. (B) Conversely, pyruvate is reduced to lactate with the addition of two hydrogens from the reaction involving the oxidation of $NADH + H^+$ to NAD^+.

radical chemistry. Recall that oxidized compounds have fewer electrons than reduced compounds, making them more reactive. An antioxidant is a compound that donates electrons to an oxidized molecule, making it more reduced and less reactive. You will learn in this section how antioxidants such as catalase, vitamin E, and vitamin C protect the cell from oxidized compounds and radicals.

Oxygen-centered free radicals, also commonly known as **reactive oxygen species (ROS)**, include the **superoxide radical** ($^{\bullet}O_2^-$), hydrogen peroxide (H_2O_2), and the hydroxyl radical ($^{\bullet}OH$). Although the ground state of diatomic oxygen (O_2) is classified as a radical species, it is only mildly reactive. Oxygen has two unpaired electrons in different orbits that have matching spins. Therefore, O_2 can react only with other atomic species that have two unpaired electrons with spin antiparallel to that of the unpaired oxygen electrons, a rare occurrence in nature. As a result, the **reduction** of O_2, a vital process in aerobic metabolism, must occur one electron at a time. One-electron reduction of oxygen results in a highly reactive species with a single unpaired electron (remember what you read above about the loss of electrons and reactivity), the superoxide radical ($^{\bullet}O_2^-$); an additional one-electron reduction of $^{\bullet}O_2^-$ results in the equally destructive hydrogen peroxide (H_2O_2) molecule. Aerobic organisms have developed enzymatic mechanisms that allow for the full reduction of oxygen to harmless compounds such as H_2O **(Figure 4.15)**. Reactive oxygen species and their reduction can occur in different parts of the cell, including the mitochondria, nucleus, and cytosol.

Mitochondrial ATP synthesis produces the majority of superoxide ions

The mitochondria of aerobic organisms **(Figure 4.16)** use 95% of all oxygen consumed by the body for the process of synthesizing **adenosine triphosphate (ATP)**. This molecule, shown in **Figure 4.17**, is the primary molecule that supplies chemical energy for cellular reactions. In its simplest form, oxidative metabolism converts the energy stored in the carbon-carbon bonds of energy nutrients—carbohydrates and fats (proteins are rarely used for energy in aerobic organisms)—into potential usable energy in the form of ATP. The conversion of ATP to

Figure 4.15 General scheme of the enzymatic reduction of oxygen (O_2) to water (H_2O). During normal aerobic metabolism, oxygen is reduced by one electron to form the superoxide radical ($^{\bullet}O_2^-$). Almost instantaneously, enzymes catalyze the reduction of $^{\bullet}O_2^-$ to hydrogen peroxide (H_2O_2). In turn, other enzymes catalyze a two-electron reduction of H_2O_2 to form H_2O.

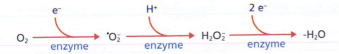

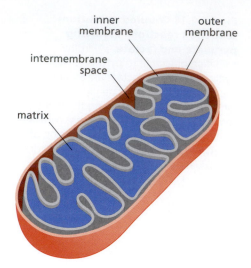

inner
membrane

outer
membrane

intermembrane
space

matrix

Figure 4.16 The mitochondrion.
The outer membrane separates the
mitochondrion from the cytosol and
contains enzymes associated with fat
metabolism. The inner membrane is folded
in order to increase the surface area for the
reactions associated with electron transport
and ATP synthesis. The matrix contains
enzymes for oxidative phosphorylation,
as well as the mitochondrial DNA. The
intermembrane space contains the
enzymes necessary for the transport of ATP
out of the mitochondrion.

adenosine diphosphate (ADP) releases the energy that drives many
biochemical reactions.

The transfer of energy from nutrient substrates to ATP occurs in two
distinct but coupled systems within the mitochondria. The first system
generates electrons through a series of oxidation-reduction reactions
as part of the **tricarboxylic acid cycle (TCA cycle)**, also known as the
Krebs cycle and the citric acid cycle. The TCA cycle is located in the
matrix of the mitochondria. The TCA cycle releases electrons from the
carbon-carbon bonds through a cascade of enzyme-catalyzed oxida-
tions. These electrons are then "shuttled" to the second system, the
electron transfer system (ETS), located in the mitochondrion's inner
membrane. The ETS uses these electrons, through a series of enzyme-
catalyzed reductions, to provide the necessary energy for ATP synthesis
(Figure 4.18). The overall process is called **oxidative phosphorylation**.

As shown in **Figure 4.19**, oxidative phosphorylation begins when carbons from nutrient substrates—carbohydrates and fats—are used to
generate the molecule **acetyl-CoA**. At specific points in the TCA cycle,
compounds called electron (energy) carriers or reducing equivalents,
such as nicotinamide adenine dinucleotide (NAD^+) and flavin adenine
dinucleotide (FAD), "pick up" the electrons released in the oxidation
reactions. These electron carriers deliver the electrons to the ETS.

In understanding how the ETS aids in the synthesis of ATP, it can be
helpful to think in terms of physics and to remember three important
points: (1) the free energy associated with electrons drives the system;
(2) a proton (H^+) gradient must be established between the mitochondrial matrix and the intermembrane space; (3) the concentration of
protons in the intermembrane space must be greater than the concentration of protons in the matrix. The electrons generated in the TCA
cycle are shuttled to the ETS by the reducing equivalents, $NADH + H^+$
and $FADH_2$, and released through oxidation reactions **(Figure 4.20)**.
The protons from these reactions are "pumped" through the inner membrane into the intermembrane space by specialized proteins. The use
of these specialized proteins is necessary because the inner membrane
is impermeable to protons. The impermeability of the inner membrane
is required to establish the proton gradient. Electrons provide the free
energy that drives the pumps. As the electrons give up their free energy,
the proton gradient becomes greater. As the electrons are transferred
within the ETS, the free energy maintaining the gradient decreases and
protons flow back into the matrix. But this flow can happen at only one
site on the membrane, the enzyme ATP synthase, which catalyzes the
reaction that makes ATP.

**Figure 4.17 Structure of adenosine
triphosphate (ATP) and adenosine
diphosphate (ADP).** The release
of energy from the breaking of one
phosphate bond in the conversion of ATP
to ADP supplies the majority of chemical
energy for cellular metabolism.

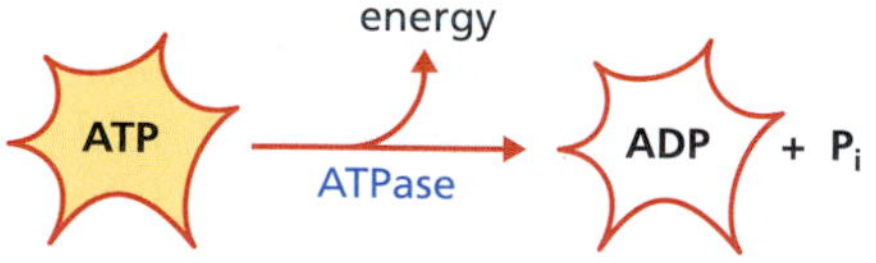

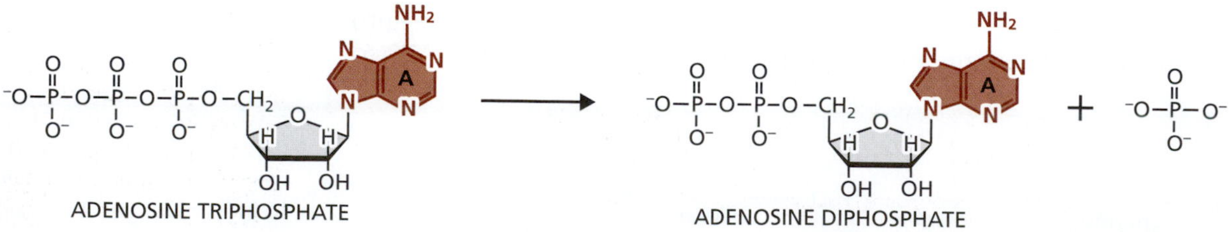

energy

ATP → ADP + P_i

ATPase

NH₂

N
N
A
N
N

⁻O–P–O–P–O–P–O–CH₂
O   O   O

O⁻ O⁻ O⁻

H   H
OH OH

ADENOSINE TRIPHOSPHATE

NH₂

N
N
A
N
N

⁻O–P–O–P–O–CH₂
O   O

O⁻ O⁻

H   H
OH OH

ADENOSINE DIPHOSPHATE

+

O
⁻O–P–O⁻
O⁻

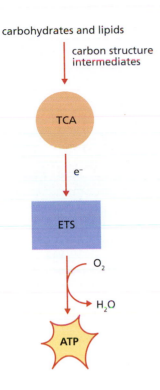

Figure 4.18 Production of cellular energy (ATP) from food energy (carbohydrates and lipids). The potential energy in the carbon-carbon bonds of carbohydrates and lipids cannot be used directly by the cell. Rather, through the reduction of carbon-containing intermediates in the tricarboxylic acid (TCA) cycle, electrons are released and react with oxygen in the electron transfer system (ETS), with the synthesis of adenosine triphosphate (ATP).

The energy state of any system will strive to reach equilibrium. Therefore, even the low free-energy state of the mitochondrial matrix must be dissipated. It is here that oxygen has its role. Oxygen acts as the final acceptor of free energy in the form of electrons and protons, producing water. While this system is very efficient in terms of the complete reduction of oxygen to water, some oxygen will be reduced by only one electron, and a superoxide radical will be generated.

The coupling of the TCA cycle to the ETS is extremely efficient with respect to the use of oxygen as the final electron acceptor during ATP synthesis. That is, very few, if any, superoxide radicals are generated when the TCA cycle is active. However, when the rate of ATP synthesis is low, during periods when the ATP-ADP ratio is high in the

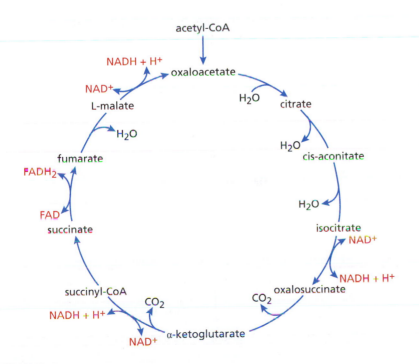

Figure 4.19 The tricarboxylic acid (TCA) cycle. Food substrates (carbohydrates and fats) are oxidized to acetyl-CoA, which can enter the TCA cycle. In a series of oxidation-reduction (redox) reactions, electrons are generated and shuttled to the electron transfer system (see Figure 4.20) by the electron carriers NADH + H+ and FADH2 (shown here in *red*).

Figure 4.20 A simplified version of the electron transfer system (ETS) and the synthesis of ATP. The high energy of electrons drives the pumping of protons by the three respiratory complexes into the intermembrane space and establishes the proton gradient. In the process, the free energy decreases and protons (H^+) flow back through ATP synthase, providing the energy needed to synthesize ATP. The free energy left after ATP synthesis is dissipated when oxygen accepts an electron and proton in a reaction that produces water.

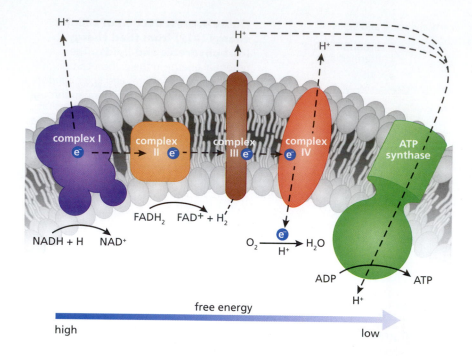

mitochondrial matrix, the conditions in the ETS are favorable for the generation of the superoxide radical, $^\cdot O_2^-$ **(Figure 4.21)**. High ATP-ADP ratios in the mitochondrial matrix inhibit the TCA cycle, and electrons are not shuttled to the ETS. Protons are not flowing through ATP synthase, and thus O_2 is not being reduced to water. The low concentration of protons in the matrix inhibits the formation of water from reduced oxygen. Therefore, oxygen can become reduced by one electron, donated from the electron-rich (reduced) ETS complexes, forming $^\cdot O_2^-$. Estimates suggest that 1–2% of all cellular oxygen consumption results in the superoxide radical.

Figure 4.21 Steps in the generation of the superoxide radical, $^\cdot O_2^-$, in mitochondria.

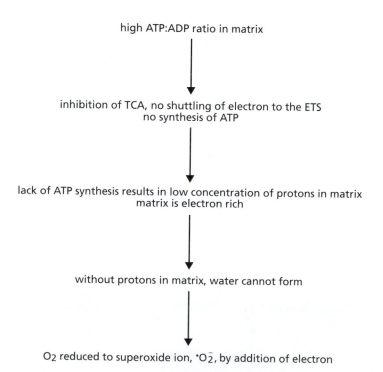

$$2 \cdot O_2^- + 2H^+ \xrightarrow{\text{SOD}} H_2O_2 + O_2 \xrightarrow{\text{catalase}} 2H_2O + O_2$$

Figure 4.22 Reduction of the superoxide radical to water in mitochondria. Superoxide dismutase (SOD) reduces two superoxide radicals to hydrogen peroxide (H_2O_2). Catalase then reduces hydrogen peroxide to water.

Enzymes catalyze reduction of the superoxide radical to water

As described above, $\cdot O_2^-$ can be generated during normal aerobic metabolism. Left unattended, these radicals react quickly with other atoms, leading to the possibility of cellular damage. Fortunately for aerobic organisms, the mitochondria contain two enzymes that catalyze the reduction of $\cdot O_2^-$ to water: **superoxide dismutase (SOD)** and **catalase**. Superoxide dismutase has a very high affinity for $\cdot O_2^-$, resulting in immediate reduction of this radical to H_2O_2 **(Figure 4.22)**. However, H_2O_2 can also cause significant harm to the cell. The second enzyme in the reduction cascade, catalase, performs a two-electron reduction of H_2O_2 to form H_2O.

Cytosolic reduction also generates free radicals

Although the oxygen concentration in the cytosol is considerably lower than that in the mitochondria, superoxide ions can still be generated from a single-electron reduction. When this occurs, cytosolic $\cdot O_2^-$ is reduced to water by superoxide dismutase and catalase, a reaction similar to that occurring in the mitochondria (shown in Figure 4.22). Cytosolic H_2O_2 can also be reduced to water with the aid of the enzyme **glutathione peroxidase (Figure 4.23)**. However, under certain cytosolic conditions, hydrogen peroxide is converted to the highly reactive hydroxyl radical, $\cdot OH$. This conversion occurs nonenzymatically when (1) ferric iron or cupric copper (Fe^{2+} or Cu^{2+}) and H_2O_2 are present together in the cytosol, a reaction known as the **Fenton reaction (Figure 4.24A)**; and when (2) Fe^{2+}, H_2O_2, and $\cdot O_2^-$ are present together in the cytosol, a process known as the **Haber-Weiss reaction** (Figure 4.24B).

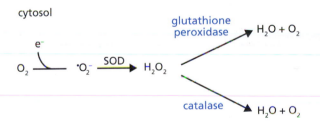

cytosol

Figure 4.23 Cytosolic reduction of the superoxide ion to water. The reduction of O_2 to $\cdot O_2^-$ is followed by reduction of the superoxide to H_2O_2 by cytosolic SOD. The hydrogen peroxide is then reduced to H_2O and O_2 by catalase or glutathione peroxidase.

(A) the Fenton reaction

$$(Fe^{2+} \text{ or } Cu^{2+}) + H_2O_2 \longrightarrow (Fe^{3+} \text{ or } Cu^{3+}) + \cdot OH + OH^-$$

(B) the Haber-Weiss reaction

$$O_2 + H_2O_2 \xrightarrow[\quad]{Fe^{2+} \quad Fe^{3+}} O_2 + OH^- + OH^-$$

Figure 4.24 Generation of the hydroxyl radical ($\cdot OH$) in the cytosol. Hydrogen peroxide can be converted to the highly reactive hydroxyl radical, $\cdot OH$, in (A) the Fenton reaction and (B) the Haber-Weiss reaction. Both reactions proceed nonenzymatically.

Oxygen-centered free radicals lead to the accumulation of damaged biomolecules

Radicals are highly reactive and cause significant alterations in the structure of many biological molecules, such as nucleic acids, lipids, and proteins. In turn, the alteration in structure leads to a reduction in the molecule's biological activity. This damage can be observed at all levels of cellular organization **(Figure 4.25)**. As you will see, the accumulation of lipid peroxides in membrane phospholipids renders the cellular and mitochondrial membranes less effective at maintaining the barrier between the extracellular and intracellular compartments. In turn, chemical reactions that are sensitive to the concentration of solutes and water are affected. Oxidative damage to components of the DNA transcription and translation process results in proteins with altered amino acid sequences. Moreover, because ROS can also affect the proteins of the DNA repair mechanisms, inappropriate base-pair substitutions may not be removed and replaced with the proper base. It seems that proteins involved in the removal of damaged proteins from cell are also affected by ROS. That is, accumulation of damaged proteins is exacerbated by a decline in the function of waste removal by the cell.

Although mitochondria generate the majority of cellular ROS as $\cdot O_2^-$, the efficiency of the mitochondrial antioxidant system (SOD and catalase) results in very few superoxide ions to cause damage. Rather, the ROS of the cytosol (H_2O_2 and $\cdot OH$) seem to be responsible for most of the cellular damage caused by free radicals, although they account for only 15% of the total cellular ROS. This reflects the fact that $\cdot OH$ reacts very quickly—in 1×10^{-12} seconds—with polyunsaturated fats, compounds found throughout the cell but primarily in biomembranes (see the discussion below).

Figure 4.25 Reactive oxygen species can lead to the cellular accumulation of damaged biomolecules at all levels of cellular organization.

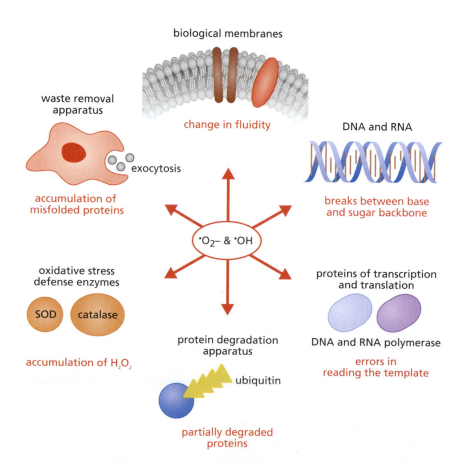

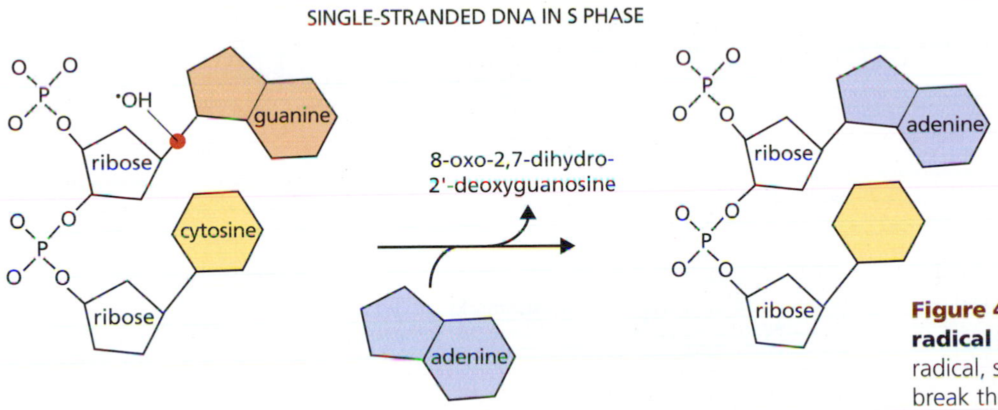

SINGLE-STRANDED DNA IN S PHASE

Figure 4.26 Effect of the hydroxyl radical (˙OH) on DNA. The hydroxyl radical, shown here as a *red dot*, can break the bond between guanine and the sugar backbone of DNA during replication. If the repair mechanism of DNA does not recognize the substitution, the replicated DNA will be out of sequence. The by-product of the base-pair break, 8-oxo-2,7-dihydro-2'-deoxyguanosine, can be used as a measure of DNA damage.

Oxygen-centered radicals can also result in an alteration in the base-pair sequence of DNA. While most research has found that in senescent cells, DNA errors due to damage from radicals are rare, the hydroxyl radical (˙OH) does have a high affinity for the bond between the sugar backbone of DNA and guanine **(Figure 4.26)**. If, for example, the error occurs during the replication process, adenine may be substituted for guanine. The substitution of adenine for guanine would result in an A-T pairing rather than a G-C pairing during DNA replication. In turn, the resulting protein would be assembled with an incorrect amino acid sequence, which could lead to decreased biological activity or accumulation of damaged proteins.

Cell membranes are susceptible to damage by reactive oxygen species

Cell membranes maintain a highly efficient barrier between the intracellular and extracellular compartments, while at the same time allowing an exchange of molecules essential for cellular functions. The unique properties of cell membranes are due to the chemical structure and physical arrangement of **fatty acids** (**lipids**) and the various molecules embedded in the membranes. The membrane lipids are known as **phospholipids**, because they contain a water-soluble phosphate head and two water-insoluble fatty acid tails **(Figure 4.27)**. Cell membranes are also known as **lipid bilayers**, because the phospholipids are arranged in a double layer, such that the phosphate ends of one layer of phospholipids point toward the extracellular space, while the phosphate ends of the other layer point toward the intracellular space. The lipid components of the phospholipids point toward each other within the interior of the membrane. This arrangement ensures that water and water-soluble molecules do not pass freely between the extracellular and intracellular spaces, an important property for maintaining proper chemical balance of the cell. Water-soluble molecules are transported into or out of the cell by specialized structures embedded in the membrane.

A highly functioning membrane is dependent, in large part, on the physical arrangement of the phospholipids and the structures embedded in the membrane. If adhesion between the structures in the membrane and the phospholipids is too tight, water-insoluble molecules cannot pass freely through the membrane. If adhesion is too loose, an inappropriate amount of water and water-soluble molecules can enter or escape from the cell. Thus, the bilayer membrane has evolved so that the lipids and proteins are in perfect balance to achieve the proper

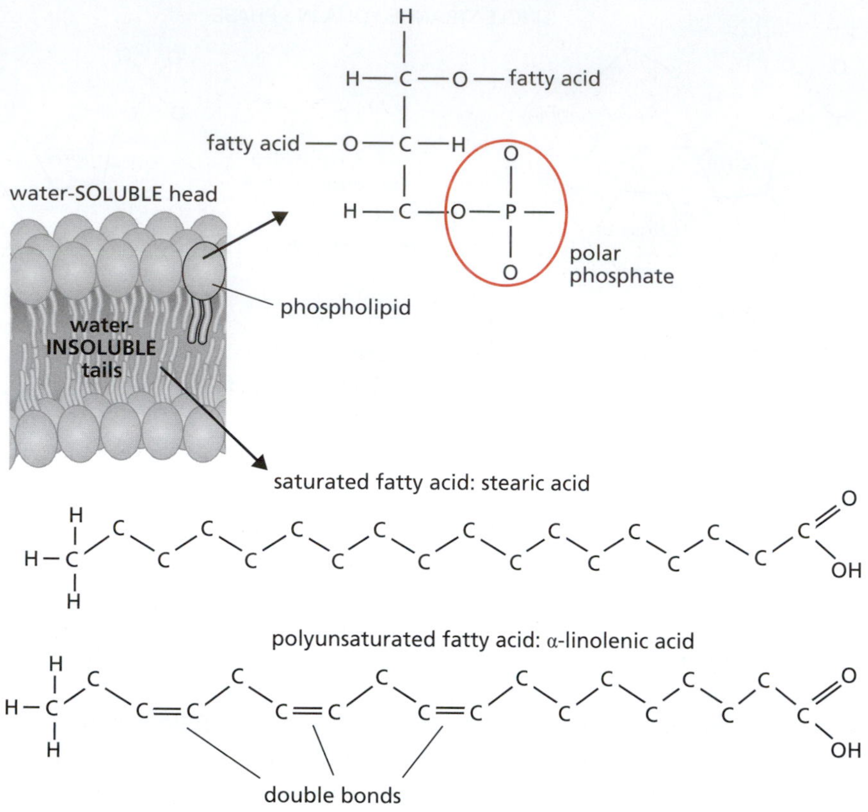

Figure 4.27 Structure of the phospholipid found in cell membranes. The polar phosphate group of the phospholipid head is water-soluble, or hydrophilic. The fatty acid tails are water-insoluble, or hydrophobic. The phospholipid can contain two saturated fatty acids, two unsaturated fatty acids, or one of each. Most phospholipids found in biological membrane have one of each: a saturated and an unsaturated fatty acid.

electrochemical composition for just the right membrane structure. Any change in that balance will alter membrane function and thus cellular function. Alterations in biomembranes caused by ROS can lead to disruption of membrane permeability, imbalance of the electrolyte gradient, inhibition of normal membrane protein mobility, and disruption in a variety of other functions critical to normal cellular activity.

The fluid nature of the cell membrane reflects, in large part, the bonding characteristics of the fatty acids contained in the phospholipids. Phospholipids are a mixture of **saturated** and **unsaturated fatty acids** that combine to give the membrane a particular viscosity. Saturated fatty acids have greater viscosity at body temperature than do unsaturated fatty acids. Thus, combining the two into phospholipid imparts a fluidity to the membrane that results in optimal function.

The double-bond structure of the polyunsaturated fats found in the lipid component of cell membranes is particularly susceptible to "attack" by •OH. The •OH "attack" on polyunsaturated fats causes a chain reaction that propagates additional free radicals and alters the polyunsaturated fat to form a new molecule, a lipid peroxide. As shown in **Figure 4.28**, the •OH donates its unpaired electron to a double bond of a polyunsaturated fat (LH), forming a lipid radical (L•) and water. This reaction marks the initiation phase of the free-radical chain reaction. Cytosolic O_2 is reduced by the L•, generating the lipid peroxyl radical (LOO•), which, in turn, attacks another LH, resulting in L• and lipid peroxide (LOOH) and the further generation of LOO•, LOOH, and L•.

If the lipid-radical reaction that produces lipid peroxide was left unregulated, cellular function would be greatly hampered, if not altogether halted. Fortunately, the cell has developed a mechanism to prevent the formation of lipid peroxides, which involves vitamin E (tocopherol) and vitamin C (ascorbic acid). Vitamin E exists in the cell close to or within

(A)

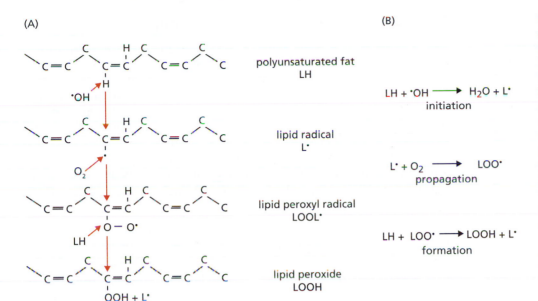

polyunsaturated fat
LH

lipid radical
L•

lipid peroxyl radical
LOOL•

lipid peroxide
LOOH

(B)

$$LH + {}^{\bullet}OH \longrightarrow H_2O + L^{\bullet}$$
initiation

$$L^{\bullet} + O_2 \longrightarrow LOO^{\bullet}$$
propagation

$$LH + LOO^{\bullet} \longrightarrow LOOH + L^{\bullet}$$
formation

Figure 4.28 Formation of lipid peroxide. (A) The hydroxyl radical "attacks" the polyunsaturated fat (LH) at a double bond, resulting in formation of the lipid radical (L•). In turn, cytosolic O_2 reacts with the unpaired electron of the lipid radical to form the lipid peroxyl radical (LOO•). The lipid peroxyl radical attacks another polyunsaturated fat, forming the lipid peroxide molecule (LOOH) and a new lipid radical, and so on. (B) Initiation and propagation of the free-radical chain reaction leading to the formation of lipid peroxide.

the cell membrane and has an affinity for the hydroxyl radical and the lipid radical that is considerably greater than that of the double bond of membrane polyunsaturated fats. As shown in **Figure 4.29**, the lipid peroxyl radical is oxidized to LOOH by reduction of α-tocopherol to the α-tocopherol radical, thereby terminating the lipid peroxide chain reaction. However, the α-tocopherol radical needs to be oxidized back to reduced α-tocopherol. This is accomplished through a multi-step process that includes vitamin C (ascorbic acid), glutathione, and NAD⁺.

(A)

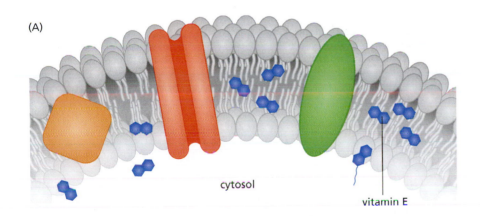

cytosol

vitamin E

Figure 4.29 Termination of the free-radical chain reaction by vitamin E (α-tocopherol). (A) Vitamin E is positioned in or near the cell membrane. (B) If a free radical arises, vitamin E is reduced to the α-tocopherol radical during oxidation of the lipid peroxyl radical. (C) The α-tocopherol is reoxidized in a multi-step process involving ascorbic acid, glutathione, and NAD⁺.

(B)

HO

......C₁₆H₃₃

α-tocopherol (vitamin E)

LOO•

LH

O•

......C₁₆H₃₃

α-tocopherol radical

(C)

α-tocopherol ← → dehydroascorbic acid ← reduced glutathione ← → NAD+

α-tocopherol radical   ascorbic acid   oxidized glutathione   NADPH

Reactive oxygen species can have beneficial effects

Our discussion so far has painted a fairly bleak picture of ROS. However, ROS can also have beneficial effects for the aerobic organism. The most widely know beneficial effect is found in the immune system. When a foreign invader, such as a bacterium, enters the bloodstream or tissues, the immune system releases a number of cells from the lymph nodes that attack and destroy the intruder. One of these cell types is the macrophages, which release enzymatically produced molecules designed to break down the cellular structure of the invading organism. Included among these molecules is $\cdot O_2^-$, whose primary job appears to be breaking down the invader's cell membrane through the generation of the lipid peroxide moiety **(Figure 4.30, top)**.

Recent research has also shown that ROS and/or by-products of ROS reduction, such as H_2O_2, may prevent oxidative damage by inducing further expression of the antioxidants superoxide dismutase and catalase (Figure 4.30, bottom). While the precise mechanism is still unclear, inducing production of the superoxide radical in cultured cells seems to activate the promoter regions of the genes for SOD and catalase.

TELOMERE SHORTENING AND REPLICATIVE SENESCENCE

In the previous section, we focused our attention on a mechanism in which the random or stochastic phenomenon of oxidative damage results in an accumulation of molecular damage that leads to cellular aging. We now turn our attention to a theory of cell longevity that involves a more programmed approach—the telomere-shortening theory. The **telomere-shortening theory** predicts that repeated

Figure 4.30 Two examples of how ROS are beneficial to an organism. Immune function (*top*): Cells of the immune system called macrophages secrete toxic chemicals, such as the superoxide radical and hypochlorite, in response to an invading organism. The superoxide ions help to break down the cell membrane of the bacterium via the lipid peroxidation pathway. Aerobic metabolism (*bottom*): The generation of ROS during normal aerobic metabolism (AM) can also stimulate the expression of superoxide dismutase (SOD) and catalase, although the precise mechanism is unknown.

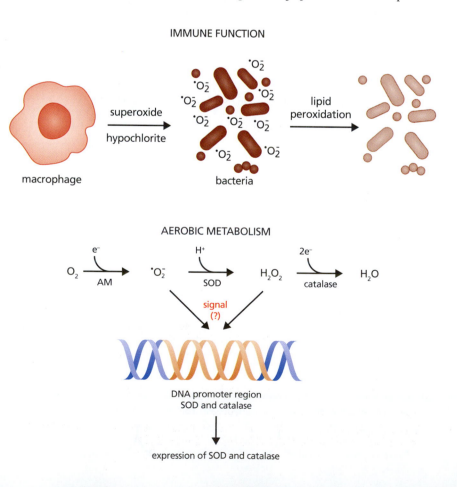

replication of a chromosome shortens the **telomeres** (the repetitive, noncoding base-pair sequences at the ends of each chromosome) to the point at which no further replication can occur without affecting the coding sequences of the DNA. According to this theory, the DNA repair mechanism "recognizes" the difference between telomeres and coding sequences and halts cell division before the transition from G_1 to S phase if genetic information is about to be used to synthesize an RNA primer. That is, the telomere is a biological clock, of sorts, that counts the number of cell replications.

Telomeres prevent the lagging strand from removing vital DNA sequences

Recall that the replication of eukaryotic linear DNA occurs in both directions from the replication origin, with each strand of the original DNA being used as a template for a new copy. Because DNA polymerase can synthesize a new strand only in the $5' \rightarrow 3'$ direction, the lagging strand template must use a backstitching maneuver on the Okazaki fragments to complete the replication of the chromosome. This mechanism requires the synthesis of an RNA primer for each new Okazaki fragment formed. Thus, when DNA polymerase reaches the end of the chromosome, there may not be enough DNA available with which to initiate the final RNA primer on the lagging strand template. This is generally referred to as the "end replication problem."

In mitotic eukaryotic cells, telomeres solve the end replication problem. Telomeres are rich in thymine (T) and guanine (G), are highly conserved across the eukaryote kingdoms, and in newly replicated DNA, have a length of 5000–10,000 base pairs. When DNA polymerase reaches the end of the coding sequences on the lagging strand template, it uses the telomere as the template for the RNA primer needed to begin the final Okazaki fragment **(Figure 4.31)**. However, a small number of

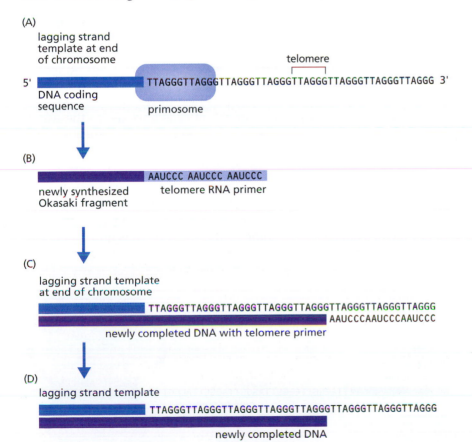

(A)

lagging strand template at end of chromosome

telomere

5' TTAGGGTTAGGGTTAGGGTTAGGGTTAGGGTTAGGGTTAGGGTTAGGGTTAGGG 3'

DNA coding sequence

primosome

(B)

AAUCCC AAUCCC AAUCCC
telomere RNA primer

newly synthesized Okasaki fragment

(C)

lagging strand template at end of chromosome

TTAGGGTTAGGGTTAGGGTTAGGGTTAGGGTTAGGGTTAGGGTTAGGG
AAUCCCAAUCCCAAUCCC

newly completed DNA with telomere primer

(D)

lagging strand template

TTAGGGTTAGGGTTAGGGTTAGGGTTAGGGTTAGGGTTAGGGTTAGGG

newly completed DNA

Figure 4.31 How the presence of telomeres solves the end replication problem. (A) The primosome reads the telomere sequence on the lagging strand template at the end of the chromosome, and (B) makes the RNA primer for the next Okazaki fragment. (C) The newly synthesized Okazaki fragment is then joined to the end of the new DNA. (D) Finally, the RNA primer made from the telomere is removed, and the new strand of DNA is complete.

(10–50) telomere base pairs are lost as a result of the backstitching mechanism. This small loss of telomere sequence during replication prevents the loss of genetic material at the end of the chromosome during replication.

Because replication on the lagging strand template results in the loss of a short segment of the telomeres, repeated cell divisions can result in the complete loss of the telomeres. In actively mitotic cells, however, telomere length is maintained through the action of the enzyme **telomerase**. Telomerase has a unique structure that contains both an RNA subunit and a protein catalytic subunit **(Figure 4.32)**. That is, telomerase is an enzyme that carries around its own RNA sequence. The telomerase RNA subunit contains the sequence of the telomeres for that species. In the case of humans, the RNA is approximately 450 nucleotides long, with the repeated sequence AAUCCC (DNA sequence, TTAGGG). The catalytic subunit is a **reverse transcriptase** (telomerase reverse transcriptase, TERT), an enzyme that reads the RNA sequence, as a template, and synthesizes the corresponding DNA sequence.

The mechanism of telomere elongation ensures that the 3' end is longer than the 5' end, a condition necessary for appropriate lagging strand replication. Note, however, that if the 3' end were left as is, the DNA repair mechanism could read the extended length at the 3' end as a double-strand break. Telomeres solve this potential problem by forming a loop, called the t-loop, at the 3' end, a procedure known as end-capping, thus "fooling" the DNA repair mechanism into reading it as a double strand **(Figure 4.33)**.

Shortening of the telomere may cause somatic cell senescence

Elongation of the telomeres by telomerase occurs in cells that require high fidelity during replication. That is, telomerase is expressed only in a limited number of cells, which include the germ-line and stem cells. Most somatic cells do not normally express telomerase, so telomere shortening occurs at each replication in these cell types. *In vitro* experimentation has shown that chromosomes of senescent cells do indeed have shorter telomeres than those observed on the chromosomes of pre-senescent cells. The lack of telomerase in somatic cells and the discovery of short telomeres in senescent cells have led to the widely accepted theory of replicative senescence known as the **mitotic clock theory**, which predicts that old cells sense short telomeres, which, in turn, causes cell cycle arrest **(Figure 4.34)**.

Although the mechanism that underlies the mitotic clock theory has not yet been described, current research suggests that short or uncapped telomeres induce inhibition of the cell cycle. The inhibition seems to follow the normal cell checkpoint, as the p53 protein is up-regulated in

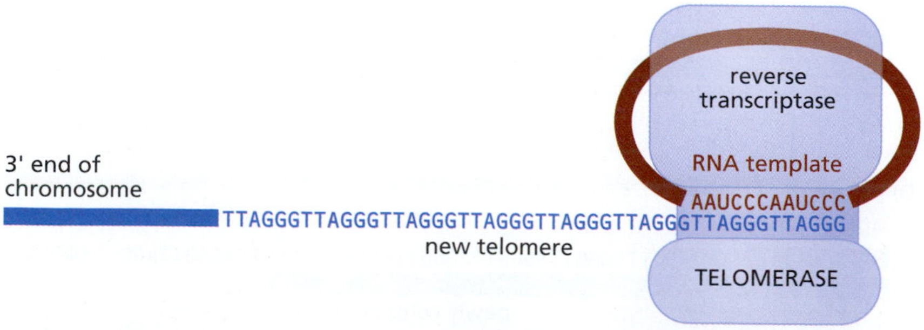

Figure 4.32 Telomerase and elongation of the telomere. Prior to the replication of DNA during the S phase, telomerase is attracted to the 3' end of the lagging strand template. Telomerase contains a reverse transcriptase, a catalytic unit that makes DNA from RNA (designated "reverse" because RNA is usually made from DNA), and the RNA template needed for DNA synthesis.

3' end of chromosome

TTAGGGTTAGGGTTAGGGTTAGGGTTAGGGTTAGGGTTAGGGTTAGGG

new telomere

reverse transcriptase

RNA template

AAUCCCAAUCCC

TELOMERASE

senescent cells. An example of how senescence may arise as a result of a short or uncapped telomere during G_1 is shown in **Figure 4.35**. Short or uncapped telomeres arise in somatic cells after several replicative cycles. The DNA damage and repair mechanism reads the telomeres as a strand break and initiates, through various signals, phosphorylation of the p53 protein. Inhibition of the cyclin-Cdk complex then halts the cell cycle in G_1. Since there is no repair mechanism that will elongate or cap the telomere—that is, no telomerase—the cell enters G_0 permanently, resulting in replicative senescence.

The mitotic clock theory may also tie predictions about the evolutionary bases of aging and longevity to cellular aging. Recall from Chapter 3 that evolutionary theories of longevity suggest that the mortality of the somatic cell arose as a result of a trade-off in the use of energy resources between the DNA repair systems of the soma and the germ line. It is easy to see how the shortening of telomeres that leads to cell senescence in somatic cells and the maintenance of telomeres in the germ line fit within the scope of such models. That is, the germ line invests a significant amount of energy in telomerase, to maintain the telomeres and the high fidelity of DNA replication needed for transfer of the genes to the next generation. Somatic cells, on the other hand, invest no resources in telomere lengthening, as they do not express telomerase. Remember that a basic tenet of the disposable soma theory is that the soma need only live long enough to support successful reproduction. It appears, although is not yet proven, that telomeres tell the cells when they have lived long enough.

(A)

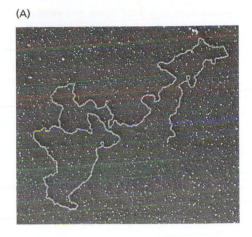

(B)

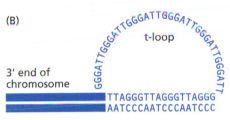

Figure 4.33 Telomere t-loop.
(A) Electron micrograph of the t-loop at the end of the chromosome. (B) Illustration of the end-capping of the telomere, a procedure used by the telomere to prevent the possible reading of the 3' elongation as a DNA double-strand break. (A, from J.D. Griffith et al., *Cell* 97:503–514, 1999. With permission from Elsevier.)

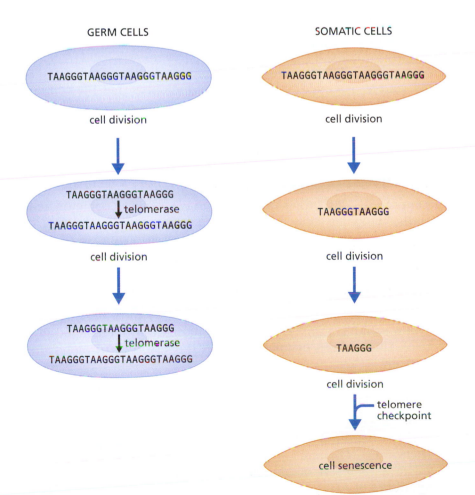

Figure 4.34 The mitotic clock theory of cell senescence. In both germ cells and somatic cells, the telomere shortens after each division. In germ cells, telomerase replaces the telomeres lost during cell replication. Somatic cells do not have telomerase, so the telomere that is lost from the lagging strand is not replaced. In a mechanism that is not well understood, the cells become senescent when the telomere checkpoint identifies critically short telomeres.

Figure 4.35 The proposed p53 protein mechanism of cell senescence. Given enough time and a sufficient number of cell divisions, telomeres in somatic cells (those without telomerase) will shorten or become uncapped. Short or uncapped telomeres are read as strand breaks by the DNA repair mechanism. In turn, signals are relayed to the p53 protein cascade, the cell cycle is halted in G_1, and the cell enters G_0 permanently.

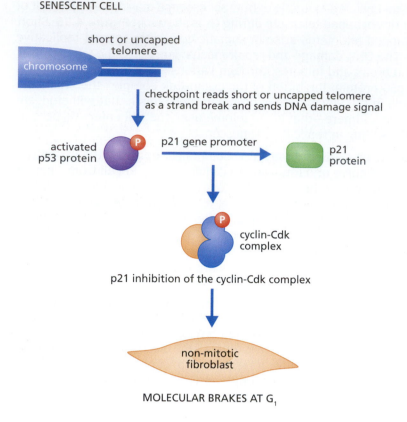

SENESCENT CELL

short or uncapped telomere

chromosome

checkpoint reads short or uncapped telomere as a strand break and sends DNA damage signal

activated p53 protein   P   p21 gene promoter   p21 protein

cyclin-Cdk complex   P

p21 inhibition of the cyclin-Cdk complex

non-mitotic fibroblast

MOLECULAR BRAKES AT G_1

ESSENTIAL CONCEPTS

- The eukaryotic cell cycle consists of four distinct phases: G_1, S, G_2, and M. An additional phase, G_0, indicates that the cell has exited the cycle.

- If a cell does not receive a mitogenic signal at the correct time, it dismantles the cell cycle control mechanisms and enters a modified version of the G_1 phase known as G_0.

- Populations of mitotic cells grown *in vitro* have a finite life span.

- Senescent cells grown in culture have several common features, including cellular enlargement, increase in the size of the nucleus, multinucleated cells, decreased synthesis of DNA, RNA, and protein, and increased secretion of inflammatory cytokines.

- Population doubling rates vary with the species and are not strongly correlated with the organism's length of life.

- The mechanism underlying the accumulation of damaged proteins that causes cellular aging is the disorder of increasing entropy.

- The oxidative stress theory of cellular aging predicts that the intracellular accumulation of damaged biomolecules results from reactions involving oxygen-centered free radicals.

- Oxygen-centered free radicals, or reactive oxygen species (ROS)—$^\bullet O_2^-$, $^\bullet OH$, and H_2O_2—arise during normal aerobic metabolism as a result of one-electron reduction of O_2 during the synthesis of ATP.

- The double-bond structure of the polyunsaturated fats found in the lipid component of cell membranes is particularly susceptible to "attack" by the hydroxyl radical. This "attack" causes a chain reaction that propagates additional free radicals and alters the polyunsaturated fat structure to form a new molecule, a lipid peroxide.

- ROS can also have beneficial effects for the aerobic organism. Macrophages of the immune system use ROS to break down bacterial membranes. ROS generated during oxidative metabolism may also induce the expression of cellular antioxidants such as SOD and catalase.

- The telomere-shortening theory predicts that repetitive replication of the chromosome shortens the telomere to the point at which no further replication can occur without affecting the coding sequences of the DNA.

- The repeating sequences of telomeres at the ends of chromosomes solve the end replication problem.

- In highly mitotic cells, telomere length is maintained by the enzyme telomerase.

- The mitotic clock theory predicts that cell senescence occurs when old mitotic cells sense short telomeres, which, in turn, causes cell cycle arrest at the G_1-S phase transition.

DISCUSSION QUESTIONS

Q4.1   Draw and explain the three growth phases of mitotic cells in culture. Describe the difference between cells in culture and a cell line. From knowledge gained in Chapter 3, which phase do you suspect would have the greatest fitness? Why?

Q4.2   List some general biological features observed in senescent cells near the end of their replicative life span.

Q4.3   The replicative life span of a cultured cell population has been widely accepted as "failure of the population to double within four weeks." Does this indicate that all cells have lost their mitotic capacity? Explain.

Q4.4   Explain why the age-related increase in entropy leads to an accumulation of damaged proteins in the cell.

Q4.5   The reduction of O_2 to water during aerobic metabolism is a vital process in the synthesis of ATP. Describe how oxygen-centered free radicals arise during the reduction of O_2.

Q4.6   Briefly explain how alterations in the structure-function relationship of proteins by ROS can lead to cellular aging.

Q4.7   The hydroxyl radical, ˙OH, can cause considerable damage to the cell. Which cell structures are particularly susceptible to attack by the hydroxyl radical? How does damage to these structures lead to cellular dysfunction and aging? How does the cell protect itself from damage by the hydroxyl radical?

Q4.8   Reactive oxygen species have no known beneficial function. True or false? Explain.

Q4.9 What is the "end-replication problem" and how do telomeres solve this problem?

Q4.10 Describe the mitotic clock theory of cellular aging and how it supports the evolutionary theory of aging. What is the major argument against the mitotic clock theory of cellular aging as a model for whole-organism aging?

FURTHER READING

THE CELL CYCLE AND CELL DIVISION AND REGULATION OF THE CELL CYCLE

Alberts B, Bray D, Hopkins K, et al. (2010) The Cell Division Cycle. In Essential Cell Biology, pp 609-650. New York: Garland Science.

Sadava D, Heller HC, Orians GH, et al. (2007) Chromosomes, the cell cycle, and cell division. In Life: The Science of Biology, p 180. Sinauer Associates.

REPLICATIVE SENESCENCE

Campisi J (2005) Senescent cells, tumor suppression, and organismal aging: good citizens, bad neighbors. *Cell* 120:513–522.

Carrel A & Ebeling A (1921) Age and muliplication of fibroblasts. *J Exp Med* 34:599–623.

Cristofalo VJ, Lorenzini A, Allen RG, et al. (2004) Replicative senescence: a critical review. *Mech Ageing Dev* 125:827–848.

Hayflick L & Moorhead PS (1961) The serial cultivation of human diploid cell strains. *Exp Cell Res* 25:585–621.

Rodier F & Campisi J (2011) Four faces of cellular senescence. *J Cell Biol* 192:547–556.

Witkowski JA (1980) Dr. Carrel's immortal cells. *Med Hist* 24:129–142.

THE CELLS OF HENRIETTA LACKS: HERS OR OURS?

Skloot R (2010) The Immortal Life of Henrietta Lacks, p 369. New York: Crown Publishers.

THE CAUSE OF CELLULAR AGING: ACCUMULATION OF DAMAGED BIOMOLECULES

Hayflick L (2007) Biological aging is no longer an unsolved problem. *Ann NY Acad Sci* 1100:1–13.

Hayflick L (2007) Entropy explains aging, genetic determinism explains longevity, and undefined terminology explains misunderstanding both. *PLoS Genet* 3:e220.

Lambert FL (2007) A student's approach to the second law and entropy. http//entropysite.oxy.edu/students_approach.html

Mitteldorf J (2010) Aging is not a process of wear and tear. *Rejuvenation Res* 13:322–326.

Toussaint O, Raes M & Remacle J (1991) Aging as a multi-step process characterized by a lowering of entropy production leading the cell to a sequence of defined stages. *Mech Ageing Dev* 61:45–64.

OXIDATIVE STRESS AND CELLULAR AGING

Alberts B, Bray D, Hopkins K, et al. (2010) Essential Cell Biology, pp 81–117, 425–452. New York: Garland Science.

Beckman KB & Ames BN (1998) The free radical theory of aging matures. *Physiol Rev* 78:547–581.

Brand MD, Affourtit C, Esteves TC, et al. (2004) Mitochondrial superoxide: production, biological effects, and activation of uncoupling proteins. *Free Radic Biol Med* 37:755–767.

Harman D (1956) Aging: a theory based on free radical and radiation chemistry. *J Gerontol* 11:298–300.

Jang YC, Perez VI, Song W, et al. (2009) Overexpression of Mn superoxide dismutase does not increase lifespan in mice. *J Gerontol A Biol Sci Med Sci* 64:1114–1125.

Landis GN & Tower J (2005) Superoxide dismutase evolution and life span regulation. *Mech Ageing Dev* 126:365–379.

Lu T & Finkel T (2008) Free radicals and senescence. *Exp Cell Res* 314:1918–1922.

Pamplona R (2008) Membrane phospholipids, lipoxidative damage and molecular integrity: a causal role in aging and longevity. *Biochim Biophys Acta* 1777:1249–1262.

Yu BP (2005) Membrane alteration as a basis of aging and the protective effects of calorie restriction. *Mech Ageing Dev* 126:1003–1010.

TELOMERE SHORTENING AND REPLICATIVE SENESCENCE

Beliveau A, Bassett E, Lo AT, et al. (2007) p53-dependent integration of telomere and growth factor deprivation signals. *Proc Natl Acad Sci USA* 104:4431–4436.

Delany ME, Daniels LM, Swanberg SE & Taylor HA (2003) Telomeres in the chicken: genome stability and chromosome ends. *Poult Sci* 82:917–926.

Harley CB, Futcher AB & Greider CW (1990) Telomeres shorten during ageing of human fibroblasts. *Nature* 345:458–460.

Mather KA, Jorm AF, Parslow RA & Christensen H (2011) Is telomere length a biomarker of aging? A review. *J Gerontol A Biol Sci Med Sci* 66:202–213.

GENETICS OF LONGEVITY

5

"AGING IS LIKE THE NEWEST VERSION OF A SOFTWARE—IT HAS A BUNCH OF GREAT NEW FEATURES BUT YOU LOST ALL THE COOL FEATURES THE ORIGINAL VERSION HAD."

-CARRIE LATET, POET

In Chapter 3, we established the theoretical basis of how genes selected for their impact on survival to reproductive age (that is, on fitness) also determine longevity, and we showed how mathematical theories and empirical laboratory research have demonstrated the "fixing" of genes that delay the start of reproduction, which increases longevity. We did not, however, discuss which genes and/or genetic pathways are involved in increasing longevity. In this chapter we take a closer look at some specific genes and genetic pathways that can alter life span.

Our exploration into the genetic basis of longevity focuses primarily on genes and pathways for which a general consensus has been established, arrived at through repeated experimentation. Keep in mind, as you read through this chapter, that our knowledge comes largely from studies on three simple, short-lived organisms—a yeast, a worm, and a fly—that are raised and studied under highly controlled laboratory conditions. Although at least 200 genes have been identified as having some impact on the rate of aging and longevity, only a handful of these have been studied in sufficient detail to be presented here. And although we can demonstrate that the manipulation of specific genes can affect longevity, it is important to remember that all genes have been selected for their ability to enhance survival to reproductive age, not for longevity. That is, genes are fixed because they provide a reproductive advantage, not because they increase the longevity of the species.

We begin with a review of gene expression and its regulation in eukaryotes, then briefly describe some of the methods used to analyze gene expression. Then we look at some specific genes and genetic pathways that affect longevity in *Saccharomyces cerevisiae* (a yeast), *Caenorhabditis elegans* (a roundworm), *Drosophila melanogaster* (fruit fly), and *Mus musculus* (mouse).

OVERVIEW OF GENE EXPRESSION IN EUKARYOTES

The combined research efforts of many biologists, beginning with Darwin, in the mid-nineteenth century, through the mid-1940s, led to the discovery that genes, the instructions for building and maintaining organisms, reside on the chromosomes within the nucleus of eukaryotes. However, some controversy remained as to whether the genes were part of the cell's **deoxyribonucleic acid (DNA)** or were proteins. The controversy began to dissipate in the late 1940s as the structure of DNA became clearer. **X-ray crystallography** of DNA (a method that shows the atomic arrangement in a molecule), performed by Rosalind Franklin and Maurice Wilkins, helped James Watson and Francis Crick more clearly define the structure of DNA and suggest that DNA could easily be rearranged and must be the location of genes. In April and May of 1953, Watson and Crick published two papers describing the structure of DNA and how the pairings of four **nucleotide bases**, **adenine (A)** with **thymine (T)** and **guanine (G)** with **cytosine (C)**, formed the instructions for building proteins.

These 1953 publications by Watson and Crick describing the structure of DNA revolutionized biology. In the 60 years since their discovery, the basic mechanism of how the information in DNA is transferred to the amino acid sequence in a protein has been completely worked out. In this first section of the chapter we explore that process.

Transcription of DNA produces complementary RNA

Gene expression begins when DNA transfers its information to **ribonucleic acid (RNA)** in a process called **transcription (Figure 5.1)**. The synthesis of RNA from a DNA template has many similarities to the process of DNA replication that you learned about in Chapter 4. The double-stranded, double-helical DNA must open up and unwind to expose the base pairs. Then, one (but only one) of the DNA strands acts as the template for an RNA molecule, and, as in DNA replication, the nucleotides are added one by one. That is, the RNA is complementary to the DNA. There are, however, some significant differences between DNA replication and RNA transcription. Adenine in DNA pairs with the base **uracil (U)**, as opposed to thymine, in forming the base sequence in RNA **(Figure 5.2)**. That is, the four bases in RNA are adenine, uracil, cytosine, and guanine. The finished RNA molecule, known as an **RNA transcript**, is a single-stranded molecule. Another important difference between DNA replication and RNA transcription is the time needed to complete the process and the number of molecules that are produced. RNA molecules are typically only a few thousand nucleotides in length, compared with the 250 million base pairs of DNA found in a medium-size chromosome.

Whereas DNA replication times are discussed in terms of hours, RNA transcription takes place in minutes. Several molecules of **RNA polymerase**, the enzyme that carries out transcription, can work on the same gene at the same time. As one RNA polymerase is finishing a transcript, another begins. This allows for several RNA transcripts to be made in a relatively short time, and thus the rapid synthesis of proteins.

RNA polymerase begins its work when it recognizes a **promoter region** on the DNA, a specific sequence of nucleotides indicating the starting point for RNA synthesis **(Figure 5.3)**. A single RNA polymerase performs several functions in building the RNA transcript. First, it opens up the double-stranded DNA to expose the bases. Then, the active site of the enzyme catalyzes a reaction that adds to the RNA a nucleotide that is complementary to the nucleotide in the DNA template—adding nucleotides, one at a time, in the 5'→3' direction. Finally, RNA polymerase rewinds the DNA into its double-helical structure. Elongation of the RNA transcript continues until RNA

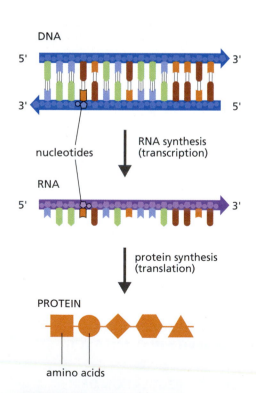

Figure 5.1 Transfer of genetic information from DNA to protein.

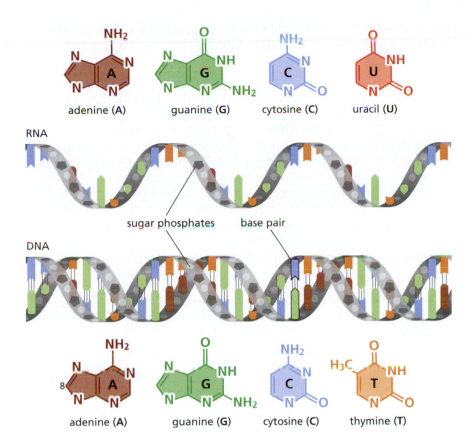

Figure 5.2 RNA and DNA. RNA (*top*) and DNA (*bottom*) consist of long chains of nucleotides, each of which consists of a nitrogenous base, a sugar, and a phosphate group. The four bases in DNA are adenine, guanine, cytosine, and thymine. RNA contains uracil instead of thymine.

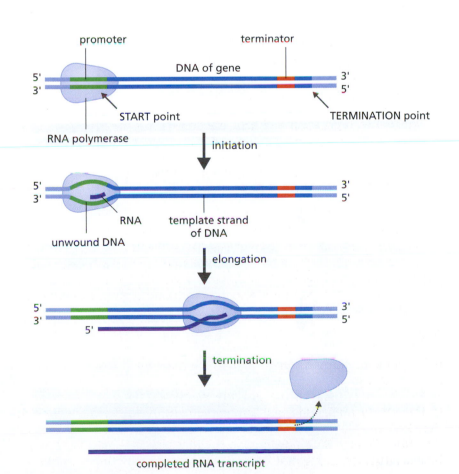

Figure 5.3 DNA transcription. RNA polymerase binds to the promoter region of DNA. During initiation, RNA polymerase unwinds the two DNA strands and initiates RNA synthesis. During elongation, the polymerase continues to assemble an RNA molecule with a nucleotide sequence complementary to that of the DNA template strand. When RNA polymerase reaches the termination site on the DNA, it unbinds from the DNA template and releases the newly made RNA transcript.

polymerase encounters another specific sequence of nucleotides on the DNA known as a **termination site**.

RNA polymerases recognize the promoter region of the DNA by its specific shape **(Figure 5.4)**, which is formed by the binding to the promoter of specialized proteins called **general transcription factors**. The general transcription factors find the promoter region of a gene by locating a sequence known as the **TATA box**, because its sequence is composed primarily of the bases thymine (T) and adenine (A). Binding of the general transcription factors to the promoter region's TATA box causes the DNA to bend outward. RNA polymerase uses this bend as the landmark for beginning the transcription process. Once RNA polymerase binds to the DNA and begins transcription, the general transcription factors are released, to be used again for another gene transcription.

Eukaryotic cells modify RNA after transcription

The RNA transcript contains the entire base-sequence of the gene including its noncoding regions known as **introns**. Another type of RNA made from the RNA transcript, called **messenger RNA (mRNA)**, is the one used by the cell as the template for protein synthesis. Messenger RNA contains only **exons**, the coding regions of the gene. Thus, the introns must be removed from the RNA transcript before the molecule leaves the nucleus. The mechanism for removing introns is known as **RNA splicing (Figure 5.5)**, which is carried out by other RNA molecules called **small nuclear RNAs (snRNAs)**. The snRNAs are packaged with other proteins to form small nuclear ribonucleoproteins (snRNPs, pronounced "snurups"). snRNPs combine with other proteins to form the spliceosome, the structure that carries out RNA splicing. The spliceosome cleaves the RNA transcript (or pre-mRNA) at the 5' end of an intron, as recognized by a few short nucleotide sequences common to most introns. The snRNAs then slide down the RNA until they find another set of unique nucleotide sequences at the 3' end of

Figure 5.4 Transcription factors and the transcription initiation complex. Transcription factor proteins bind to the TATA box in the promoter region. The transcription factors mediate the binding of RNA polymerase and the initiation of transcription.

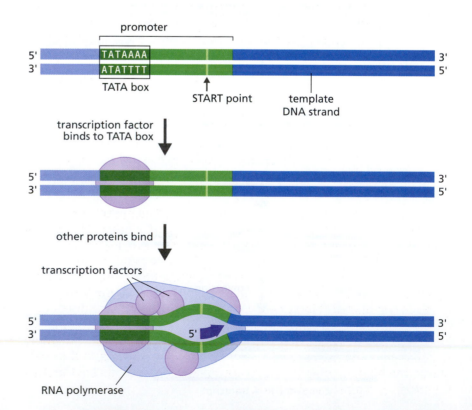

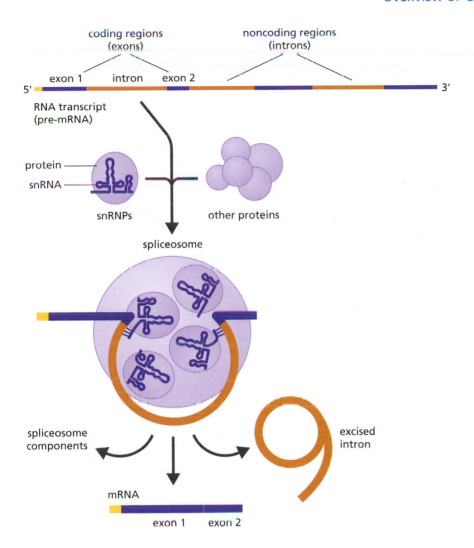

coding regions
(exons)

noncoding regions
(introns)

exon 1   intron   exon 2

5'   3'

RNA transcript
(pre-mRNA)

protein

snRNA

snRNPs   other proteins

spliceosome

spliceosome
components

excised
intron

mRNA

exon 1   exon 2

Figure 5.5 RNA splicing. The RNA transcript combines with small nuclear ribonucleoproteins (snRNPs) and other proteins to form the spliceosome. The spliceosome cuts each end of the intron sequence, removing it from the transcript. The exons are spliced together, forming messenger RNA (mRNA).

the intron, and here they make another slice, freeing the intron from the RNA. The ends of the exons are then sealed together. This process continues until all the introns are removed from the RNA, and the pre-mRNA becomes mRNA.

Synthesis of mRNA occurs in the nucleus whereas **translation** (synthesis of the protein encoded by the DNA), the primary purpose of mRNA, takes place in the cytosol. Therefore, the mRNA must be transported out of the nucleus. Transport of mRNA into the cytosol presents a potential problem for the cell, because many RNA fragments, such as excised introns, are present in the nucleus after splicing has been completed. In other words, how does the nucleus recognize only the mRNA and then transport only this molecule out of the nucleus?

The nuclear envelope contains openings called **nuclear pore complexes (Figure 5.6)**. These pores allow passage of a molecule having a specific structure, and this structure is created by binding proteins that are specific to mRNA. For example, completed and active mRNA contains unique regions of nucleotides characterized by strings of adenine bases called **poly-A tails**. The ubiquitous **poly-A-binding proteins** found in the nucleus bind to these regions, and the nuclear pore complexes recognize this complex as mRNA. Poly-A-binding proteins are just one of several types of proteins found in the nucleus that bind only to completed and active mRNA. Once the mRNA passes into the cytoplasm, the binding proteins are removed and degraded, as are the RNA fragments remaining in the nucleus.

Figure 5.6 RNA modification occurs after transcription. (A) Enzymes modify the ends of the mRNA molecule by adding a modified guanosine cap to the 5′ end and a poly-A tail to the 3′ end. (B) Two proteins—a binding protein that attaches to the poly-A tail and a cap-binding protein that binds to the 5′ end—form a complex with the completed mRNA molecule, allowing the nuclear pore complex to distinguish between RNA fragments and mRNA. Once in the cytosol, the cap-binding protein is exchanged for a protein synthesis initiation factor.

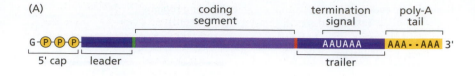

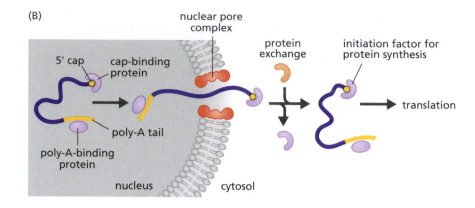

Translation is the RNA-directed synthesis of a protein

The nucleotide sequence of the mRNA is the code that is used for protein synthesis. The nucleotides are read in groups of three known as **codons**, and each group of three corresponds to a specific amino acid; this correspondence between codons and amino acids is known as the genetic code **(TABLE 5.1)**. A single amino acid can have more than one codon; for example, the codons GCU, GCC, GCA, and GCG all code for

TABLE 5.1
THE GENETIC CODE

Amino acid	Abbreviations		Codons					
Alanine	Ala	A	GCU	GCC	GCA	GCG		
Arginine	Arg	R	AGA	AGG	CGU	CGC	CGA	CGG
Asparagine	Asn	N	AAU	AAC				
Aspartate	Asp	D	GAU	GAC				
Cysteine	Cys	S	UGU	UGC				
Glutamate	Glu	E	GAA	GAG				
Glutamine	Gln	Q	CAA	CAG				
Glycine	Gly	G	GGA	GGC	GGG	GGU		
Histidine	His	H	CAU	CAC				
Isoleucine	Ile	I	AUU	AUC	AUA			
Leucine	Leu	L	UUA	UUG	CUU	CUC	CUA	CUG
Lysine	Lys	K	AAA	AAG				
Methionine	Met	M	AUG[1]					
Phenylalanine	Phe	F	UUU	UUC				
Proline	Pro	P	CCU	CCC	CCA	CCG		
Serine	Ser	S	AGU	AGC	UCU	UCC	UCA	UCG
Threonine	Thr	T	ACU	ACA	ACC	ACG		
Tryptophan	Trp	W	UGG					
Tyrosine	Tyr	Y	UAU	UAC				
Valine	Val	V	GUU	GUC	GUA	GUG		
Stop			UAA	UAG	UGA			

[1] This also acts as an initiation codon.

the amino acid alanine (Ala). The codon AUG codes for methionine and the start of the protein coding message, and three codons, UAA, UAG, and UGA, known as **stop codons**, end the protein-coding message.

The mRNA does not directly bind to the amino acids. Rather, two specialized molecules known as adaptors—**transfer RNA (tRNA)** and **aminoacyl-tRNA synthetase**—are responsible for the process of translating the nucleotide message in the mRNA into the correct amino acid sequence of a protein. The tRNAs are small molecules, about 80 nucleotides in length. They have a site for linkage to the codon in the mRNA at one location and a site unique to the corresponding amino acid at another **(Figure 5.7)**. Aminoacyl-tRNA synthetase is the enzyme that covalently couples the tRNA to its corresponding amino acid. A tRNA coupled to its amino acid is called a **charged tRNA**.

So far, we've seen that mRNA contains the genetic code for protein synthesis and that tRNA, along with an aminoacyl-tRNA synthetase, brings the amino acid to the mRNA. The actual process of bonding the amino acids together to form the protein is a much more complex process that requires 50–80 different proteins and a physical location that holds all the different molecules, including enzymes and other proteins, together. The protein-manufacturing complex that houses all these proteins and provides the space for the process to take place is called the **ribosome**. Each cell contains millions of ribosomes, so that many different proteins or several copies of the same protein can be synthesized simultaneously and with great speed **(Figure 5.8)**.

Proteins can be modified or degraded after translation

When the ribosome releases the completed protein, noncovalent interactions among the protein's amino acids cause the protein to fold upon itself, creating its **tertiary structure**. The tertiary structure is essential to the active state of the protein. Some proteins require modification after the translation process before they are completely active. There are several types of post-translational modification, but the two most common are **phosphorylation**, addition of a phosphate group, and **glycosylation**, addition of a glucose molecule. In general, proteins subject to post-translational modification are used as intracellular messengers and signals. That is, they signal other proteins what to do by the presence or absence of non–amino acid molecules bound to the protein. As you'll learn later in this chapter, the state of post-translational modification of a protein known as DAF-16 has a significant role in the longevity characteristics of *C. elegans*.

A properly, efficiently functioning cell depends, to a large degree, on its ability to regulate the amount and structure of its constituent proteins. Complex mechanisms have evolved in the eukaryotic cell to ensure that improperly folded or damaged proteins are removed and that the concentration of an individual protein stays at the optimal level. These mechanisms involve thousands of proteins participating in hundreds of biochemical pathways. Here we discuss briefly some generalities common to most protein-degradation pathways **(Figure 5.9)**.

Maintenance of the correct tertiary structure of a protein—its active form—is assisted by a group of regulatory cytoplasmic proteins known as **chaperones**. Chaperones perform two major functions: (1) they help the protein to fold correctly, and (2) if a protein has not folded correctly or is damaged, they mark it for degradation with another protein called **ubiquitin**. A large complex of enzymes and other proteins called the **proteasome** recognizes ubiquitin-marked proteins. The cylinder-shaped proteasome contains enzymes called **proteases** that break the

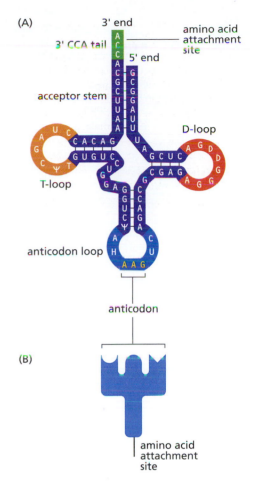

Figure 5.7 Structure of tRNA.
(A) tRNA consists of a single strand of RNA that forms several double-stranded regions in which segments of the RNA base pair with each other. A single-stranded loop at the end of the molecule contains a triplet of bases called the anticodon. A short single strand at the opposite end is the site where an amino acid attaches to form the charged tRNA. (B) Schematic representation of the tRNA molecule.

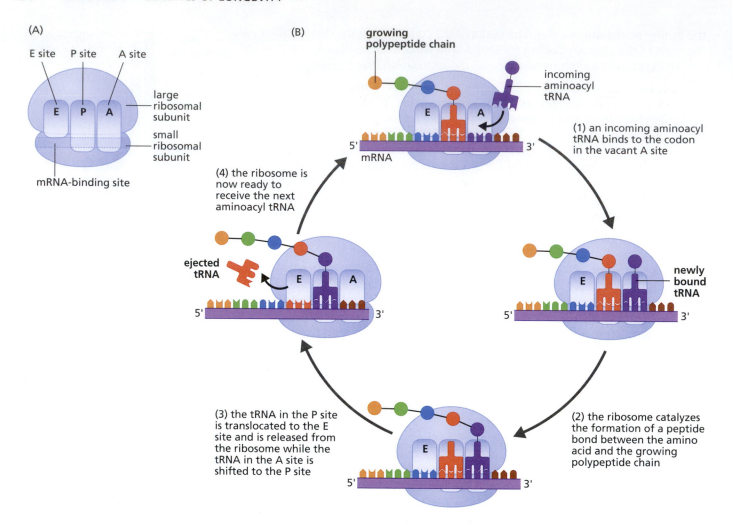

Figure 5.8 Protein synthesis on the ribosome. (A) A ribosome consists of two subunits—the large subunit and the small subunit. The ribosome has an mRNA-binding site and three tRNA-binding sites, known as the P, A, and E sites. (B) Translation takes place in a four-step cycle. (1) The anticodon of an incoming charged tRNA pairs with the mRNA codon in the A site of the ribosome. (2) The charged tRNA donates its amino acid to the growing protein, using the energy of the high-energy bond created during the charging process. (3) The large subunit translocates (*to the right, as shown here*), moving the tRNAs from the P and A sites to the E and P sites, respectively, leaving the A site open for the next charged tRNA. The tRNA in the E site is enzymatically removed from the ribosome. The small subunit shifts (*to the right*) to match the large subunit, pulling the mRNA along. (4) The ribosome is now ready to receive the next charged tRNA in the A site, and the process starts again with step 1. This cycle continues until the ribosome encounters a stop codon in the mRNA, at which point the completed protein is released from the ribosome.

bonds, called peptide bonds, between the amino acids of proteins. The amino acids released by this breakdown are returned to the intracellular amino acid pool to be used in another round of protein synthesis. You will learn in Chapter 9 that a dysfunction in the protein-degradation pathways that involve chaperones and ubiquitin may be the underlying biochemical cause of Alzheimer's and Parkinson's diseases.

REGULATION OF GENE EXPRESSION

Expressing the correct gene, at the correct time, and in the correct amount is critical to the proper functioning of a cell. Disruption in the regulation of gene expression can lead to serious dysfunction within the cell and possible altered states of metabolism, which, in turn, can affect the organism's longevity. A problem facing biogerontologists in

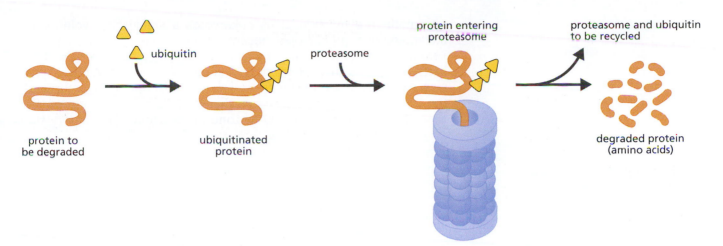

determining which of the 24,000 genes of the human **genome** affect longevity is that the type of gene expressed in a cell differs with the cell type. Hepatocytes (liver cells) and cardiac myocytes (heart cells) have radically different gene expression profiles for radically different functions, even though each cell contains the identical sequence of genomic DNA and uses the same transcription and translation processes to convert the genetic code of the DNA into proteins. Nonetheless, as you learned in the previous chapter, because both cell types are somatic, both are mortal. These two facts—different gene expression in different cells and the universal mortality of somatic cells—have led many biogerontologists to suggest how unlikely it is that longevity reflects differences in a single gene, or even in groups of genes. Rather, many biogerontologists now believe that longevity reflects a complex interaction between gene expression and the processes that regulate gene expression, or lack of expression.

In this section, we examine the basic mechanisms that regulate gene expression. The processes that govern gene expression occur at virtually every step in conversion of the genetic code into a functioning protein. However, study of the genetic control of longevity is a very young science and has had time to elucidate only a few regulatory pathways in gene expression in sufficient detail to allow for discussion here. These studies have focused primarily on transcriptional regulation. Post-transcriptional control, the regulatory process occurring after RNA polymerase has bound to a gene's promoter, has been less well described in terms of its impact on the rate of aging and longevity. Therefore, we begin with a description of transcriptional regulation, then give a brief overview of post-transcriptional mechanisms—for which research has only just begun, or remains to be done—to describe their role in modulating longevity.

Gene expression can be controlled by changing nucleosome structure

In eukaryotes, an efficient organization has evolved to deal with the enormous size of the DNA; if left in its linear form, DNA would easily occupy the entire volume of a cell **(Figure 5.10)**. The DNA, RNA, and various proteins make up the material called **chromatin**. DNA is further organized into the **chromosomes**, bound to proteins called **histones**. Unfolded chromatin has the appearance of beads on a string. Each "bead" consists of DNA wound around a histone protein called a **histone octomer**. Together, the wrapped DNA and histone proteins make up the **nucleosome**. The short segments or strings of DNA between nucleosomes are known as linker DNA.

Figure 5.9 Degradation of a protein by a proteasome. Small ubiquitin proteins are attached to the protein to be degraded. The proteasome recognizes the ubiquitin-tagged (ubiquitinated) protein, unfolds it, and stores it in its central cavity. Enzymes within the proteasome then break apart the protein into small peptides and amino acids, which are released back into the cytosol.

Chromatin organization gives eukaryotes a significant evolutionary advantage over prokaryotes, because it protects the genes from inadvertent chemical reactions. However, the nucleosomes can also be a barrier to gene transcription, since they can block the TATA boxes (promoter regions) associated with genes. For transcription to go forward, the histones must be modified so that they loosen their grip on the DNA. This is accomplished by **histone acetylation**, a regulatory step

Figure 5.10 DNA packing in a eukaryotic chromosome.

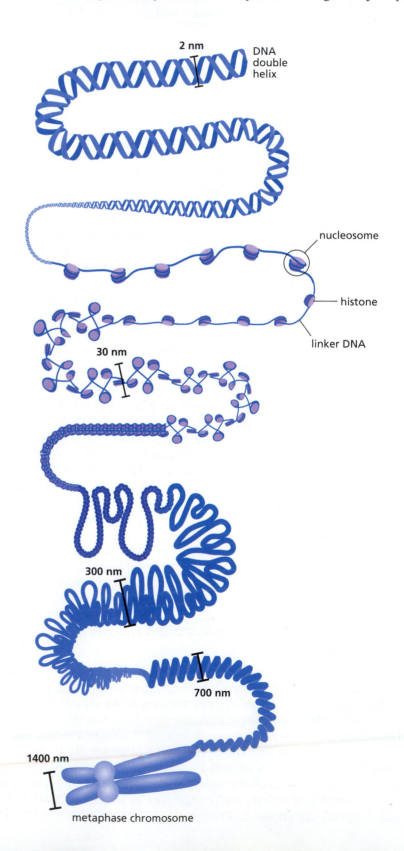

2 nm — DNA double helix

nucleosome

histone

linker DNA

30 nm

300 nm

700 nm

1400 nm

metaphase chromosome

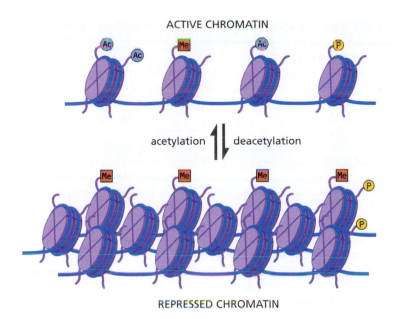

ACTIVE CHROMATIN

acetylation ⇅ deacetylation

REPRESSED CHROMATIN

Figure 5.11 Histone acetylation and deacetylation. Before the initiation of gene transcription, the nucleosomes are tightly packed together. During histone acetylation, the addition of acetyl groups to the tails of the histones changes the charge on the DNA and "opens up" the pattern of nucleosome packing. The promoter regions now become more accessible to the binding proteins that initiate gene transcription. During deacetylation, the acetyl groups are removed from the histone tails, causing the nucleosomes to move closer together.

that attaches acetyl groups (–COCH$_3$) to certain amino acids in the histone. When histones are acetylated, they change shape so that they grip the DNA less tightly, particularly in the promoter region. As a result, transcription proteins can access the genes in the acetylated region **(Figure 5.11)**. The opposite of histone acetylation, histone deacetylation, is discussed later in the chapter, as this mechanism appears to be important in the regulation of genes associated with longevity.

Gene expression is controlled by binding of proteins to DNA

One of the more important findings in genetics and cell biology was the discovery that the expression of genes can be turned on or off by the interaction of gene regulatory proteins with specific segments of noncoding DNA known as **control elements (BOX 5.1)**. A eukaryotic protein-coding gene can contain two types of control elements: **enhancers**, which increase RNA transcription, and **silencers**, which decrease RNA transcription. A transcription factor that binds to an enhancer and stimulates transcription of a gene is called an **activator**. A transcription factor that binds to a silencer and inhibits transcription is called a **repressor**. Both activators and repressors may be hundreds or thousands of nucleotides away from the promoter. The regulation of transcription results mainly from binding of activators and repressors to enhancers and silencers and transmission of their effect to the proteins bound to the promoter regions. Control elements seem to have a significant role in determining the longevity of some species, as described later in this chapter.

Activator proteins are expressed in response to hormones, metabolic products of hormones, amino acids, individual nutrients, and thousands of other types of molecules. When an activator protein binds to the enhancer site, the promoter region of the gene bends to expose the TATA box for the binding of general transcription factors **(Figure 5.12)**. DNA bending brings the bound activators closer to the promoter. The physical distortion in the DNA strand attracts a type of RNA polymerase called RNA polymerase II, as well as additional transcription factors and proteins that aid in transcription (well over a 100 proteins are involved in this process). One such protein, known as a **mediator**, helps bind the

BOX 5.1 CONSERVATION OF GENETIC SWITCHES: FROM FRUIT FLY TO HUMAN

One of the greatest questions in biology after Watson and Crick's description of DNA structure was, "How can two cells of one animal contain identical DNA yet differ so radically in their functions?" For example, consider the differences between a human liver cell (hepatocyte) and muscle cell (myocyte). Among many other functions, hepatocytes detoxify harmful compounds in the blood by using a specialized set of enzymes. Myocytes also contain these enzymes, but the low level of expression of detoxifying proteins in these cells makes this pathway almost nonfunctional. On the other hand, actin and myosin, the major contractile proteins of the body, are expressed in muscle cells at levels millions of times greater than that observed in the liver—their function in the hepatocyte is limited to structures that hold the cell together.

Prior to 1960, the differences in expression patterns among cells were thought to reflect a loss of genes during the differentiation process. This view was consistent with the observation that most metazoan cells lose the ability to divide after differentiation, although no empirical evidence describing a loss of specific genes existed. In the early 1960s, however, the pioneering work of François Jacob and Jacques Monod solved at least part of the mystery surrounding the regulation of gene expression. These Nobel laureates (along with André Lwoff) observed that, in the bacterium *Escherichia coli*, the enzyme β-galactosidase, which breaks the bond connecting glucose to galactose in the sugar lactose, was expressed only when lactose was present in the bacterium's growth medium. Using different mutants of *E. coli*, Jacob and Monod discovered that lactose influenced the binding of a protein to DNA. When lactose was absent, the protein was tightly bound to the DNA and prevented expression of β-galactosidase. Conversely, when lactose was present, the binding protein fell off the DNA and β-galactosidase was expressed, leading to breakage of the bond between glucose and galactose. Jacob and Monod had discovered the first genetic switch, the *lac* repressor, a finding demonstrating that differences in gene expression among cells reflect a regulatory process rather than a difference in genes per se. Tens of thousands of genetic switches in virtually all species examined have been identified since Jacob and Monod's initial studies in bacteria.

More importantly, subsequent research demonstrated that the DNA sequences in the binding sites of the genetic switches are highly conserved. For example, genes that determine the shape of the human body at the time of fetal development, known as the *Hox* genes, differ by less than one or two base pairs from the same genes found in the fruit fly. That is, the genes responsible for ensuring that our legs, arms, and fingers are in the proper position are not just similar but almost identical to the genes that determine the position of the wings, legs, and antennae of the fruit fly. Until this discovery, most scientists accepted the concept that differences between species reflected differences in the genes. That is, humans had very different genes than did the fruit fly. The discovery of the *Hox* genes changed this idea and suggested that, even though flies and humans took very different evolutionary paths almost 500 million years ago, both species retained identical base-pair sequences that regulate protein expression. Differences between species do not reflect different genes; rather, they reflect differences in when and how much of a protein is expressed. From a biochemical perspective, the fly and the human are identical. Indeed, the radical change in our understanding of protein expression led Monod to state (and rightfully so), "What is true for *E. coli* is true for the elephant."

enhancer and promoter regions. Together, RNA polymerase II, general transcription factors, and the mediator form the **transcription initiation complex**.

Repressor proteins can be as important as activator proteins to the regulation of gene transcription. Of the multiple mechanisms that eukaryotes use to repress gene transcription, four are most common: repressor proteins may (1) have a higher affinity for the enhancer site than does the activator protein; (2) mask the enhancer site by binding to a repressor site close to the activation site; (3) prevent formation of the transcription initiation complex; and (4) prevent formation of the gene activation complex altogether. This last is done indirectly through the process of histone acetylation-deacetylation. The first three mechanisms are shown in **Figure 5.13**. Later in this chapter, we discuss how the repression of gene transcription, also known as **gene silencing**, plays a significant role in the genetics of longevity.

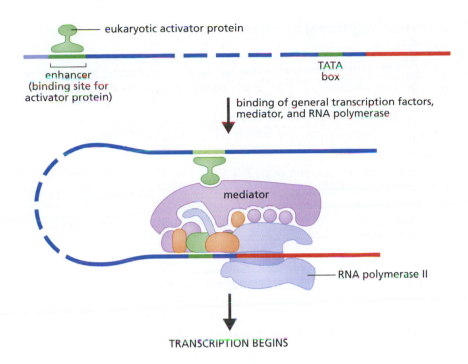

Figure 5.12 The action of activators and enhancers on gene transcription. Activator proteins bind to enhancer sequences in the DNA. The enhancer site, which may be thousands of nucleotides away from the promoter region, loops out and binds to the promoter region. Once the enhancer site comes within the proximity of the promoter, a mediator protein helps the enhancer bind to the promoter region. The promoter region, bound enhancer site, and RNA polymerase form the transcription initiation complex.

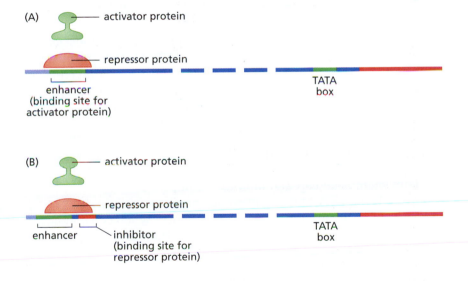

Figure 5.13 General mechanisms of gene transcription repression. (A) Competition between repressor and activator proteins for the same binding site. The repressor has a greater affinity for the binding site than does the activator. (B) Repressor proteins bind to a region near the enhancer site and prevent binding of the activator protein by masking the enhancer site. (C) Repressor proteins prevent formation of the transcription initiation complex.

Post-transcriptional mechanisms can also control gene expression

While most gene regulation occurs at the transcriptional level, before RNA polymerase binds to the gene's promoter region, significant regulation occurs post-transcriptionally. Here we look briefly at two types of post-transcriptional control, **alternative RNA splicing (Figure 5.14)** and **translational initiation**, which will prove useful in evaluating the genetic control of the rate of aging and longevity.

Recall that introns are removed from the pre-mRNA transcript before mRNA becomes active, by the process of RNA splicing. The snRNAs that perform the splicing recognize the start and end of an intron by regions known as **consensus sequences**. However, these consensus sequences have significant plasticity and allow the snRNAs to begin splicing the intron at slightly different locations—the process of alternative RNA splicing. This results in changes in the nucleotide sequence of the finished mRNA, resulting in slightly different proteins with slightly different activity. That is, alternative RNA splicing allows the cell to make several variations of a common protein from one gene, increasing the coding potential of the genome.

Regulation of alternative RNA splicing occurs by the binding of regulatory proteins to the pre-mRNA and can vary with cell type or developmental stage. For example, the *Drosophila* gene *dsx* encodes a protein that is important to secondary sexual characteristics. If the RNA splicing results in six exons being translated, male sex characteristics are expressed. Conversely, if RNA splicing results in four exons in the mRNA, female characteristics are expressed. Thus, one gene results in two proteins, suggesting different genotypes.

Cells can also control the amount of protein being synthesized by regulating the initiation of translation. Before mRNA leaves the nucleus, RNA polymerase "caps" the 5′ end with a methyl group attached to a guanosine (G) nucleotide (see Figure 5.6). This capping is another way of distinguishing mRNA from other RNA fragments in the nucleus so that only mRNA moves through the nuclear pore complexes. The methylguanosine cap also serves as a location of inhibition of translation. The cap is normally found very close to the first AUG (start) codon. Repressor proteins expressed in response to cellular signals recognize the methylguanosine cap and bind to the mRNA just before the AUG. Binding of the repressor protein prevents the ribosome from translating the base sequence into the amino acid sequence. When cell conditions change, the repressor protein is degraded and translation begins.

Figure 5.14 Alternative splicing can produce different mRNAs from the same gene.

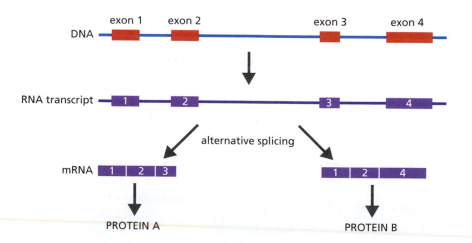

ANALYZING GENE EXPRESSION IN BIOGERONTOLOGY

One of the most important advances in biology occurred during the 1970s, when scientists were able to remove a piece of DNA from a chromosome and determine its sequence. The ability to extract and characterize a DNA segment led to the development of technologies that allowed scientists to manipulate a gene sequence or develop entirely new genes, then to study how these treatments affected the functions of cells and organisms. These techniques, known collectively as **recombinant DNA technology**, would change the direction of all research in biology. From ecology to molecular genetics, scientists could now identify specific genes and gain a more complete understanding of the fundamental mechanisms underlying the why and the how of life.

Recombinant DNA technology also spawned the boom in the commercial biotechnology industry. Techniques that were once the domain of only the most highly trained and skilled scientists were now automated and commercialized so that the nonspecialist could apply sophisticated gene analysis techniques to specific research interests. Biogerontologists began using these techniques to identify possible genes that influence the rate of aging and longevity.

You learned in Chapter 1 how biogerontologists use simple organisms such as yeast, worms, and flies to study the genetics of aging. These organisms provide the biogerontologist with several advantages, including (1) their genomes have been sequenced—that is, the sequence and locations of all genes are known; (2) the long history of general genetics research using these organisms has resulted in hundreds if not thousands of readily available mutants with a variety of genotypes and phenotypes; and (3) the life spans of these organisms are short, allowing for several longevity studies within a month (yeast) or a year (worms and flies). You also learned that any genes identified as modulating the rate of aging and longevity in these simple eukaryotes will have **gene orthologs**—genes having similar functions—in more complex organisms, given the phylogenetic relationships. Once a gene has been shown to affect longevity in simple eukaryotic organisms, a similar gene can be identified in a more complex organism, such as a mouse, and its possible relevance to the rate of aging or longevity in humans evaluated.

Using the enormous amount of accumulated knowledge about the genetics of these three model organisms, biogerontologists have identified a few genes that seem to directly modulate the rate of aging and longevity. The details of these discoveries are presented in the sections that follow. First, we explore the general techniques and methods used in the discovery of these longevity genes.

The general process leading to the identification of genes that modulate the rate of aging and longevity requires several steps **(Figure 5.15)**. The complexity of this process is such that no single researcher or laboratory could possibly perform all the steps required. The gene discovery process reflects a community of researchers throughout the world, relying on one another's scientific ability to perform the critical steps in the process. Although the techniques and methods described briefly here are presented as a single, seamless procedure, keep in mind that each step entails, for the most part, independent laboratories completing work over several years.

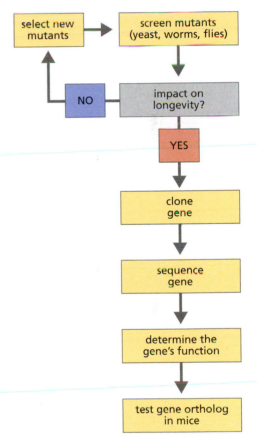

Figure 5.15 General scheme for identifying genes that affect longevity. Researchers consult computer databases or the research literature to choose a mutant strain that has a genotype or phenotype that may affect longevity. Once the mutant strain has been selected, the next step is to clone the gene. Cloning is the process of making several identical copies so as to have enough genetic material to sequence the gene. Once the gene has been sequenced, its function is determined. The final step of the process is to identify the gene's ortholog in mice and evaluate whether it has a similar effect on longevity.

Genetic analysis in biogerontology begins with the screening of mutants

By far the greatest advantage of using *S. cerevisiae*, *C. elegans*, and *Drosophila* for genetic analysis in aging research is that their entire genomes have been sequenced. Sequencing of a gene, however, does not provide information about the function of that gene. To gain that knowledge, geneticists start by creating mutant strains of these organisms in which a single gene has been altered to produce a specific genotype. Looking at the differences between the genotypes of the mutant strain and the **wild type** (that is, the organism in its normal or natural condition, as distinct from an atypical or mutant type) gives scientists the first information about the function of that gene. A biogerontologist with a particular hypothesis about how genes affect the rate of aging and longevity can then select mutant strains showing genotypes consistent with a particular hypothesis. (Because most research is performed with public money, researchers are obligated to make their mutant strain available to other scientists.) In general, biogerontologists select strains with mutations affecting growth and/or reproduction. (Recall, from Chapter 3, the close relationship between growth, reproduction, and longevity.) Since many genes affect growth and reproduction, the biogerontologist must select several different mutants and then compare their longevity characteristics to determine which one best suits his or her needs.

The comparing of mutant strains, known as **genetic screening**, may seem relatively easy and straightforward. However, these beginning steps of genetic analysis are the most critical and the most difficult. Most genes are pleiotropic (a single gene producing more than one phenotype), and mutant strains often show several phenotypes associated with the mutation of a single gene. Even if one of the mutant strains has a desirable longevity characteristic, it may also have a phenotype that could confound life span and cause significant variation in results. Thus, the genetic biogerontologist may have to refine the mutation so that only the longevity phenotype is present. Refining the genome so that there is only one genotype or phenotype resulting from one gene mutation requires meticulous research, with lots of trial and error that ends in failure. It can take years for a researcher to succeed in developing a mutant that provides a suitable and reliable experimental system for testing a hypothesis about the impact of genes on longevity.

Identification of gene function requires DNA cloning

Even after a scientist has developed a reliable mutant strain with a phenotype having extended longevity, the function and regulation of the gene, as well as the structure of the protein it encodes, will be unknown. To determine the function of a gene, it must be **cloned**—that is, identical copies of the gene must be created. By cloning a gene, the researcher produces enough DNA material to determine the gene's base-pair sequence. Knowing the base-pair sequence, the researcher can determine the amino acid sequence of the protein it encodes and thus the protein's possible function.

The cloning process was once a very laborious task requiring several different procedures, perhaps taking days to complete. Today, these procedures are automated and combined into a single process known as **polymerase chain reaction (PCR)**, which takes only hours to complete **(Figure 5.16)**. The process begins with the isolation of either DNA or mRNA from tissue or cells, procedures that are fairly easy. DNA is

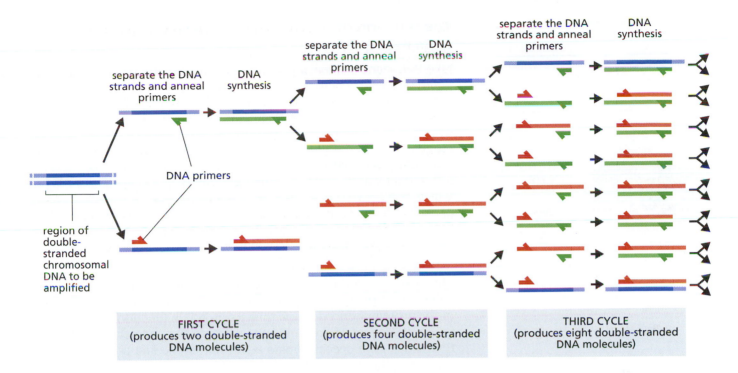

FIRST CYCLE
(produces two double-stranded
DNA molecules)

SECOND CYCLE
(produces four double-stranded
DNA molecules)

THIRD CYCLE
(produces eight double-stranded
DNA molecules)

used as the starting sequence if the scientist needs to clone the entire genomic sequence of the gene, since genomic DNA includes both exons (the mRNA-coding sequences) and introns. If determining the amino acid sequence of a protein is the objective, only the exon sequence of the gene is needed, and the starting material is mRNA (recall that mRNA is the complementary product of only the exons). Here we describe the process using mRNA as the starting material.

Before beginning the PCR process, the researcher must design and chemically create a set of oligonucleotides (*oligo-* meaning "many") that are exclusively complementary to the 3′ end of the gene being isolated. The 3′ end of the gene is known because the sequence of the gene being used to develop the mutant is known. In everyday practice, the researcher designs the oligonucleotides and sends those plans to a company for synthesis. When the oligonucleotides, or "primers," are added to a solution containing the mRNA, they direct an enzyme called a **reverse transcriptase** (which makes DNA from mRNA, the reverse of the usual process in cells) to start replicating the gene being amplified. The deoxyribonucleotides necessary to make DNA are also added to the solution. The reverse transcriptase uses the mRNA template to make a new, complementary DNA (cDNA) strand containing only the exons from the gene being isolated.

At this point in the process, the mRNA and the new cDNA are bound together by hydrogen bonds in a double strand and must be separated so that only the cDNA is amplified. The PCR machine heats the solution to around 90°C to break the hydrogen bonds. DNA polymerase and a second set of primers, complementary to the 3′ end of the cDNA, are then added, and a new strand of cDNA complementary to the first cDNA is made. The PCR machine slowly cools the solution, and the two complementary strands of cDNA form a double helix. The amplification process is now a simple matter of heating and cooling the solution in the presence of deoxyribonucleotides and DNA polymerase. Each cycle doubles the number of identical genes. It takes about 30 cycles to produce enough material for sequencing the gene.

Figure 5.16 Cloning a gene using PCR. After identification of the segment of DNA to be cloned, the DNA is heated to separate the strands, and short complementary sequences known as primers are attached to the 3′ end of the sequence of interest. The primers direct a reverse transcriptase to the start of the gene. The DNA is then mixed with a DNA polymerase and deoxyribonucleoside triphosphates, and the DNA is replicated to create two daughter DNA molecules. Repeating the process multiple times, using the newly synthesized fragments as templates, quickly amplifies the number of copies.

The function of the gene can be partially determined from its sequence

Like PCR, gene sequencing is now automated. The newly synthesized gene copies are added to a solution that breaks the cDNA into fragments terminating at specific bases. Because there are literally billions of copies of the new gene, cDNA fragments are generated that terminate at every position on the gene. The robotics of the sequencing machine mix reagents that add a fluorescent dye to the terminal base of the cDNA fragment. Each base is labeled with a different color. The fragments are separated by atomic weight, using **gel electrophoresis (Figure 5.17)**, which is also automated by the sequencing machine. Based on the presence of the fluorescent dyes, detectors in the sequencing machine read the position of each fragment on the gel relative to the other fragments, and with the aid of computers, the sequencer determines the sequence of the gene.

The final step in this process involves applying the genetic code (see Table 5.1) to the gene sequence to determine the amino acid sequence of the protein it encodes and thus its probable function. However, the raw gene sequence does not tell us where the first set of three bases that codes for a specific amino acid (that is, the first codon) starts. At this stage, the researcher can only make an educated prediction about the function of the gene, based on the collective knowledge of previous work in genetics. The National Institutes of Health maintains a large computer database of gene sequences with known functions. (Once a gene's function is determined, researchers submit their gene sequence to this database.) Computer software has been developed that allows researchers to compare their sequence to sequences of known genes and receive information on the probable function of their gene. Researchers can use this information to focus their research and, using recombinant technology, can determine more precisely certain characteristics of their gene, including the start and stop codons.

In situ hybridization can reveal a gene's function

Determining the function of a gene often starts with an evaluation of where and when the gene is expressed. To accomplish this, a technique call *in situ* **hybridization** (*in situ* is Latin for "in place") can be used. Since the sequence of the gene is known, segments of the gene can be labeled with a detectable and measurable compound, such as a radioactive isotope or fluorescent dye, and introduced into cells, tissues, or whole organisms. When these labeled nucleic acid probes encounter a complementary sequence of mRNA, they bind to that sequence and create a hybrid molecule. The hybrid molecules can then be isolated and measured using the appropriate detection machines.

In situ hybridization is one more technique that can be used to evaluate the expression pattern of a gene. For example, suppose we determined, using computer databases, the most likely function of a gene extracted from *C. elegans*. Our analysis suggests that the gene's sequence has high homology with genes that express their mRNA in response to certain ambient temperatures. Because our gene is unique, we don't know what that temperature might be. We can determine the temperature, and thus the function of the gene, by injecting *C. elegans* with labeled DNA probes and then expose the animals to different temperatures. The greatest *in situ* hybridization will occur when a temperature is reached that causes the greatest gene expression. In addition, this experiment provides information about which cells or tissues have the greatest expression of the gene. As you'll see later in the chapter,

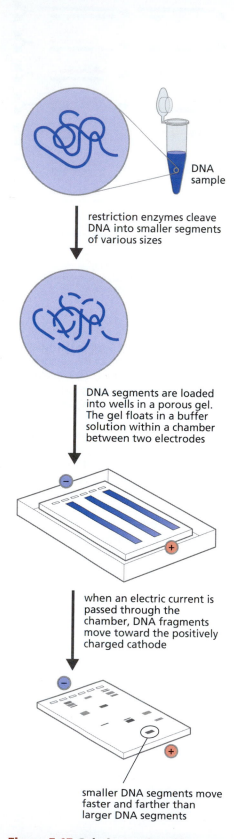

DNA sample

restriction enzymes cleave DNA into smaller segments of various sizes

DNA segments are loaded into wells in a porous gel. The gel floats in a buffer solution within a chamber between two electrodes

when an electric current is passed through the chamber, DNA fragments move toward the positively charged cathode

smaller DNA segments move faster and farther than larger DNA segments

Figure 5.17 Gel electrophoresis. Electrophoresis on a porous gel physically separates DNA molecules on the basis of their size and electrical charge.

an experiment similar in design to this hypothetical one was used to determine which cells in *C. elegans* express a protein associated with enhanced longevity.

Genetically altering organisms helps evaluate a gene's impact on human longevity

As you'll recall, one of the many reasons that a genetic biogerontologist begins genetic analysis with simple organisms is that they have shorter life spans than more complex organisms. This means that scientists can screen for longevity genes in a relatively short amount of time. Once the gene has been cloned and sequenced, and its function determined in simple organisms, mutant strains of more complex animals with physiology closer to that of humans can be developed. In general, the mouse has been the mammal of choice for genetic analysis of the rate of aging and longevity.

A common method for evaluating a gene's influence on the rate of aging or longevity is to develop mutant strains of mice that either have an extra copy of the gene or do not have the gene at all. Comparing these mutants with the wild-type mouse demonstrates whether the gene alters the rate of aging or longevity. Having determined which mouse gene to modify, it is necessary to know whether gene expression or gene silencing caused the increase in longevity, knowledge that is gained from analysis of the simpler organisms. If repression (gene silencing) causes the longevity effect, then the gene ortholog is removed from the mouse's genome. This type of mutant is called a **gene knockout**. If expression of the gene causes an increase in longevity, adding an extra gene results in greater expression than is observed in the wild-type mouse. This type of mutant is called a **transgenic organism. Figure 5.18** illustrates how transgenic and knockout genes are produced for use in mice. Once one of these genes has been created, it is amplified and inserted into mouse embryos and will then be incorporated into the mouse's genome.

DNA microarrays are used to evaluate gene expression patterns at different ages

Our discussion to this point has focused on recombinant DNA technology used to determine a single gene's function(s). However, in the real world, a cell may express hundreds and maybe thousands of genes in response to a single stimulus or cellular event. Until the mid-1990s, determining the gene expression pattern in response to a single stimulus was impossible. The advent of hybridization techniques led to the development of a tool called **DNA microarrays**, which allow researchers to determine the gene expression pattern of a cell at a single point in time.

DNA microarrays are pieces of glass, about the size of a postage stamp, that have been embedded with thousands of single-stranded DNA sequences from specific genes with known or unknown functions. Robots are used to precisely place the DNA sequences on the glass so that their position is known. Then, mRNA is extracted from the cell of interest and converted to cDNA, which is labeled with fluorescent probes. The microarray and the cDNA probes are incubated together, allowing hybridization to occur. The microarrays are then washed to remove unbound probes, and the hybridized molecules are identified by their fluorescence, using automated microscopes. **Figure 5.19** shows this process, using a hypothetical example of analyzing mRNA isolated from the muscle cells of mice at two different ages.

Figure 5.18 Making a transgenic or knockout gene for use in mice. (A) Gene addition. After cloning the gene ortholog, a fragment of the DNA with the original gene is exposed to restriction nucleases. The restriction enzymes cleave the DNA at specific sequences. The cloned gene is then added to a mixture containing DNA ligase, an enzyme that stitches DNA together at specific sites. The final product is a DNA fragment that has two copies of the same gene. (B) Gene removal. Restriction enzymes and DNA ligase are used to remove the identified gene, leaving the mouse genome without the gene of interest (such as a suspected longevity gene).

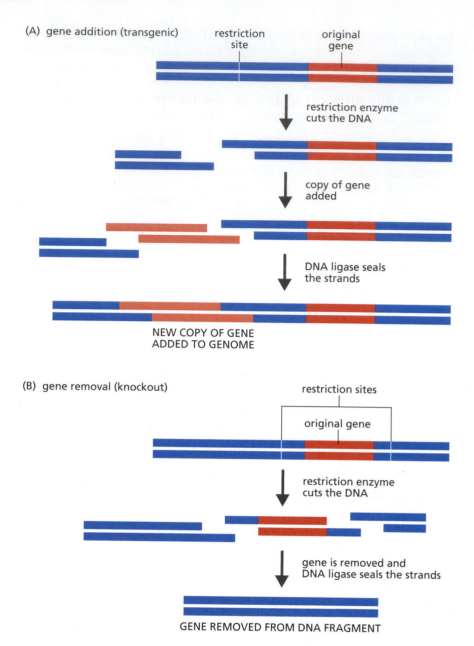

(A) gene addition (transgenic)

restriction enzyme cuts the DNA

copy of gene added

DNA ligase seals the strands

NEW COPY OF GENE ADDED TO GENOME

(B) gene removal (knockout)

restriction enzyme cuts the DNA

gene is removed and DNA ligase seals the strands

GENE REMOVED FROM DNA FRAGMENT

Experiments similar to the hypothetical example shown in Figure 5.19 have been carried out. The results were mixed. Differences in gene expression between young (3–6 months) and old (26–30 months) mice were extremely modest and depended on the organ from which the mRNA had been isolated. For none of the organs did the difference

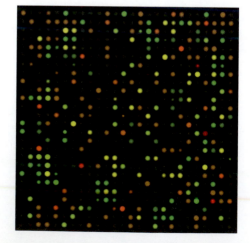

Figure 5.19 DNA microarrays reveal gene expression patterns. DNA microarrays are often used to evaluate gene expression patterns in two samples treated with a different stimulus or obtained under different conditions. In this hypothetical example, mRNA was isolated from muscle cells of mice at two different ages. After converting the mRNA into cDNA, the fragments are labeled with a red (young mice) or a green (old mice) fluorescent dye and mixed together. The solution is allowed to hybridize to the cDNA fragments on the microarray. After the unbound fragments are washed off, the fluorescence is scanned by an automated detector. Green dots indicate greater gene expression in the old mice; red dots, greater expression in the young mice; yellow dots, expression in both young and old mice; and brown dots, no expression of the gene by either group. (Courtesy of Alila Medical Images/Shutterstock.)

between young and old mice exceed 3% of the total genes represented on the microarray, with most organs showing only about a 1.0–1.5% difference. These values are largely considered statistically insignificant. Moreover, the gene expression pattern of the old mice was not consistent across organs. That is, the age-related gene expression pattern in the heart was not the same as that observed in the brain, which was not the same as that in the muscle, and so on. These results are consistent with the proposition that aging is not a result of a genetic program but more closely reflects random stochastic events that vary considerably among tissue types.

The results of investigations designed to evaluate age-related gene expression patterns suggest that the usefulness of microarray technology to biogerontology may be in determining how interventions that delay aging alter the gene expression pattern. For example, we now know that maintaining a healthy weight by reducing caloric intake can significantly slow the rate of aging of many organ systems in humans. The knowledge that this effect may be directly related to an age-related alteration in the gene expression pattern arises from microarray studies performed using cDNA from mice undergoing calorie restriction. Restricting the caloric intake to 70–80% of calories normally consumed by mice given free access to food results in a significant reduction in functional loss in all physiological systems (this is further described in Chapter 10). Comparison of gene expression profiles between free-eating and calorie-restricted aged mice shows dramatic differences. Although differences in the gene expression profile vary from organ to organ, and how these differences directly affect the rate of aging has yet to be determined, these results provide solid evidence that genes can affect the rate of aging and the life span. The identification of specific genes that respond to interventions will greatly advance our knowledge of the factors contributing to inherited differences in the rate of aging.

GENETIC REGULATION OF LONGEVITY IN *S. CEREVISIAE*

The single-cell eukaryotic organism *Saccharomyces cerevisiae* has been used for many years in aging research. It is a model organism for this research because it is short-lived, is inexpensive to grow and maintain, and has well-characterized genetics and physiology. Commonly known as brewer's or bread yeast, *S. cerevisiae* was the first eukaryotic organism to have its genome sequenced (completed in 1996). The 12,156,590 base pairs in the DNA sequence are contained in a haploid set of 16 chromosomes. There are approximately 6600 known genes with an average size of 490 codons (1.47 kbp). The genome of this yeast is compact, with 75% of the DNA sequence as exons. Although researchers have identified many yeast strains having mutations that influence longevity, only a few of these have been screened in sufficient detail to warrant discussion here. Genetic mechanisms related to longevity in *S. cerevisiae* include genomic instability, gene silencing, and mutations in nutrient-responsive pathways. We take a look at each of these in more detail in this section, but first we need to briefly discuss how the yeast reproduces, because its reproductive cycle affects its longevity.

S. cerevisiae reproduces both asexually and sexually

Yeast can exist in two chromosomal states, one **diploid** (containing two sets of chromosomes) and one **haploid** (containing only one set of chromosomes). The haploid cells reproduce mitotically, creating a

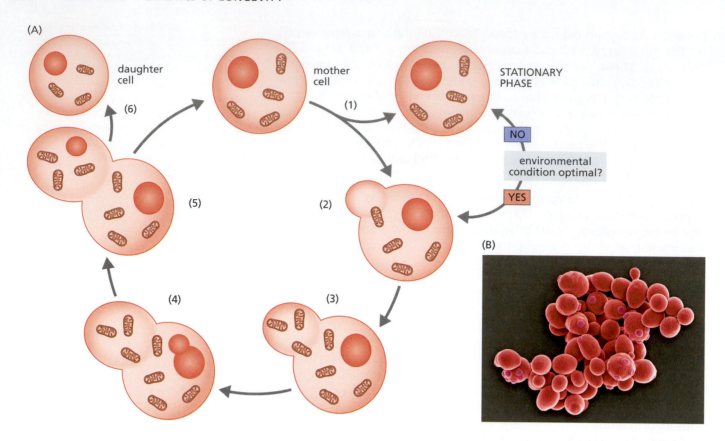

Figure 5.20 The asexual reproductive cycle of *S. cerevisiae*. (A) Asexual reproduction occurs by budding. (1) The mother cell can sense whether environmental conditions are optimal for reproduction. If they are suboptimal, the mother cell may enter a stationary phase until conditions improve. (2) If environmental conditions are suitable for reproduction, a bud forms as an extension of the mother cell's cytoplasm. (3) New cytoplasmic organelles are synthesized and migrate into the forming bud. (4) Mitosis begins, the genetic material is duplicated, and a new nucleus splits from the mother nucleus. (5) The newly formed nucleus migrates into the bud. (6) Cytokinesis occurs, and mother and daughter cells separate. (B) Scanning electron micrograph of new yeast cells (daughter cells), forming as buds on mother cells. The number of daughter cells produced by a mother cell, as indicated by the number of scars left by daughter cells after separation, provides a measure of reproductive senescence. Note the smaller size of the newly separated daughter cells. (B, courtesy of S. Gschmeissner/Getty Images.)

new cell from the parent cell by budding. During budding, the nucleus of the mother cell undergoes mitosis to form a daughter nucleus, which migrates into the forming bud **(Figure 5.20A)**. **Chitin**, a tough polymer of nitrogen-containing sugars, forms between the bud and the mother cell. When mitosis is complete, the new daughter cell separates from the mother cell (Figure 5.20B). After separation, a circular chitin residue known as the **bud scar** remains permanently on the surface of the mother cell.

Diploid reproduction occurs as a matter of chance and is regulated, at least in part, by environmental conditions. Budding results in two types of daughter cells, a and α. If the two opposite mating types meet, they can fuse and enter the diploid phase of the cell cycle. In environmental conditions that are favorable for the survival of daughter cells, the diploid yeast reproduces by mitosis, or budding. If, however, environmental conditions are such that the survival of daughter cells is at risk, the diploid yeast produces four gametes called **ascospores**, two with a and two with α chromosomes. When environmental conditions are more favorable, the spores germinate and produce four haploid yeast cells.

The life history of budding yeast can be described by the Gompertz mortality equation, which distinguishes replicative senescence in the single-cell eukaryotic yeast from mitotic cells derived from the tissue of multicellular organisms. As shown in **Figure 5.21**, the number of bud scars and the size of the mother cell provide convenient markers for estimating the reproductive age of the mother cell. Individual *S. cerevisiae* produce an average of 20–30 new daughter cells (20–30 bud scars) before entering replicative senescence. However, neither the number of bud scars nor the size of the mother cells is the direct cause of senescence. Mutations that increase the number of bud scars and/or cell size do not affect longevity.

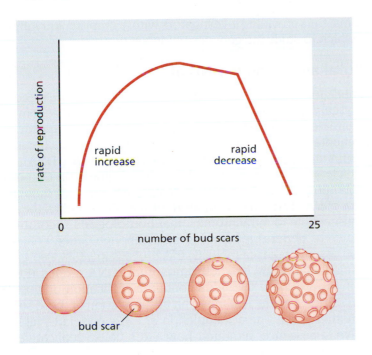

Figure 5.21 Bud scars and replicative senescence of *S. cerevisiae*. A young yeast cell is small, has few bud scars, and undergoes a rapid rate of budding. As shown in the graph, the initial burst of reproduction is followed by an extended period with a stable reproductive rate. As the yeast approaches its maximum number of buddings, reproduction rapidly declines and the organism dies.

Environmental conditions influence reproduction and life span

In Chapter 3, we showed that the proximal environment has a significant influence on the life history pattern and reproductive success in populations of single-cell organisms such as *S. cerevisiae*. The life history of *S. cerevisiae* shows the typical pattern associated with wild, single-cell populations: (1) a rapid increase in population when resources (food, space, etc.) are abundant; (2) stabilization of population size when resources match the intrinsic rate of growth (r = reproduction rate – death rate = 0); (3) decrease in population size (decrease in reproduction) as resources begin to decline; and (4) replicative senescence and death of the population when resources are exhausted.

S. cerevisiae has two reproductive strategies that help extend its reproductive life span and prevent the death of the population during times of limited or depleted environmental resources. Sexual reproduction of the diploid yeast produces ascospores, which can survive several years and can withstand many different types of environmental conditions. This provides a long-term solution for extending reproduction during times of environmental stress. The budding of haploid yeast provides a short-term solution for extending the reproductive period. When a budding yeast detects that environmental conditions are detrimental to the survival of a daughter cell, asexual reproduction (budding) is turned off. The budding yeast enters a nondividing state, called the stationary phase, but remains metabolically active. If the environmental conditions improve within 1–2 weeks, the yeast resumes reproductive activity; otherwise, the cell dies.

In the stationary phase, the yeast has significantly elevated levels of oxygen radicals toward the end of its life span. In this phase, mutant yeast cells without genes for superoxide dismutase (SOD) have shorter life spans, whereas mutants that overexpress SOD have significantly longer post-reproductive life spans. The increase in oxygen radicals in the nondividing yeast seems to stimulate the p53 checkpoint protein, which, in turn, results in an apoptotic-like death. The nondividing yeast has senescence characteristics similar to those of more complex organisms, thus providing investigators with a simple eukaryotic organism in which to study the genetics of longevity.

Structural alteration in DNA affects life span

The genetic mechanisms that account for replicative senescence in the budding yeast remain, in large part, unknown. However, genomic instability, characterized by a structural alteration in the DNA, appears to play a fundamental role in the yeast life span. The genomic instability associated with aging yeast occurs in the **ribosomal DNA (rDNA)**, located in the nucleolus. The yeast rDNA locus is a segment on Chromosome XII, containing 100–200 repeats of 9 kbp, arranged in tandem, that codes for rRNA. The repetitive nature of the rDNA makes these areas highly susceptible to intrachromosomal **homologous recombination**, also known as **general recombination**. Although general recombination is an essential part of meiosis, this process in yeast also gives rise to potentially damaging **extrachromosomal rDNA circles (ERCs)**, as shown in **Figure 5.22**. The ERCs contain a replication origin site and thus can self-propagate, leading to an accumulation of these structures in the mother yeast. The accumulation of ERCs correlates highly with replicative senescence, although how ERCs cause replicative senescence has yet to be demonstrated. Studies in *S. cerevisiae* suggest that the production of ERCs and the subsequent replicative senescence may be related to the acetylation-deacetylation mechanism of gene expression regulation.

The ERCs are a "biological clock" of sorts, associated with the life span of *S. cerevisiae*. This conclusion is the result of two findings. First, budding yeast containing a mutation that prevents the formation of ERCs live significantly longer than wild-type yeast. Second, daughter cells that arise from old mother yeast cells receive ERCs and have significantly shorter life spans than ERC-free daughter yeast.

The SIR2 pathway is linked to longevity

Silencing in the transcription of rDNA into rRNA appears to be regulated by a highly conserved gene named *silent information regulator 2* (*sir2*). Overexpression of sir3 results in decreased rDNA expression and reduced formation of ERCs. Overexpression of *sir2* doubles the yeast's replicative life span. The means by which the *sir2* gene and its encoded protein (SIR2) cause extension of the yeast replicative life span has not been completely worked out. We do know, however, that the SIR2 protein is a histone deacetylase, which causes compaction of the chromatin and gene silencing.

SIR2-induced gene silencing seems to be connected directly to the pathways that convert food into energy. As described in Chapter 4, electrons arising from the oxidation of glucose are shuttled to the ATP-synthesizing apparatus by electron carriers. One of these carriers, oxidized nicotinamide adenine dinucleotide (NAD^+), activates or "turns on" the SIR2 protein. Thus, the cell concentration of NAD^+ directly affects the level of SIR2 activity; low concentrations of NAD^+ deactivate

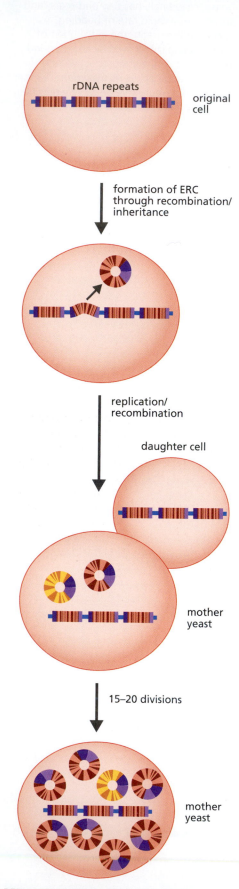

REPLICATIVE SENESCENCE AND DEATH

Figure 5.22 Formation of extrachromosomal rDNA circles (ERCs) and yeast replicative senescence. The budding yeast contains highly repetitive sequences of rDNA, occurring in tandem, that form ERCs through general recombination. ERCs form in the mother cell after each round of budding. The accumulation of ERCs is associated with replicative senescence and death of the yeast.

the SIR2 protein; high concentrations of NAD+ activate the SIR2 protein. The NAD+ concentration fluctuates with the amount of food in the yeast's environment. Times of plenty lead to increased reproduction and relatively high rates of metabolic activity. In turn, the metabolic pathways generate increased electron flow and a lower ratio of NAD+ to NADH + H+ **(Figure 5.23)**. Thus, the activity of the SIR2 protein declines, rDNA is transcribed, and reproduction (budding) proceeds. Conversely, when food becomes scarce, metabolism decreases and the resulting diminished electron flow increases the ratio of NAD+ to NADH + H+. In turn, NAD+ binds to SIR2, causing activation of the histone deacetylase and compaction of the chromatin, silencing of the rDNA locus, reduction of the budding process, and an extended replicative life span for the yeast.

Loss-of-function mutations in nutrient-responsive pathways may extend the life span

The amount of food in the yeast's environment controls, in large part, the rate of reproduction and thus the length of life. An abundance of food results in high rates of reproduction and shorter life span. Food shortages delay reproduction and extend life span. The way in which the yeast recognizes the amount of food in its environment and then alters reproduction appears to be related to specific nutrients.

The biochemical or physiological link among extracellular nutrients, reproduction, and life span may involve highly conserved cellular

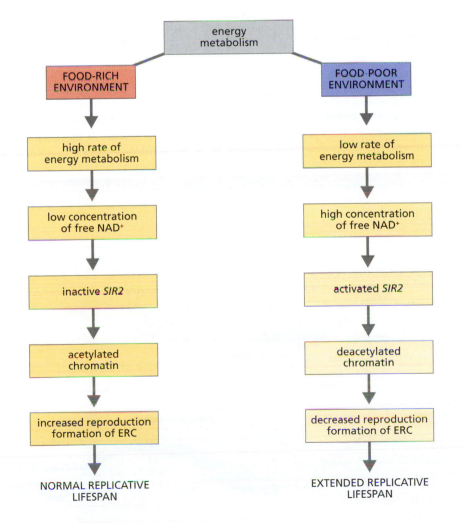

Figure 5.23 A proposed mechanism that links energy metabolism in *S. cerevisiae* to an extended life span. Times of plenty lead to increased reproduction and relatively high rates of metabolic activity (*left*). During periods of scarce food supply and low reproduction, energy metabolism in the yeast is reduced (*right*), leading to decreased rDNA transcription, reduced budding, and inhibition of ERC formation.

signaling pathways. These pathways in yeast begin when a receptor on the cell membrane binds nutrients or nutrient metabolites. At least one of the signaling pathways responds to the uptake of glucose and may involve the regulation of sir2, but the details of this pathway are not clear. Other pathways are initiated through **ligands**, molecules that bind specifically to receptor sites on other molecules. Certain ligands bind to **G-protein-coupled receptors**, a type of receptor known to initiate signaling cascades linking nutrient availability to the expression of genes important to growth and reproduction in all eukaryotes.

Two general nutrient-responsive pathways have been identified that alter life span in yeast: **target of rapamycin (TOR)** and **protein kinase A (PKA)**. Mutations in these two pathways extend life span by inhibiting enzymes that catalyze the phosphorylation of proteins, although each pathway is independent. While the details of how TOR and PKA extend longevity are unclear, the mechanisms undoubtedly involve components of the highly conserved G-protein signaling process.

GENETIC REGULATION OF LONGEVITY IN *C. ELEGANS*

The nematode *C. elegans*, just 1 mm long, lives in the soil in several climate zones; it has been used as a model organism for the genetics of longevity for more than 30 years. This roundworm feeds primarily on bacteria, but it can be grown in the laboratory on several different growth media.

There are two sexual forms of *C. elegans*, hermaphrodite and male. Hermaphrodites produce both sperm and eggs and reproduce by self-fertilization **(Figure 5.24)**, whereas males produce only sperm. Males arise spontaneously at very low frequency (1 in 500) and can fertilize hermaphrodites; hermaphrodites cannot fertilize other hermaphrodites.

All 959 cells of the adult hermaphrodite are post-mitotic; the adult male consists of 1031 post-mitotic cells. Being completely post-mitotic and having a significant post-reproductive life span makes *C. elegans* an exceptional model for studying the genetic regulation of longevity in multicellular organisms. Another important advantage in using *C. elegans* is that the lineage and function of every cell are known. Interestingly, approximately one-third of the cells are nerve cells, which makes *C. elegans* especially useful for describing cell regulatory and signaling pathways.

Figure 5.24 The two sexual forms of *C. elegans*: hermaphrodite and male.

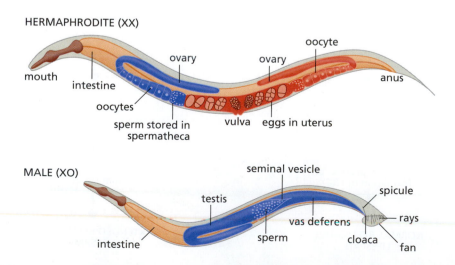

The genome of *C. elegans* has approximately 100 million bp, eight times the size of the yeast genome and about three-quarters the size of the *Drosophila* genome. *C. elegans* has six chromosomes, five pairs of **autosomes** and one **sex chromosome**. Hermaphrodites have two sex chromosomes (designated XX). Males have one X chromosome (designated XO). Males cannot produce progeny on their own. However, they can cross-fertilize hermaphrodites. The sequence of the genome was completed in 1998 and has approximately 20,000 protein-coding genes. In this section, we examine the dauer formation gene (*daf-2*), along with other genes associated with *daf-2*, and the clock genes.

Regulation of dauer formation extends life span

Development from egg to adult in *C. elegans* consists of four larval stages and takes about 3–4 days to complete. Adults are reproductively active for the first 4 days of adult life and may live an additional 10–15 days after this 4-day reproductive period. When *C. elegans* finds itself within an environment that will support the survival of offspring, the worm progresses through the four larval stages to adulthood in the normal 3- to 4-day period. However, if environmental conditions are less than optimal for reproduction, development is halted at larval stage 3 and a metabolically active but reproductively silent larva called the **dauer** forms **(Figure 5.25)**.

Figure 5.25 Developmental stages of *C. elegans*. Eggs are fertilized in the adult hermaphrodite and laid a few hours later—at about the 30- to 40-cell stage. After the eggs hatch, the worms pass through four larval stages (L1, L2, L3, and L4), each of which ends in a molt. On reaching adulthood, each *C. elegans* produces about 300 progeny. Its life span is about 2 weeks. If environmental conditions are less than optimal in stage 1, growth is arrested at stage 3 and a dauer forms. *C. elegans* can remain in this arrested state of development for several months. When environmental conditions improve, the dauer resumes development into the adult worm.

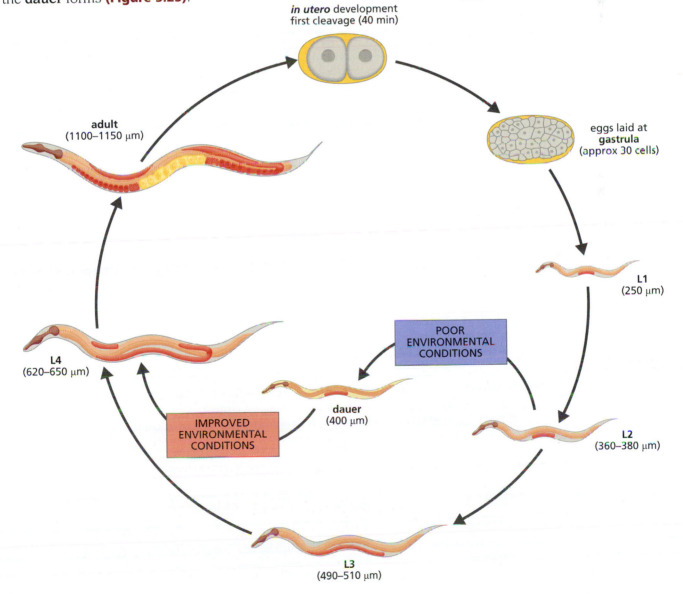

in utero development
first cleavage (40 min)

adult
(1100–1150 μm)

eggs laid at
gastrula
(approx 30 cells)

L1
(250 μm)

POOR
ENVIRONMENTAL
CONDITIONS

L4
(620–650 μm)

IMPROVED
ENVIRONMENTAL
CONDITIONS

dauer
(400 μm)

L2
(360–380 μm)

L3
(490–510 μm)

The sexually immature dauer is smaller than its stage-3 counterpart in the normal growth cycle, and it can survive for several months without food and in temperatures and soil conditions that are less than optimal for reproduction. It accomplishes this feat by reducing its metabolic rate, limiting protein synthesis, and surviving on stored fat. Anatomical changes to the dauer increase its resistance to stress and thereby increase its chance of survival. These changes include (1) increasing the thickness of its cuticle; (2) closing the buccal (mouth) cavity; and, in some cases, (3) increasing the concentration of endogenous antioxidants. So effective are these morphological alterations that the *C. elegans* dauer can survive several hours of insults from detergents, radiation, and a host of other noxious agents. Nonetheless, the dauer can rapidly resume development when environmental conditions improve. Genetic and biochemical mechanisms allow the dauer to re-enter the larval stages within one hour of placement of food in its environment; molting to larval stage 4 occurs within just eight hours after feeding.

Genetic pathways regulate dauer formation

Several proteins have been identified as important in formation of a *C. elegans* dauer, as shown in **TABLE 5.2**. The genetic pathways involved in formation of the dauer and its ability to survive under extreme stress have provided some important insights into at least one mechanism that can extend life span.

Genetic breeding of *C. elegans* mutants has identified two genes in the same signaling pathway—***age-1*** (so named for its aging phenotype) and the **dauer formation gene, *daf-2***—that are required for normal growth and reproduction. Without these genes, *C. elegans* forms dauers. Cloning of the *age-1* gene shows its protein product (AGE-1) to be a member of the highly conserved phosphatidylinositol-3-kinase (PIK-3) family. These proteins are important intracellular intermediates between membrane receptors that initiate a signal and the desired intracellular action. Cloning of the *daf-2* gene reveals that it encodes a transmembrane receptor protein (DAF-2) that has homology with an **insulin/insulin-like growth factor (IGF-1) receptor** found in other species. A highly conserved gene, *daf-2*, like its genetic orthologs, is important for normal growth and reproduction.

During times of environmental plenty, signals transferred to the worm result in expression of a DAF-2-binding protein, for which the

TABLE 5.2

BRIEF DESCRIPTION OF PROTEINS IDENTIFIED AS IMPORTANT IN FORMATION OF A DAUER AND EXTENSION OF LIFE SPAN IN *C. ELEGANS*

Protein name	Type of protein	General function
DAF-2	Insulin/IGF-1 receptor	Receptor for insulin/IGF-1 protein
AGE-1	Phosphatidylinositol-3-kinase (PIK-3)	Transduction of signal from receptor proteins to intracellular pathways; primarily involved in the transduction of mitogen signals from hormones
PDK1	Phosphatidylinositol-dependent protein kinase	Signal transduction intermediate between AGE-1 and AKT-1; phosphorylates AKT-1
AKT-1, 2, 3, etc. (also known as PKB)	Protein kinase B	Group of proteins involved in signal transduction pathways of cell growth and apoptosis
DAF-16	Forkhead transcription factor	Repression of genes involved in growth and development
Clk-1	Demethoxyubiquinone mono-oxygenase	Enzyme involved in biosynthesis of ubiquinone (coenzyme Q)

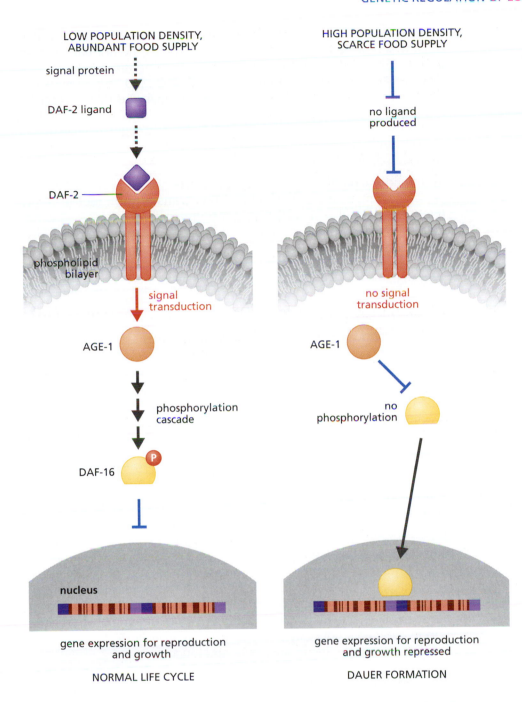

Figure 5.26 Regulation of gene expression in adult reproduction and dauer formation in *C. elegans*. Optimal environmental conditions stimulate the binding of an insulin-like ligand to the DAF-2 receptor (*left*). The signal propagates through the membrane and attracts intracellular AGE-1, a phosphatidylinositol-3-kinase, to the membrane, where the kinase initiates a phosphorylation cascade. The cascade ends with phosphorylation of DAF-16, a class of proteins involved in regulating reproduction. The phosphorylation of DAF-16 prevents its entry into the nucleus, and in the absence of DAF-16, genes involved in reproduction and growth can be expressed. Poor environmental conditions prevent the phosphorylation of DAF-16, which enters the nucleus and represses genes involved in reproduction (*right*); this results in dauer formation.

precise structure, kinetics, and specific ligand remain largely unknown **(Figure 5.26)**. Binding of the DAF-2-binding protein to the DAF-2 receptor causes the AGE-1 protein to migrate from the interior of the cytoplasm to the cell membrane. This finding provides an important clue that dauer formation is regulated by the repression of genes tied to an endocrine signal. That is, phosphatidylinositol-3-kinases are activated largely in response to hormones.

The next major advance in our understanding of dauer formation and its relationship to life-span extension in *C. elegans* came when researchers discovered that phosphorylation of DAF-16 was required for normal growth and reproduction. As shown in Figure 5.26, phosphorylation of DAF-16 prevents its migration from the cytoplasm into the nucleus. As a result, DAF-16 cannot exert its inhibitory effect on genes involved in growth and reproduction; *C. elegans* progresses normally through the four larval stages into adulthood and exhibits the normal life span of 10–15 days. If, on the other hand, environmental conditions are inadequate to support reproduction, the DAF-2-binding protein is not synthesized and no signal is transduced through the DAF-2 receptor to AGE-1. The DAF-16 protein is not phosphorylated, and the protein enters the nucleus, where it suppresses genes that regulate growth and reproduction. If the repression of genes by DAF-16 occurs during larval stage 3, a dauer forms.

In addition, cloning of the *daf-16* gene has shown that the DAF-16 protein is a member of the **forkhead box transcription factor family (FOXO)**, and that this particular transcription factor represses genes that govern growth and reproduction in *C. elegans* in response to low-food conditions. FOXO is a highly conserved group of proteins that are involved in reproduction in many different species, including several mammals.

Weak mutations in the DAF-2 receptor extend life span

Weak mutations in critical genes regulating dauer formation result in an extended adult life span. **Weak mutations** are alterations in a gene that result in expression being reduced rather than completely eliminated. When weak mutations are induced in *daf-2*, phosphorylation of DAF-16 is reduced but not prevented. Stage-3 larvae have slightly reduced metabolism compared with the wild type and do not commit to dauer formation. Rather, the reduction in DAF-16 phosphorylation allows *C. elegans* to progress normally through the four stages of larval development into the adult worm. While the progression to adulthood may be normal, the effects of the mutation are extraordinary: The adult worm with the weak mutation in *daf-2* lives 50–300% longer than the wild-type *C. elegans*.

The *C. elegans* extended-life-span mutant appears and behaves, for the most part, just like a normal-life-span adult. The adult worm remains fertile, and the hermaphrodite produces only slightly fewer offspring, although some slight modification to the mutation can result in infertility, indicating a close relationship between reproduction and extended life span. Unlike the dauer, the adult worm containing the weak mutation in *daf-2* eats normally and responds vigorously to changes in the environment, including temperature and touch.

Life extension is linked to neuroendocrine control

Cloning of the *daf-2* gene and identification of the DAF-2 protein as an insulin/IGF-1-like receptor indicates that gene regulation leading to extended life span of the adult worm is under neuroendocrine control. Further evidence of neuroendocrine regulation of *daf-2* came when the signal transduction pathway was found to involve a PIK-3 (AGE-1), a phosphorylation mechanism often found in hormonal regulation of the cell cycle. Elegant experimentation in which the *daf-2* mutation was localized to cells of different lineages—neural and muscle cells—confirmed that the environmental sensing and signal transduction involved in the *daf-2* mutation leading to extended life span were limited to neural tissue. When the mutation was limited to only muscle cells, there was no increase in longevity. However, limiting the mutation to neuroendocrine cells increased longevity. That is, neuroendocrine mechanisms seem to regulate dauer formation and extended life span in *C. elegans*. The hormonal control of longevity is discussed in detail in the next two sections, for *D. melanogaster* and *M. musculus*.

Mitochondrial proteins may be the link between extended life span and metabolism

A direct link between energy metabolism and life-span extension in *C. elegans* has been identified with another gene regulatory pathway, found in mitochondria—the **clock genes** (*clk-1*, *clk-2*, *clk-3*, and *gro-1*). Clock genes are a highly pleiotropic set of genes, so named because they regulate the timing and synchronization of several mitochondrial functions. Loss-of-function mutation in *clk-1* affects several systems and results in a "slowing down" of many functions, including larval development, egg production, and brood size. The slowdown in function of the *clk-1* mutant also results in a slightly smaller worm that lives 15–30% longer than the wild-type *C. elegans*. Because of its pleiotropic nature, the *clk-1* genetic pathway most likely to be associated with the increase in longevity has not yet been identified.

The most commonly accepted theory concerning the clock genes links a slowdown in metabolic rate to extended longevity. The Clk-1 protein has been identified as a demethoxyubiquinone mono-oxygenase (DMO), an enzyme required for the biosynthesis of ubiquinone, also known as coenzyme Q (CoQ). Coenzyme Q is found in high concentration in the inner membrane of mitochondria, where its acts as an electron transfer protein between the flavoproteins and cytochrome *b*. The loss-of-function mutation in *clk-1* that results in diminished levels of CoQ slows, but does not stop, the transfer of electrons through the ETS. In turn, energy reserves fall and physiological functions must slow down.

The pathway that regulates the gene expression of DMO from *clk-1* has not been described, nor has the mechanism for life-span extension. However, two theories have been advanced that center on the function of CoQ in mitochondria. One theory suggests that a reduction in CoQ level decreases the amount of ETS activity, which, in turn, reduces the production of oxygen-centered free radicals. A decrease in free radicals would reduce cellular damage and possibly extend life span. This explanation, however, would not account for the pleiotropic effect of the *clk-1* gene in the slowing of several, diverse systems. The second theory predicts that the regulatory process in the nucleus "senses" the chronic low energy state of the cell—that is, low ATP—resulting in reduced gene expression of proteins involved in various physiological functions, a kind of energy thermostat. This theory explains the pleiotropic nature of the *clk-1* gene, but falls short of explaining the extended life span. As with many genes shown to extend life span in *C. elegans*, a more precise explanation of why the clock genes extend life span awaits characterization of the pathways that regulate gene expression.

GENETIC REGULATION OF LONGEVITY IN *D. MELANOGASTER*

The mutation of individual genes in simple organisms such as *S. cerevisiae* and *C. elegans* has shown that longevity can have a highly regulated genetic component. The use of these simple organisms is aided by the fact that most of their genes have a single function and are expressed in response to a single biochemical pathway or environmental stimulus. That is, the gene regulatory pathway resulting in an alteration in life span can be isolated rather easily in these simple organisms, relative to more complex life forms. As we move up the evolutionary ladder, the levels of anatomical and physiological complexity are the result of an increased level of regulation of gene expression. Genes in complex organisms tend to be highly pleiotropic—they affect more than one physiological function—and the expression of any one gene can be regulated by several pathways. Thus, isolating and describing pathways of the genetic regulation of longevity tends to be more difficult in complex organisms than in simple organisms such as *S. cerevisiae* or *C. elegans*.

The description of genetic regulation of longevity in complex organisms such as *D. melanogaster* has been limited, for the most part, to life-span analysis in genetically selected strains. These results have clearly shown that mutations in many different gene loci can extend longevity **(TABLE 5.3)**. Such results, while supporting the possibility of genetic regulation of longevity, cannot conclusively demonstrate that a specific gene encoding a specific protein is directly responsible for extending life span. Given our new understanding of how genes are regulated (see the earlier section "Regulation of Gene Expression"), it is probable that the mutated gene that imparts alterations in longevity may be just one part of a more complex pathway. Therefore, to minimize unnecessary speculation and uncertainty with regard to our discussion of the effects of genes on life span, our exploration into the genetic regulation of longevity in complex organisms focuses primarily on pathways that appear to be highly conserved, from yeast to human. As you'll see, these pathways have a connection to neurohormonal signaling, similar to that found in experiments with *C. elegans*. We begin with a brief exploration of *Drosophila* mutants that have extended longevity, to set the stage for a more in-depth analysis. Then we discuss three pathways that demonstrate, in *Drosophila*, the connection between neurohormonal signaling and longevity.

Drosophila has a long history in genetic research

Most of what we know about classical and molecular genetics has come from studies of *Drosophila melanogaster*. The first successful breeding of a *Drosophila* for a specific trait was reported more than 100 years ago. Using *Drosophila*, Thomas Hunt Morgan (1866–1945) showed that genes are contained within the chromosome, for which he won the Noble Prize for Medicine in 1933, the first ever for a geneticist. The discovery that the genes regulating the placement of body parts, the *Hox* genes (see Box 5.1), are virtually identical in all multicellular organisms came, in large part, from studies of *Drosophila*.

The genome of *Drosophila* contains approximately 165 million base pairs, of which about 20% make up 14,000 genes. The DNA is contained in four diploid chromosomes, three autosomes and one sex chromosome. The *Drosophila* life cycle consists of six stages: embryogenesis, three larval stages, a pupal stage, and an adult stage **(Figure 5.27)**. Development time, from egg to adult, takes approximately nine days

TABLE 5.3
SOME GENE MUTATIONS IN *DROSOPHILA* THAT EXTEND LIFE SPAN

Gene name	Function or protein product	Type of mutation	Increase in longevity
dsir2	Histone deacetylase	Overexpression	57%
Dts3	Hormone receptor	Knockout	19%
MnSOD	Oxidative defense	Overexpression	33%
hsp70	Heat shock protein; protein repair	Overexpression	4%
chico	Signal transduction receptor	Loss-of-function	48%
dFOXO	Transcription factor	Overexpression	56%
InR	Insulin/IGF-1 receptor	Loss-of-function	85%
mth	Signal transduction receptor	Knockout	35%
mei-41	DNA repair	Overexpression	22%

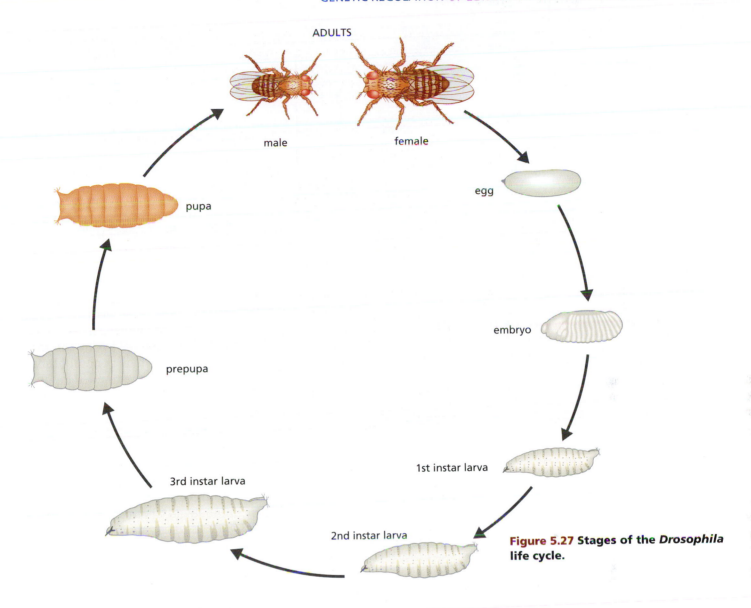

Figure 5.27 Stages of the *Drosophila* life cycle.

under ideal temperature conditions (25°C) and with sufficient food, but can take longer if temperature and food are less than optimal. Males are smaller than females. Females are receptive 12 hours after emergence and can remain fertile for 15–25 days. The average life span of wild-type *Drosophila* is 40–50 days.

Genes that extend longevity are associated with increased stress resistance

Table 5.3 lists some genes that have been identified as extending longevity in *Drosophila*. Many of these genes have a common function, and that function is anti-stress. For example, the chaperone protein encoded by *hsp70* (heat shock protein) belongs to a group of proteins that mark misfolded or damaged proteins for degradation through the ubiquitin pathway (see "Proteins Can Be Modified or Degraded after RNA Translation," earlier in this chapter). Extensive work has been completed in *Drosophila* and mice to show that overexpression of *hsp70* can lead to extended longevity. Another group of proteins involved in anti-stress functions and repeatedly shown to have a significant influence on the rate of aging and longevity in *Drosophila* is the endogenous antioxidants, proteins that degrade oxygen-centered free radicals. Overexpression of superoxide dismutase, one enzyme in the pathway

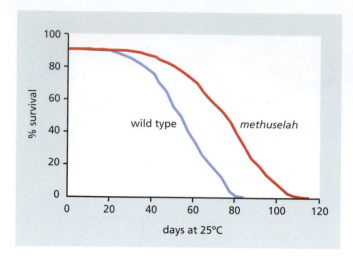

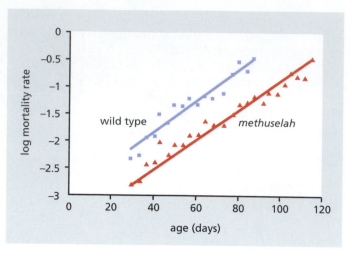

Figure 5.28 Survival curve and log of mortality rate for wild-type and *mth* transgenic *Drosophila*. Note that the log of mortality for the *mth* transgenic strain is shifted to the right, indicating that the rate of aging has been slowed. (From Y.J. Lin, L. Seroude, and S. Benzer, *Science* 282:943–946, 1998. With permission from AAAS.)

that reduces the superoxide radical to water, has consistently, but not always, extended longevity in *Drosophila*. While it is clear that heat shock proteins and endogenous antioxidants are involved in extended longevity in *Drosophila*, they are expressed in response to other intracellular signals. That is, some gene(s) upstream from the genes for these two proteins must be the factor that accounts for the extended longevity. Those genes await identification.

The connection between anti-stress genes and longevity became clearer with the discovery and cloning of the **Methuselah (mth) gene**. (Methuselah is a biblical character who was reported to live 969 years.) Careful screening of *Drosophila* mutants identified a long-lived strain that was heat resistant. The transgenic *mth* fly was shown to live significantly longer than the wild type **(Figure 5.28)**, and it was more effective at resisting the stress of paraquat (a chemical that induces overproduction of oxygen-centered radicals), starvation, and high temperatures **(Figure 5.29)**. Cloning of the *mth* gene revealed significant

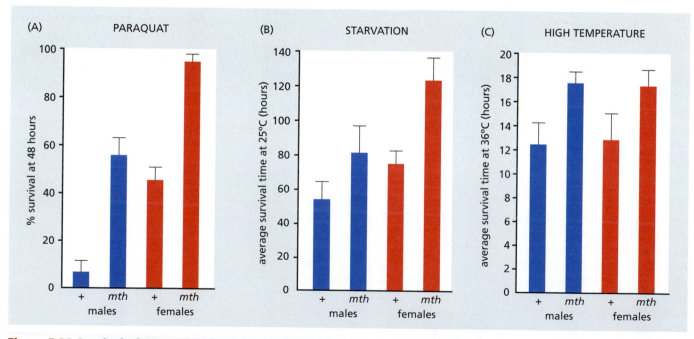

Figure 5.29 Survival of *Drosophila* in response to stress. These graphs show the survival of male and female wild-type (+) and *mth* transgenic *Drosophila* in response to (A) paraquat, (B) starvation, and (C) high temperature. In all treatments, male and female flies carrying the *mth* mutation survived longer than the control, wild-type flies. (From Y. J. Lin, L. Seroude, and S. Benzer, *Science* 282:943-946, 1998. With permission from AAAS.)

homology to a G-protein-coupled receptor. Recall that, in yeast, a G-coupled receptor has been identified with extended longevity. The central role of G-coupled receptors in longevity was confirmed when the binding of ligands specific to G-coupled receptors extended the life span of the transgenic *mth Drosophila* mutant.

Genes controlling *Drosophila*'s growth also extend life span

You have learned that extended longevity in *S. cerevisiae* and *C. elegans* appears to be tightly coupled to environmental conditions, which, in turn, affect reproduction. Environmental conditions that place stress on the animal, such as scarcity of food, repress gene expression and halt growth and reproduction until the environment is more favorable to the survival of offspring. The signaling pathways that repress growth and reproduction in *S. cerevisiae* and *C. elegans* also extend longevity. You have also learned from experiments in *C. elegans* that the signal initiating dauer formation and extended longevity is of neuroendocrine origin. *Drosophila*, like many insects, has also developed a strategy for delaying reproduction during times of poor environmental conditions. This strategy, called **diapause**, seems to be heavily influenced by neurohormonal signals and is characterized by reproductive silence, reduced energy metabolism, and resistance to stress. Biogerontologists have taken advantage of diapause in *Drosophila* to evaluate genes that affect longevity.

A mutant strain of *Drosophila* known as *chico* (Spanish for "small boy") has provided further evidence to suggest that insulin-linked gene regulatory pathways, those associated with growth and reproduction, may play a significant role in the rate of aging in species more complex than *C. elegans*. The *chico* flies are about half the size of wild-type *Drosophila*; the size difference is the direct result of this mutant having fewer and smaller cells **(Figure 5.30)**. The *chico* mutant contains a loss-of-function mutation (knockout) that reduces the expression of an insulin receptor substrate, CHICO. The *chico* gene is highly homologous to the *daf-2* mutant of *C. elegans*, and the reduced expression of *chico* has been shown to extend mean life span in *Drosophila* **(Figure 5.31)**. Moreover, the CHICO protein stimulates growth through a pathway similar to that described in *C. elegans*: the inhibition of a forkhead transcription factor, dFOXO.

The possibility that an evolutionarily conserved mechanism involving an insulin pathway may participate in the regulation of life span

Figure 5.30 Female *chico* mutant (left) and wild-type (right) adult *Drosophila*. (From M.D. Piper et al., *J. Intern. Med.* 263:179–191, 2008. With permission from Wiley.)

Figure 5.31 Life span of female and male flies with and without the *chico* gene. Increased survival of the *chico* knockout was limited to mean life span. This may indicate that *chico* plays a larger role in slowing the rate of aging than in directly increasing longevity. (From D.J. Clancy et al., *Science* 292:104–106, 2001. With permission from AAAS.)

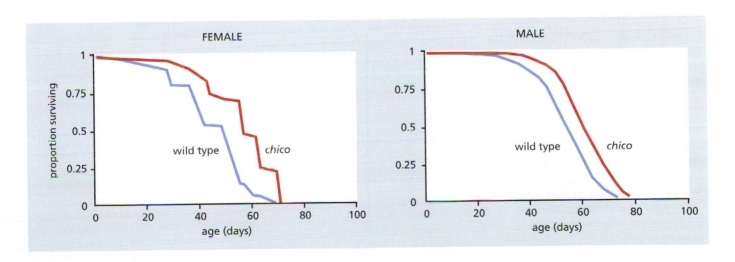

gained further support with the finding that a loss-of-function mutation in an insulin pathway receptor, InR, also increased life span. This mutation resulted in a significant decrease in kinase-like activity. A unique characteristic of this loss-of-function mutant is that egg production decreases significantly compared with wild-type flies, a finding linking this mutation to diapause **(Figure 5.32)**. The link between reproductive diapause and longevity grew even stronger when it was found that administration of a hormone involved in larval development, **juvenile hormone** (JH), restores the function of InR in the adult fly and decreases life span.

GENETIC REGULATION OF LONGEVITY IN *M. MUSCULUS*

The simple organisms *S. cerevisiae* and *C. elegans* have provided valuable information to support the possibility that reproduction, development, and longevity are linked, which in turn supports the evolutionary theories of senescence. These insights into the biology of

Figure 5.32 Relationship between reproductive diapause, juvenile hormone (JH), and life span. *Drosophila* in reproductive diapause does not lay eggs. Egg laying resumes when environmental conditions improve, and life span is extended relative to the wild-type fly. Loss-of-function mutation in an insulin-like receptor (InR) reduces egg production and extends life span (*center*). Treating the InR-mutated flies with JH can restore normal egg production (*right*), but this treatment decreases life span.

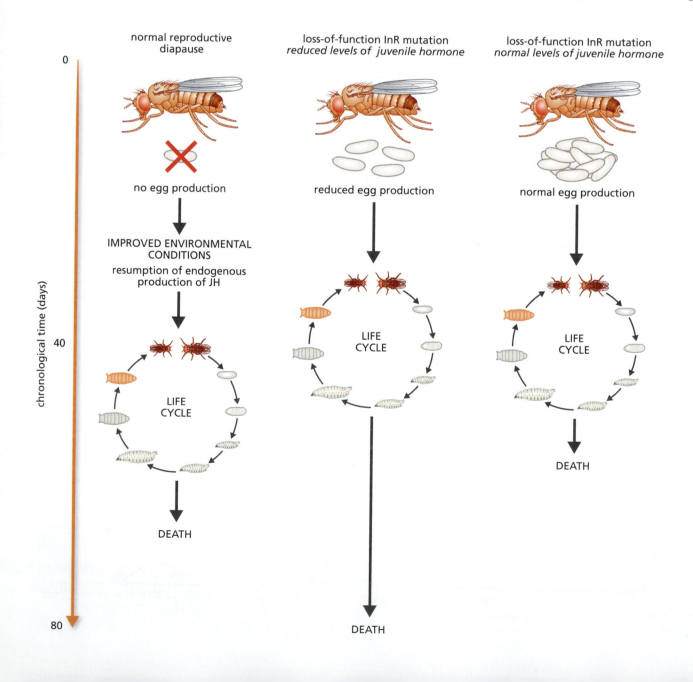

aging and longevity have been further supported in studies of the significantly more complex *Drosophila*. Now we turn our attention to the common laboratory mouse. *Mus musculus* has a genome of 2.5 billion base pairs contained in 19 diploid autosomes and 2 diploid sex-linked chromosomes. The number of protein-coding genes currently stands at 21,839 and represents about 4% of the entire base-pair sequence.

The use of genetically modified mice has added significantly to our understanding of the genetic pathways that affect longevity. In general, the findings from longevity studies using genetically modified mice are consistent with the findings reported for *S. cerevisiae*, *C. elegans*, and *Drosophila*. That is, mice genetically altered to inhibit neuroendocrine pathways important to normal growth live significantly longer that the wild-type animals. This section explores the results of those investigations. First, we take a brief look at the challenges faced by biogerontologists in sorting out which genes affect longevity and which genes may have a greater role in slowing the rate of aging and delaying age-related disease.

Many *M. musculus* genes have been reported to affect longevity

With the automation and commercialization of recombinant DNA technology, non-geneticists are able to construct knockout and transgenic mice with relative ease. Many of these genetically modified mice are being evaluated for their usefulness in biogerontological research. Between 30 and 40 different genes have been identified as having some impact on longevity in mice **(TABLE 5.4)**. Several genetically modified mice have been developed to evaluate whether the longevity genes or the signaling pathways associated with these genes identified in *S. cerevisiae*, *C. elegans*, and *D. melanogaster* also extend longevity in a mammalian species. Other genetically modified mice have been developed to evaluate whether a gene known to have a role in a specific disease process also plays a role in differences in longevity.

In recent years, the virtual explosion of research evaluating the impact of a single gene on longevity in mice has provided significant advances

TABLE 5.4
SOME GENE MUTATIONS IN *M. MUSCULUS* THAT EXTEND LIFE SPAN

Gene or mutant name	Function or phenotypic effect	Type of mutation	Approximate increase in longevity
Ames dwarf (*df/df*)	Deficient in growth hormone, prolactin, and thyroid hormone	Genetically bred	40%[1]
FIRKO	Fat-specific insulin receptor	Knockout	18%[1]
GHR/GP (also known as GHRKO)	Growth hormone receptor (GHR) combined with growth hormone-binding protein (GP)	Knockout	25–30%[2]
GPx4	Glutathione peroxidase	Knockout	10%[2]
Irs1[-/-]	Insulin receptor substrate	Knockout	32%[1] (females only; males no difference)
Irs2[-/-]	Insulin receptor substrate	Knockout	18%[1]
PAPP-A	Pregnancy-associated plasma protein A; inhibition of IGF-1 binding	Knockout	30–40%[1]
Snell dwarf (*Pit-1*)	Transcription factor associated with growth hormone	Knockout	23%[1]
S6K1	Ribosomal protein kinase	Knockout	17%[1] (females only; males no difference)

[1] Mean and maximum life span   [2] Mean life span

in the field of biogerontology. It has also brought significant challenges in terms of interpretation and in determining the relevance of this research to the study of longevity. Many investigations suggesting that overexpression of a gene (transgenic strains) or elimination of a gene (knockouts) enhances longevity have failed to follow the standards for animal care and use procedures for biogerontology, as discussed in Chapter 2 (see Box 2.1). These failings include an insufficient number of animals for proper statistical analysis, studies performed on only one gender, insufficient details on housing conditions, and lack of pathology data. The absence of pathology data and probable cause of death may present the most significant challenge in determining the relevance of a gene in enhancing longevity. Since all the genes associated with increased longevity are also highly pleiotropic, it is possible that the overexpression or elimination of a gene has an effect on immunity and disease. For example, one study that lacked pathology data listed eight different phenotypic differences between the knockout mouse and the control. Without a description of pathology and/or probable cause of death, it is impossible to determine whether the gene affected longevity per se, a genetic effect, or whether the differences were due to an alteration in the rate of aging or prevention of disease, a stochastic effect.

Another challenge facing scientists trying to determine the relevance of longevity results from studies of genetically manipulated mice is the lack of replication of the findings. Repeating experiments to confirm the accuracy of the data is a cornerstone of the scientific method; without replication, results must be considered preliminary. In species that have a short life span—that is, *S. cerevisiae*, *C. elegans*, and *Drosophila*—hundreds of experiments on longevity can be completed within a couple of years, with minimal investment in housing and care of the organisms. In contrast, performing just two longevity experiments in mice would take nearly 10 years, at a cost of more than $130,000 for housing and care. The mean life span of the most commonly used strain of non-genetically altered mouse—the C57BL/6—is between 26 and 28 months; the maximum life span can be as much as 40 months. Genetically altering a gene of the C57BL/6 mouse has typically resulted in an increase in life span of about 25%. This means that the maximum life span may be as great as 50 months (4 years). In addition, the average cost of housing one mouse in a facility designed for longevity studies can be upward of $400. Consider a longevity investigation of mice, with 4 groups of 40 each (considered by most statisticians to be the minimum number needed for a significant longevity analysis); the cost of housing alone would be $64,000.

These methodological issues associated with many longevity investigations that use genetically manipulated mice do not lessen their importance to biogerontology. They simply make them less applicable to the study of longevity. Many transgenic and knockout strains have already provided significant insight into mechanisms underlying age-related disease. However, here we are concerned with the genetics of longevity, a process significantly different from that leading to age-related disease. For this reason, we discuss here only longevity-enhancing genes that have been shown not to be affected by or to affect specific pathology and have been replicated, either in mice or in other species.

Decreased insulin signaling links retarded growth to longevity

Recall that knockout mutations of *C. elegans* and *Drosophila* that target insulin/IGF-1-like signaling pathways retard development and/or

TABLE 5.5
THE EFFECT OF INSULIN ON PROTEIN METABOLISM AND GROWTH

Function	Significance to growth
Increases amino acid uptake into cells	Increases availability of amino acids for protein synthesis
Increases rate of DNA transcription to RNA	Promotes protein synthesis
Increases mRNA translation on ribosomes	Stimulates synthesis of new proteins
Inhibits protein catabolism	Prevents excessive release of amino acids from the cell; affects synthesis of new proteins

growth and increase longevity. These results were important because insulin signaling pathways that affect growth are highly conserved phylogenetically and are found in both vertebrates and invertebrates **(TABLE 5.5)**. Therefore, it is likely that inhibition of insulin signaling in mice may also extend longevity.

Deleting genes that code for an intracellular signaling protein and an insulin receptor increases longevity in mice, supporting similar findings in *C. elegans* and *Drosophila*. And like the knockouts in *C. elegans* and *Drosophila*, both of these genetically altered mice are considerably smaller than their wild-type counterparts. This once again suggests a close relationship between blunted growth and longevity. In one line of investigations, a gene encoding a major intracellular **effector** (an agent causing an effect) of insulin receptor binding called insulin receptor substrate 1 (*Ins1*) was eliminated. The increase in longevity observed in the knockout mouse was seen only in females. The gender difference in longevity of genetically altered mice is not an uncommon finding, the reason for which remains to be seen.

The other genetically altered mouse showing increased longevity lacks a gene that codes for an insulin receptor specific to **adipose tissue**, the *fat-specific insulin receptor knockout*, or **FIRKO**. The increased life span of the *FIRKO* mouse **(Figure 5.33)**, although demonstrating once again the close relationship between insulin signaling pathways and longevity, also suggests that a reduction in fat storage may affect life span. In addition to its role in growth, insulin increases the amount of glucose that enters the fat cell. While only minuscule amounts of glucose are converted to fat, glucose provides the carbon backbone for the compound glycerol, a major component of triglycerides, the storage

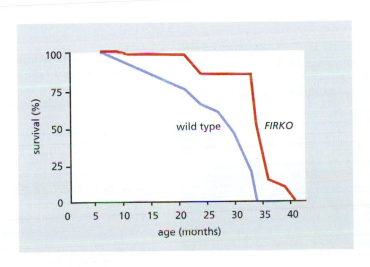

Figure 5.33 Attenuation in insulin signaling pathways increases longevity. Survival curves for male and female mice, wild type and *FIRKO* (knockout mutant lacking the gene for a fat-specific insulin receptor). (Adapted from M. Bluher, B.B. Kahn, and C.R. Kahn, *Science* 299:572–574, 2003. With permission from AAAS.)

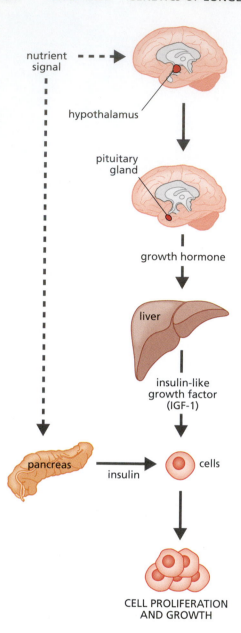

Figure 5.34 Simplified description of the growth hormone/insulin-like signaling pathway in mammals. The hypothalamus sends signals to the pituitary gland, located under the hypothalamus, causing secretion into the blood of growth hormone (somatotropin). Growth hormone stimulates the synthesis of insulin-like growth factor 1 (IGF-1) by the liver, and this factor is released into the general circulation. At the same time, the nutrient signal causes the pancreas to secrete insulin into the general circulation. Together, insulin and IGF-1 induce mitosis and growth of the organism.

form of lipid found in adipose tissue. Since the adipose tissue in the *FIRKO* mutant lacks an insulin receptor, these mice have only limited amounts of adipose tissue, and the lack of adipose tissue explains their smaller size. You will learn in Chapters 9 and 10 that obesity has a major impact on mortality, and a healthy body weight reduces a person's risk for many age-related diseases.

Diminished growth hormone signaling links insulin-like signaling pathways to increased longevity

The reduction in body size caused by a disruption in insulin or insulin-like signaling pathways and leading to increased longevity is highly conserved in *C. elegans*, *Drosophila*, and *M. musculus*. Insulin and insulin-like signaling pathways in mammals are part of a much larger regulatory system that affects growth and development, involving the secretion of growth hormone from the pituitary gland **(Figure 5.34)**. It is not surprising, therefore, that mutations causing a disruption in growth hormone signaling and thus stunted growth in mice result in increased longevity.

Three dwarf mutants—the Ames dwarf, the Snell dwarf (*Pit-1*), and GHR/GP (also known as GHRKO, growth *hormone receptor knock-out*)—all have a disruption in growth hormone-initiated intracellular signaling that causes a smaller body size and increased longevity compared with the wild-type strain **(Figures 5.35 and 5.36)**. The mutations in the Ames dwarf and the Snell dwarf are highly pleiotropic, resulting in a reduced blood concentration of growth hormone, prolactin, and thyroid hormone. The GHR/GP mutant does not show alteration in other hormones, but like the Ames dwarf and Snell dwarf, it has a reduced blood concentration of IGF-1.

It also appears that the dwarf mutation slows the rate of aging and delays the onset of certain age-related diseases. For example, the time required for collagen to denature has proved to be an indication of the rate of aging in mice. This is because, once synthesized, collagen in the tail remains there for life and becomes more difficult to denature over time (the reason for this is explained in detail in Chapter 8). Thus, the faster collagen can be denatured, the biologically younger the protein is. Collagen takes almost four times as long to denature in 19-month-old *Pit-1* control mice as it does in mutants. In addition, while the incidence of neoplastic (cancer) diseases is virtually identical in the wild type and the dwarf mutant, the average age of appearance is significantly older in the mutant. Furthermore, for any given age, the

Figure 5.35 GHR/GP knockout mice versus wild-type and heterozygous mice. Note that the GHR/GP knockout mouse (–/–) is smaller than its normal (+/+) and heterozygous (+/–) littermates. The difference in size among these three genotypes is similar to that seen in the other dwarf mouse strains, Ames and Snell. (From K.T. Coschigano et al., *Endocrinology* 141:2608–2613, 2000. With permission from the Endocrine Society.)

dwarf mutant appears biologically younger. Although the mechanism underlying the life-extension properties that seem to delay age-related loss of function has yet to be identified, these mutations are providing biogerontologists with a method to examine longevity (genetic) and the rate of aging (stochastic) simultaneously.

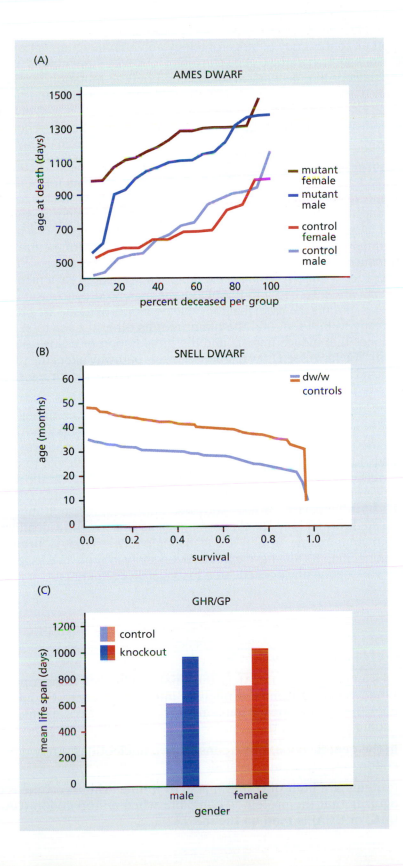

Figure 5.36 Longevity of three dwarf genotypes of the mouse. (A) Ames dwarf, (B) Snell dwarf, and (C) GHR/GP mutant. In the Ames dwarf and the GHR/GP mutant, females have significantly greater longevity than males. For the Snell dwarf, only males were evaluated. (A, adapted from H.M. Brown-Borg et al., *Nature* 384:33, 1996. With permission from Macmillan Publishers; B, adapted from K. Flurkey et al., *Proc. Natl Acad. Sci. U.S.A.* 98:6736–6741, 2001. With permission from PNAS; C, data from K.T. Coschigano et al., *Endocrinology* 141:2608–2613, 2000. With permission from the Endocrine Society.)

Genetic regulation of longevity demonstrated in mice has implications for human aging

The increased longevity of genetically manipulated mice with an inhibited growth hormone/insulin-like signaling pathway is important because this demonstrates a phylogenetically conserved mechanism for longevity. Such consistency in results strengthens the evolution-based prediction that longevity is the by-product of genes selected for survival to reproductive age. The mouse data also confirm several other findings important to the study of biogerontology and human aging. For example, a mutation in a single gene can affect longevity in mammals. While the details of the mechanism underlying how the mutation increases longevity remain to be elucidated, the rate of aging in humans clearly can be evaluated at the gene level. In the years ahead, it is likely that research will identify drugs that target specific genetically regulated pathways, with the possibility of altering the rate of human aging.

The mouse data also confirm that increased longevity reflects neuroendocrine regulation, a finding that has major implications for humans. The success of humans as a species reflects, in part, our ability to survive and thrive in virtually every environment. The ability to survive under harsh and sometimes stressful conditions is the result of a highly integrated system of neural pathways recognizing environmental cues that induce secretion of hormones that maintain homeostasis—that is, the neuroendocrine system. For example, when children eat a meal, the nutrients are absorbed into the blood; the increased concentration of nutrients informs the brain that the building blocks of tissue are present and ready to be used; and the brain informs other tissues to secrete hormones that induce mitosis and the proliferation of tissue—that is, growth. You have seen how disruption in this pathway alters the rate of aging and longevity. It is a good bet that other neuroendocrine systems will have similar effects.

One final note on the genetic regulation of longevity. Although the highly conserved growth hormone/insulin-like signaling pathway has been the main focus of this chapter, it is not the only pathway that extends longevity. Its description in eukaryotes, from simple to complex, reflects the application of good scientific procedures in the study of biogerontological genetics. Other conserved pathways are certainly involved in longevity. To discover these pathways, research must be completed in precisely the same manner as that used in identifying the growth hormone/insulin-like signaling pathway. That is, begin by screening mutants that have enhanced longevity in short-lived species, then work up the phylogenetic tree to longer-lived mammals.

ESSENTIAL CONCEPTS

- Messenger RNA (mRNA) is the final product of transcription. When released from the nucleus into the cytosol, it contains only the nucleotides complementary to the DNA's exons, the coding portions of the gene.

- In the genetic code, three consecutive nucleotides, known as a codon, specify an amino acid.

- Proteins are built one amino acid at a time. The ribosome recognizes that the protein is complete when a stop codon (UAA, UAG, or UGA) is read on the mRNA.

- Histone acetylation, the binding of acetyl groups to certain amino acids on the histone tails, opens up the tightly pack histones so that the activator sites and promoter regions of the DNA are exposed. Histone deacetylation removes the acetyl groups and prevents gene expression.

- A gene can be cloned (identical copies made) in a quick, automated process called polymerase chain reaction (PCR). Once a gene has been cloned, its function can be predicted from the nucleotide sequence by applying the genetic code; this process has also been automated.

- Segments of a gene with unknown function can be labeled with fluorescent dyes and injected into cells, tissues, or whole organisms to determine where and when gene expression occurs. This method is called *in situ* hybridization.

- Evaluation of a gene's function in complex eukaryotes is commonly done by either removing the gene or inserting an extra copy of the gene. A mutant with the gene removed is called a knockout; a mutant with an extra copy of a gene is referred to as transgenic.

- Budding yeast dies after 20–30 cycles and can be used as a model for cellular replicative senescence. Budding yeast can also enter a stationary phase when environmental conditions are not favorable from reproduction. This stationary phase can be used as a model for chronological aging.

- Gene silencing in yeast appears to be the primary mechanism that extends longevity. A few other gene-silencing pathways have been identified that are associated with low levels of food in the environment.

- The roundworm *C. elegans* has a mechanism that prolongs development and delays reproduction. A larva in an environment that is not favorable to reproduction can develop into a dauer, which can survive in this environment for months. When environmental conditions improve, the dauer re-enters the larval stage and develops into a reproductively active adult.

- Biogerontologists have identified two genes in the dauer formation pathway, *age-1* and *daf-2*, that also extend longevity. Both genes encode proteins that are highly conserved.

- Weak mutations in *daf-2* extend life span but do not repress reproduction.

- Another *C. elegans* gene, the clock gene *clk-1*, also extends longevity. The pleiotropic nature of *clk-1* has made identification of the underlying longevity-extending mechanism challenging. Because *clk-1* regulates many functions in mitochondria, the most commonly accepted theory of the association between the gene and longevity is related to energy production.

- Several genes in *Drosophila* have been identified as influencing the rate of aging and longevity. In general, these genes code for proteins that have a role in protecting cells from stress that leads to cell damage and death.

- A long-lived *Drosophila* mutant, the *Methuselah* (*mth*) mutant, has extended longevity and exceptional stress resistance. Cloning of the *mth* gene revealed significant homology to a G-protein-coupled receptor.

- A mutant strain of *Drosophila* known as *chico* contains a loss-of-function mutation (knockout) that reduces the expression of an insulin receptor substrate, CHICO. The *chico* gene is highly homologous to the *daf-2* mutant described in *C. elegans*.

- The extended longevity of the *chico* fly is associated with growth arrest through a neuroendocrine mechanism. A deficiency in juvenile hormone, which is needed for proper growth and development, has been shown to be a phenotype of the *chico* fly.

- Inhibition of insulin-like signaling that leads to increased longevity has been demonstrated in *M. musculus*. The enhanced longevity of these mutants reflects stunted growth caused by a disruption in the growth hormone/insulin-like signaling pathway.

- The conservation of genes that affect longevity, from simple to complex eukaryotes, has several implications for human aging, including (1) a mutation in a single gene can influence longevity; (2) longevity reflects, at least in part, neuroendocrine regulation; and (3) other genetic pathways associated with longevity must exist and await elucidation.

DISCUSSION QUESTIONS

Q5.1 Cells often require protein synthesis to occur quickly in response to a changing intracellular environment. Briefly list and describe the factors that contribute to the rapid pace of protein synthesis.

Q5.2 You have performed experiments to show that inhibition of gene expression is the mechanism by which a particular cellular event occurs. You also know that this inhibition occurs in the presence of a repressor protein. However, the repressor is not associated with preventing binding of an activator to an enhancer site. Speculate on what gene expression inhibition process is most likely taking place. Briefly explain the process.

Q5.3 As a specialist in geriatric medicine, you have noticed that patients with short stature seem to show signs and symptoms of premature aging. Since this fast rate of aging occurs simultaneously in several organs, you hypothesize that the relationship between stature and rate of aging may be under genetic regulation. Describe the steps you might take to determine whether a gene that codes for stature also affects longevity.

Q5.4 Yeast (*S. cerevisiae*) has developed two strategies for increasing its reproductive life span. Briefly describe the two strategies. What is the evolutionary rationale as to why these two strategies make yeast a powerful model for studying the genetics of longevity?

Q5.5 Provide evidence to support the following statement: "Reducing food availability to budding yeast increases its life span."

Q5.6 The study of *C. elegans* has great importance to farmers, because this nematode feeds on certain bacteria that live in the soil. While eating the bacteria, *C. elegans* also damages the roots of plants. From what you have learned in this chapter, suggest a way to limit the damage caused by *C. elegans*.

Q5.7 Explain briefly why the experiments using weak *daf-2* mutations (as opposed to knockout mutations) demonstrated the importance of this pathway to overall biogerontology.

Q5.8 The *Methuselah* (*mth*) gene extends life span in *Drosophila* by protecting the animal from environmental stress. For what other species has it been suggested that a similar mechanism extends life span? Discuss briefly the significance of this finding.

Q5.9 Both the *Irs1* and the *Pit-1* knockout mice are dwarf strains, although the mutations are at different locations on the genome and each has a different function. Briefly explain why these different genes both result in a dwarf mouse.

Q5.10 This chapter has emphasized the importance of a particular method in genetic research to identify genes that affect longevity. List the important findings, from yeast to mice, that have confirmed a highly conserved signaling pathway that enhances longevity. Also include why a particular model may be used to study longevity.

FURTHER READING

OVERVIEW OF GENE EXPRESSION IN EUKARYOTES

Alberts B, Bray D, Hopkin K et al (2010) Essential Cell Biology. New York: Garland Science, pp 231–267.

Watson JD (1968) The Double Helix: A Personal Account of the Discovery of the Structure of DNA. New York: Simon and Schuster.

REGULATION OF GENE EXPRESSION

Alberts B, Bray D, Hopkin K et al (2010) Essential Cell Biology. New York: Garland Science, pp 269–296.

Carroll SB (2005) Endless Forms Most Beautiful. New York: WW Norton and Company.

Jacob F & Monod J (1961) Genetic regulatory mechanisms in the synthesis of proteins. *J Mol Biol* 3:318–356.

Kornberg RD (2007) The molecular basis of eukaryotic transcription. *Proc Natl Acad Sci USA* 104:12955–12961.

ANALYZING GENE EXPRESSION IN BIOGERONTOLOGY

Alberts B, Bray D, Hopkin K et al (2010) Essential Cell Biology. New York: Garland Science, pp 327–362.

Mullis KB (1990) The unusual origin of the polymerase chain reaction. Sci Am 262(4):56–61, 64–65.

Park SK & TA Prolla (2005) Lessons learned from gene expression profile studies of aging and caloric restriction. *Ageing Res Rev* 4:55–65.

Shalon D, Smith SJ & Brown PO (1996) A DNA microarray system for analyzing complex DNA samples using two-color fluorescent probe hybridization. Genome Res 6:639–645.

GENETIC REGULATION OF LONGEVITY IN *S. CEREVISIAE*

Goffeau A, Barrell BG, Bussey H et al (1996) Life with 6000 genes. *Science* 274:546, 563–567.

Jazwinski SM, Chen JB & Sun J (1993) A single gene change can extend yeast lifespan: the role of Ras in cellular senescence. *Adv Exp Med Biol* 330:45–53.

Kaeberlein M (2010) Lessons on longevity from budding yeast. *Nature* 464:513–519.

Sauve AA, Wolberger C, Schramm VL & Boeke JD (2006) The biochemistry of sirtuins. *Annu Rev Biochem* 75:435–465.

Sinclair DA & Guarente L (1997) Extrachromosomal rDNA circles—a cause of aging in yeast. *Cell 91:1033-1042.*

GENETIC REGULATION OF LONGEVITY IN *C. ELEGANS*

C. elegans Sequencing Consortium (1998) Genome sequence of the nematode *C. elegans*: a platform for investigating biology. *Science* 282:2012–2018.

Friedman DB & Johnson TE (1988) A mutation in the *age-1* gene in *Caenorhabditis elegans* lengthens life and reduces hermaphrodite fertility. *Genetics* 118:75–86.

Guarente L & Kenyon C (2000) Genetic pathways that regulate ageing in model organisms. *Nature* 408:255–262.

Johnson TE (2008) *Caenorhabditis elegans* 2007: the premier model for the study of aging. Exp Gerontol 43:1–4.

Kenyon C, Chang J, Gensch E et al (1993) A *C. elegans* mutant that lives twice as long as wild type. *Nature* 366:461–464.

Lakowski B & Hekimi S (1996) Determination of life-span in *Caenorhabditis elegans* by four clock genes. *Science* 272:1010–1013.

Riddle DL, Blumenthal T, Meyer BJ & Priess JR (eds) (1997) *C. elegans* II, 2nd ed. Cold Spring Harbor, NY: Cold Spring Harbor Laboratory Press.

Tissenbaum HA & Johnson TE (2008) In Molecular Biology of Aging (Guarente L, Partridge L & Wallace DC eds). Cold Spring Harbor, NY: Cold Spring Harbor Laboratory Press, pp 153–183.

van der Horst A & Burgering BM (2007) Stressing the role of FoxO proteins in lifespan and disease. *Nat Rev Mol Cell Biol* 8:440–450.

GENETIC REGULATION OF LONGEVITY IN *D. MELANOGASTER*

Adams MD, Celniker SE, Holt RA et al (2000) The genome sequence of *Drosophila melanogaster*. *Science* 287:2185–2195.

Clancy DJ, Gems D, Harshman LG et al (2001) Extension of life-span by loss of CHICO, a *Drosophila* insulin receptor substrate protein. *Science* 292:104–106.

Lin YJ, Seroude L & Benzer S (1998) Extended life-span and stress resistance in the *Drosophila* mutant methuselah. *Science* 282:943–946.

Orr WC & Sohal RS (1994) Extension of life-span by overexpression of superoxide dismutase and catalase in *Drosophila melanogaster*. *Science* 263:1128–1130.

Sun J & Tower J (1999) FLP recombinase-mediated induction of Cu/Zn-superoxide dismutase transgene expression can extend the lifespan of adult *Drosophila melanogaster* flies. *Mol Cell Biol* 19:216–228.

Tatar M, Kopelman A, Epstein D et al (2001) A mutant *Drosophila* insulin receptor homolog that extends life-span and impairs neuroendocrine function. *Science* 292:107–110.

GENETIC REGULATION OF LONGEVITY IN *M. MUSCULUS*

Barbieri M, Bonafe M, Franceschi C & Paolisso G (2003) Insulin/IGF-signaling pathway: an evolutionarily conserved mechanism of longevity from yeast to humans. *Am J Physiol Endocr Metab Physiol* 285:E1064–1071.

Bluher M, Kahn BB & Kahn CF (2003) Extended longevity in mice lacking the insulin receptor in adipose tissue. *Science* 299:572–574.

Brown-Borg HM, Borg KE, Meliska CJ & Bartke A (1996) Dwarf mice and the ageing process. *Nature* 384:33.

Coschigano KT, Clemmons D, Bellush LL & Kopchick JJ (2000) Assessment of growth parameters and lifespan of GHR/BP gene-disrupted mice. *Endocrinology* 141:2608–2613.

Flurkey K, Papaconstantinou J, Miller RA & Harrison DE (2001) Lifespan extension and delayed immune and collagen aging in mutant mice with defects in growth hormone production. *Proc Natl Acad Sci USA* 98:6736–6741.

Ladiges W, Van Remmen H, Strong R et al (2009) Lifespan extension in genetically modified mice. *Aging Cell* 8:346–352.

Selman C, Lingard S, Choudhury AI et al (2008) Evidence for lifespan extension and delayed age-related biomarkers in insulin receptor substrate 1 null mice. *FASEB J* 22:807–818.

PLANT SENESCENCE

6

"IT'S SAD TO GROW OLD, BUT NICE TO RIPEN."

-BRIGITTE BARDOT, ACTRESS (1934–)

Plants make up more than 90% of the Earth's biomass and were some of the first multicellular organisms to inhabit land. Plants provide our atmosphere with oxygen while removing the waste product of animal respiration (CO_2). All food, including all animal products, begins as plants. Only plants have the ability to combine inorganic and organic elements, using the energy in particles of light (photons), to synthesize food. The aging and death of plants is critically important to the health of our soils (compost) and to plants' next generation (seeds). Plant senescence has a major impact on our daily lives as well as on the ecology of the planet in ways often not considered by animal-centric human beings.

Let's use a familiar example to demonstrate the importance of plant senescence to the ecology of the Earth. Consider a field of corn at different times in its life cycle. For the majority of the cycle, the plants progress through the normal steps of plant development—seed germination, growth, and fruit set (reproduction). All stages take place while the plant is green. Late in the season, the appearance of the corn stalk changes rapidly. Leaves begin to change color to a golden brown as photosynthesis wanes. You might also notice that while the leaves appear to be dying, the ears of corn continue to grow and ripen. The ripening of the ear and, more importantly, the development of the kernels occur precisely because the leaves are dying. That is, by breaking down their own organic components and transferring the resulting nutrients to the reproductive organs, the dying leaves become a source of life for the next generation of corn.

As the growing season comes to an end, the leaves begin to fall from the corn stalk. The **abscission** of leaves—their regulated and purposeful separation from the stalk (as also occurs for fruits and flowers)—takes place almost simultaneously with the end of seed development. Seed maturity marks an important transition for the plant. Whereas the leaves were once the source of food for seed development, they have now become a sink, collecting nutrients from other plant organs. The leaves are now preparing for next year's crop. The abscised leaves will degrade (compost) and release residual nitrogen- and carbon-containing (and other) molecules back into the ground. The composting of the leaves maintains the soil so that the next generation of plants can grow. If plant senescence did not occur, the soil would soon be depleted of nutrients and there would be no crop the following year—and, for that matter, no animal kingdom.

IN THIS CHAPTER . . .

BASIC PLANT BIOLOGY

THE BIOLOGY OF PLANT SENESCENCE

INITIATING PLANT SENESCENCE

In this chapter we consider plant senescence. We begin with a brief discussion of basic plant biology and physiology, emphasizing topics important to senescence. We then examine the mechanism of senescence in plants and look at factors that initiate senescence.

BASIC PLANT BIOLOGY

Basic biological processes in plants do not differ, to any great degree, from those in animals. Plants function within the laws of chemistry and physics and use the same general scheme of transcription, translation, cellular transport, and energy production as animals. Plants are also subject to the same evolutionary forces that affect animals—namely, adapting to their proximal environment. Therefore, in this section, we focus primarily on aspects of plant biology that differ from animal biology.

Plant cells have a cell wall, a central vacuole, and plastids

Plant cells can be distinguished from animal cells by three unique structures: the cell wall, plastids, and the central vacuole **(Figure 6.1)**. **Cell walls** come in two types, primary and secondary. **Primary cell walls** develop at the same time as other parts of the cell and tend to be very thin and elastic. The elasticity of the primary wall allows proper cell growth and elongation, processes that are important for overall plant growth. Cells communicate and exchange molecules with each other through small, tunnel-like structures in the primary cell wall called **plasmodesmata** (singular, **plasmodesma**).

Secondary cell walls, found mainly in plants that have a **xylem** vascular system (a woody vascular system made up of dead cells), become more prominent as the plant cell dies and loses its organelles. This wall is much thicker and considerably more rigid than the primary wall. The secondary wall serves as a structural support for many plants. Both types of cell walls are located outside the cell's lipid bilayer membrane (plasma membrane) and are made of **cellulose**, a molecule consisting of linear chains of glucose units.

Figure 6.1 Anatomy of the plant cell.

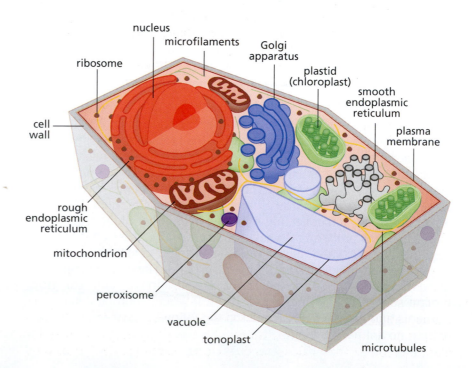

(A)

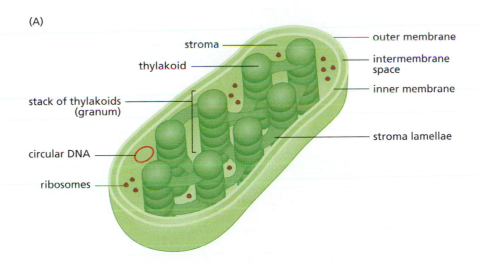

(B)

Figure 6.2 Chloroplasts and the structure of chlorophyll.
(A) Chloroplasts are bounded by two membranes that enclose the stroma, the fluid content of the chloroplast. The thylakoids are arranged in stacks called grana. Chloroplasts appear green due to their high concentration of chlorophyll. (B) Chlorophyll has a "head" called a porphyrin ring, with a magnesium ion at its center. Attached to the head is a hydrocarbon "tail," called the phytol tail, which interacts with proteins in the thylakoid membranes. Degradation of the chloroplast and of chlorophyll are important steps in plant senescence.

The **central vacuole** is a water-filled compartment that contains dissolved salts and is surrounded by a semi-permeable membrane called the **tonoplast**. Volume changes in the vacuole exert or remove pressure on the primary cell wall, and the amount of pressure exerted on the elastic wall can alter the size of the cell. These vacuole-induced changes in cell volume have great importance to the growth and health of the plant. Because plants are immobile and have primitive vascular conducting systems, each cell needs a large surface area to maximize its ability to collect water, minerals, CO_2, and light. However, physical laws governing the relation between the cell's energy requirements and its size (ratio of volume to surface area), known as **allometric scaling**, limit the size the cell can attain and still efficiently supply its cytoplasm with the nutrition needed to maintain function. Therefore, the plant uses changes in the water volume in the vacuole to change its surface area. The typical central vacuole occupies about 50% of the cell volume, but it can expand to 95% of this volume. The wilting of plants during low-water conditions reflects the loss of volume in the vacuole.

The vacuole performs several other functions, including the storage of dissolved materials needed for normal metabolism in the cell. Vacuoles contain pigments that produce the colors observed in many flowers. In addition, vacuoles are part of the **autophagic system**—the cellular digestive system—containing enzymes that digest various "used" cellular parts and transfer the products into the **phloem**, the second vascular system in plants.

Plastids are major organelles found in the cytoplasm of plant cells. They function as sites for the synthesis and storage of the chemicals necessary for photosynthesis. Plastids have a double membrane and an intermembrane space in which chemical reactions take place. There are three general types of plastids. **Leucoplasts** are colorless plastids that store starch or oil. **Chromoplasts** store pigments other than chlorophyll. For example, chromoplasts in tomatoes contain the red pigment lutein, and chromoplasts in carrots contain the yellow-orange pigment carotene, a **carotenoid**. **Chloroplasts**, the third type of plastids, contain the green pigment chlorophyll and are the sites of photosynthesis **(Figure 6.2)**.

Photosynthesis takes place in the chloroplast

Plants, unlike animals, obtain their energy from food created internally, a process called **autotrophy**. The primary energy sources for plant metabolism are glucose and fructose, which are synthesized by

Figure 6.3 Overview of photosynthesis. Photosynthesis, the synthesis of glucose from simpler precursors, using the energy from light, begins with the light reactions (see Figure 6.4) taking place in the thylakoid membranes of the chloroplast. The light reactions use solar energy to make ATP and NADPH. Water is oxidized by light energy, resulting in the release of O_2 into the atmosphere. Simultaneously, CO_2 absorbed from the atmosphere is converted to sucrose through a series of reactions in the stroma known as the Calvin cycle or dark reactions (see Figure 6.5). Sucrose is transported out of the cell and into the phloem for delivery to other plant organs.

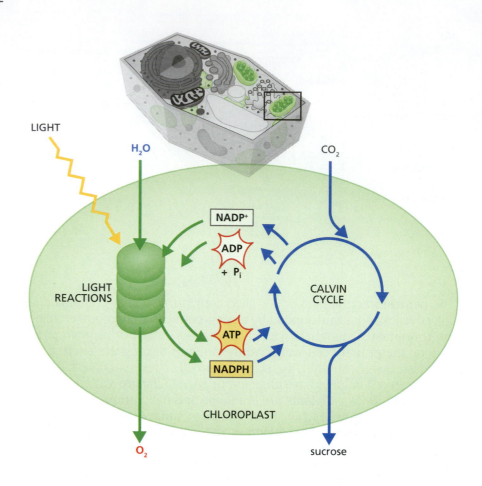

the process of **photosynthesis (Figure 6.3)**, occurring in chloroplasts. Glucose and fructose synthesized during photosynthesis combine to form sucrose. Sucrose enters the phloem to be delivered to other plant organs, where it is converted back to glucose and fructose and used for energy.

Photosynthesis has traditionally been divided into two phases: the **light reactions** (the "photo" part of photosynthesis) and the dark phase or **Calvin cycle** (the "synthesis" part of photosynthesis). The latter is named for the scientist who described the pathway, Melvin Calvin (1911–1997). The light phase consists of two interconnected light reactions **(Figure 6.4)**. Both reactions convert solar energy to chemical energy. Chlorophyll molecules attached to proteins on the thylakoid (Greek for "pouch") membranes of the chloroplasts trap photons, exciting electrons to a higher energy level. The excited electrons oxidize water, causing separation of the hydrogens from the oxygen; oxygen is released into the atmosphere. The protons generated from the oxidation reaction "drive" the formation of ATP, in a reaction very similar to that occurring in respiration-linked oxidative phosphorylation in mitochondria (see Chapter 4). The newly formed ATP is the energy source that drives the reactions in the dark phase. In the other photo-dependent reaction, the excited electrons provide the energy for the reduction of **$NADP^+$** to NADPH (Figure 6.4).

The ATP and NADPH generated in the light reactions are used in the Calvin cycle to convert CO_2 into the three-carbon structure of a **triose phosphate**, the precursor to glucose. The reactions in the Calvin cycle occur in three steps: carboxylation, reduction, and regeneration **(Figure 6.5)**. During **carboxylation**, each CO_2 molecule is attached to a sugar called ribulose 1,5-bisphosphate (RuBP), in a reaction catalyzed by the

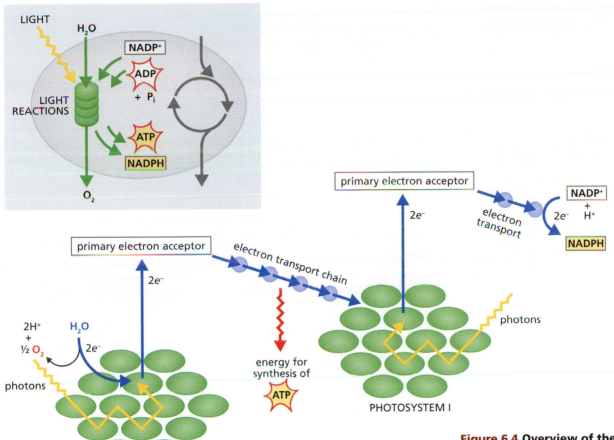

Figure 6.4 Overview of the light reactions. The thylakoid membranes contain two types of photosystems, photosystems I and II, which contain light-harvesting complexes of pigment molecules. Light is absorbed by photosystem II, and the energy excites electrons, providing the energy for the oxidation of water to O_2 and H^+. The O_2 is released into the atmosphere, and the protons (H^+) are used to drive the synthesis of ATP. Light absorbed by photosystem I provides the energy for the reduction of $NADP^+$ to NADPH. The NADPH shuttles electrons to the Calvin cycle, where they are used in the conversion of CO_2 to triose phosphate. (Adapted from J.B. Reece et al., Campbell Biology, 6th ed., San Francisco: Benjamin Cummings, 2002. With permission from Pearson.)

enzyme ribulose 1,5-bisphosphate carboxylase/oxygenase (RuBisCo). The carboxylation of RuBP generates the three-carbon compound 3-phosphoglycerate (3-PGA). During the **reduction** phase, each molecule of 3-phosphoglycerate receives an additional phosphate group from ATP, forming 1,3-bisphosphoglycerate, then electrons donated from NADPH reduce the 1,3-bisphosphoglycerate to the triose phosphate, glyceraldehyde 3-phosphate (G3P). The G3P diffuses through the chloroplast membrane into the cytosol and serves as the beginning substrate for the synthesis of glucose and, to a lesser degree, fructose.

Not all of the G3P generated in the Calvin cycle leaves the chloroplast. The third step in the Calvin cycle, **regeneration**, begins when triose phosphate isomerase converts some of the G3P into dihydroxyacetone phosphate (DHAP). Several further steps convert the DHAP to ribulose 5-phosphate, and addition of a phosphate group, from ATP, produces RuBP, thus ending one turn of the Calvin cycle and preparing for another. For each molecule of G3P synthesized, the Calvin cycle consumes nine molecules of ATP and six molecules of NADPH, which are then regenerated by the light reactions.

Plant hormones regulate growth and development

Virtually all the functions of plant hormones, or phytohormones, relate to some aspect of growth and development, such as cell division or apoptosis. Moreover, plant hormones, unlike their animal counterparts, can be synthesized in the same organ that contains their eventual target. There are five main groups of hormones: abscisic acid, auxins,

Figure 6.5 Overview of the Calvin cycle. During carboxylation, the two-step carboxylation/oxidation of ribulose 1,5-bisphosphate (RuBP), catalyzed by the enzyme RuBisCo, produces 3-phosphoglycerate. During reduction, ATP, generated in the light reactions, is used to convert 3-phosphoglycerate to 1,3-bisphosphoglycerate. Reduction of 1,3-bisphosphoglycerate by a proton donated by NADPH then produces the triose phosphate, glyceraldehyde 3-phosphate (G3P). The G3P either is transported out of the chloroplast and converted into glucose or enters the regeneration phase to form ribulose 1,5-bisphosphate, ready to undergo another round of the cycle. (Adapted from J.B. Reece et al., Campbell Biology, 6th ed., San Francisco: Benjamin Cummings, 2002. With permission from Pearson.)

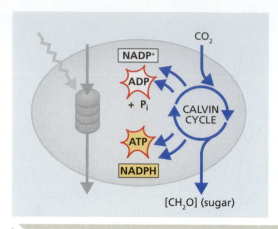

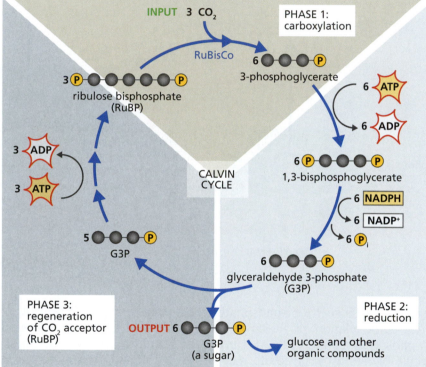

cytokinins, ethylene, and the gibberellins **(Figure 6.6)**. In addition, several other hormones are known to regulate plant growth, including salicylic acid and jasmonic acid (jasmonates). Cytokinins, abscisic acid, ethylene, salicylic acid, and jasmonates have all been shown to be important in plant senescence, and their roles in senescence are discussed later in the chapter.

Naturally occurring **auxins** are chemical variants of **indoleacetic acid**. Auxins promote root formation and growth. They also exert an inhibitory effect on lateral bud development by stimulating the growth of plant tissue at the apical (top) bud on the primary growth stem. Removal of the apical bud, or of auxin, causes lateral (auxiliary) bud development and the growth of lateral shoots. Auxins also seem to inhibit the decay of chloroplasts, which, in turn, inhibits leaf senescence.

All of the more than 90 **gibberellins** that have been identified arise from a basic molecular skeleton consisting of a four-ringed structure (*ent*-gibberellane). Gibberellins have three primary functions in the established, growing plant: promotion of plant elongation, stimulation

plant hormone	structure	type of hormone	function
abscisic acid (ABA)	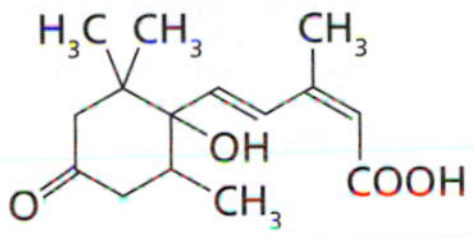	stress hormone	stimulates closure of stomata inhibits shoot growth stimulates synthesis of α-amylase promotes leaf senescence
auxins	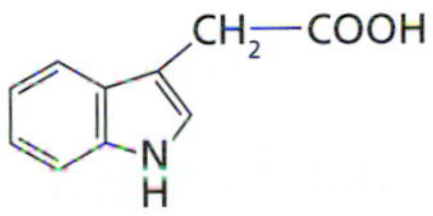	growth hormone	stimulates cell elongation stimulates cell division important to phototropism delays fruit ripening
cytokinins	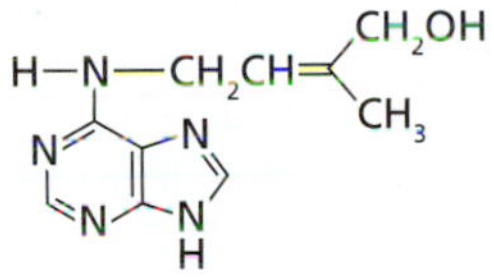	growth hormone	stimulates cell division stimulates leaf expansion stimulates growth of lateral shoots delays leaf senescence
ethylene	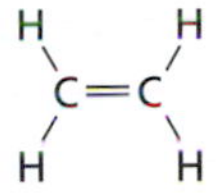	senescence hormone	leaf and fruit abscission stimulates flower opening stimulates leaf and fruit abscission/senescence stimulates fruit ripening
gibberellins	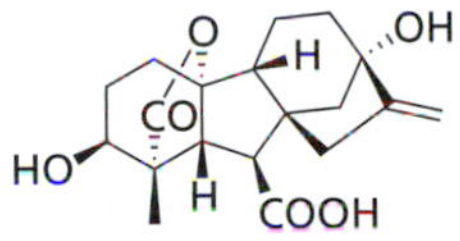	growth hormone	stimulates cell division stimulates flowering/bolting breaks seed dormancy can delay senescence in leaves
jasmonate	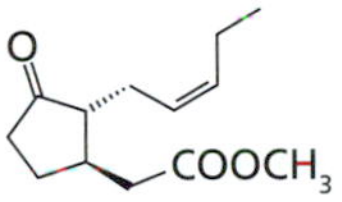	stress hormone	promotes production of defense proteins stimulates seed germination affects root growth promotes leaf senescence
salicyclic acid	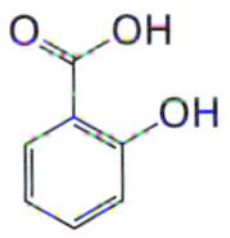	senescence hormone	aids in defense against pathogens

Figure 6.6 Functions of selected plant hormones.

of germination in dormant seeds and buds, and induction of flowering. Although the mechanisms leading to these effects have not been completely worked out, most evidence suggests that gibberellins promote the hydrolysis (bond breakage by addition of a water molecule) of starch and sucrose, which can be used in respiration as sources of energy. In general, auxins exert their effect on young plants, whereas gibberellins are more likely to enhance the growth of established plants.

THE BIOLOGY OF PLANT SENESCENCE

You learned in Chapter 3 that animal longevity evolved through genes selected for survival to reproductive age. Plant senescence evolved somewhat differently. Plant senescence, although causing the death of the plant or plant parts, is not a by-product of genes selected for reproduction or a chance event that leads to age-related decline in physiological function. Rather, it is an intrinsic part of the reproductive and development functions of the plant. Genes regulating plant senescence have been selected specifically for that process. As much as 25% of the entire plant genome participates in senescence. Here we consider how senescence and/or death of various parts of the plant support reproduction and ensure the continuation of the species.

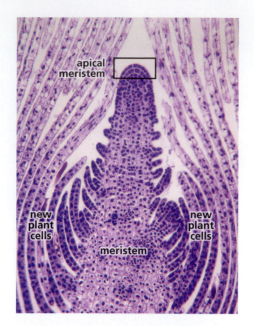

Figure 6.7 The apical meristem and the proliferation of new plant cells. (Courtesy of J. Harshaw/Shutterstock.)

Mitotic senescence occurs in cells of the apical meristem

As you learned in Chapter 4, mitotic cell division in animals has a limit, a process named replicative senescence. Plant cells also have a type of mitotic cell senescence, although the process and the type of cells affected differ fundamentally from those in animals. The difference between mitotic cell senescence in animals and plants can be traced to dissimilarities in the development and growth phases in the two phylogenetic kingdoms. An animal's body plan, meaning the anatomical location of its body parts and organs, develops exclusively during embryogenesis; the adult form is determined by the genotype. In animals, postembryonic development reflects the growth of tissues through cell division or the addition of tissue from stem cells. In addition, several different types of animal tissues have cells that retain their mitotic capacity throughout the organism's life span. In contrast, in plants, the adult body plan is determined postembryonically and reflects the plant's interaction with its environment. Roots and shoots, the body plan of a plant, are anatomically determined, and growth through cellular proliferation is what gives the plant its best opportunity for obtaining sunlight, water, and nutrition. Postembryonic cellular proliferation is carried out exclusively by the apical meristem, a numerically small group of undifferentiated mitotic cells found in only one type of plant tissue, called meristem **(Figure 6.7)**. Active apical meristem cells cause the plant to grow by laying down post-mitotic cells in a root or shoot, behind the meristem. Because the mitotic activity of plant cells occurs only during growth, the end of mitotic cell division in an apical meristem should be referred to as proliferative, rather than replicative, senescence.

All plants experience proliferative senescence. In annual, **monocarpic** plants, plants that live through only one season and reproduce only once (corn, for example), proliferative senescence occurs immediately before the whole plant begins to die. Perennial plants, plants that die back during winter and begin growth again in the spring, and deciduous trees, those that lose their leaves in the fall, undergo many cycles of apical meristem proliferation and proliferative senescence. Apical meristem proliferation may occur continuously in evergreen plants, but even in species that may live for thousands of years, proliferative senescence and whole-plant death will eventually occur **(BOX 6.1)**.

Apical meristem proliferation is completely under the regulation of hormones (Figure 6.6), and several investigations have demonstrated that proliferative senescence can be reversed by exposing the meristem to growth-promoting hormones. However, the mechanisms in apical meristem cells that respond to the absence of growth-promoting hormones by entering proliferative senescence are unknown—unlike what we know about their animal counterparts, for which five decades of research have identified possible cellular mechanisms underlying replicative senescence (see Chapter 4). Technical issues might explain, at least in part, our limited knowledge about the cellular mechanisms underlying proliferative senescence in plants. Cultures of apical meristem cells have been developed, but these systems are difficult to start and to maintain because of the small number of cells in an active apical meristem. (As is evident in Figure 6.7, it is possible to count the number of cells in an apical meristem, which ranges from 20 to 30.)

Plant biogerontologists have found at least one mechanism of replicative senescence, telomere shortening, that does *not* have a role in proliferative senescence. Recall that in animal cells lacking telomerase, the enzyme that maintains telomere length, telomeres shorten with each cycle of cell division, eventually causing coding DNA to be used

BOX 6.1 HOW OLD ARE THEY?

A bristlecone pine in the White Mountains of eastern California is estimated to be more than 4500 years old; however, bristlecone pines contain a significant amount of dead wood supporting their living core, branches, and needles. Does the true age of the tree reflect only the oldest of the living cells, estimated to be less than 200 years of age? Or the age of the oldest part of the tree itself? Clonal plants, such as the quaking aspens of the Rocky Mountains, sprout from a continuous root system that can support thousands of trees. These root systems span several miles and are estimated to be more than 10,000 years old. Individual trees that are cloned from this root system reach ages of only 150–200 years. Is the cloned aspen tree 200 years old? Or is it the age of the root system, 10,000 years?

Bristlecone pine

The White Mountains of eastern California are home to what many believe to be the oldest organism on Earth. Here, in these arid, windswept mountains, just above the tree line, a bristlecone pine (*Pinus longaeva*) **(Figure 6.8)** named Methuselah has stood for more than 4500 years.

Figure 6.8 A bristlecone pine in the White Mountains of California. (Courtesy of M. Norton/123RF.)

The longevity of the California grove of bristlecone pines was documented in 1954 by Edmund Schulman (1908–1958), a tree-ring researcher from the University of Arizona. Schulman was searching for old trees containing many growth rings, so that he could determine weather patterns throughout history. Thick rings indicate wet years, whereas thinner rings suggest drought. Schulman and a colleague, after conducting research in Idaho, traveled to the White Mountains before returning to Tucson. They had heard rumors of trees in the White Mountains that were several thousand years old. They did find a grove of trees that were 1000–1600 years old. Old, yes, but not significantly older than many species they had previously described. Nonetheless, Schulman was intrigued by the ability of a tree to last up to 1600 years in such harsh conditions. He and his colleague returned the following year to find several trees, including Methuselah, that were 3000–5000 years of age.

Schulman's discovery and subsequent report in *National Geographic* led more researchers to the White Mountains, as well as to other locations throughout the west that have similar topography. Within a few years, several sites of long-lived bristlecone pines were found in Nevada, Utah, and Colorado. A grove on Wheeler Peak in Nevada contained trees as old as those in the White Mountains. One tree, Prometheus, was found to contain 4862 rings, making it older than any tree in California. Unfortunately, the Forest Service allowed the tree to be cut down for study. The cutting down of a bristlecone pine is now a felony. The groves of bristlecone pines in Wheeler Peak are now part of the Great Basin National Park.

The longevity of the bristlecone pines reflects some unique anatomical and physiological properties. Needles turn over on a 30-year cycle, which provides a photosynthetic safety net during successive years of drought.

There also appear to be survival advantages in having a large amount of dead wood supporting the smaller portions of living tissue. The dead sections of the bristlecone pine provide protection against lightning strikes, improve water absorption, and prevent attack by several species of insects. In addition, the chemical composition of the living tissue is a natural deterrent to bacteria and other microscopic predators.

Clonal colonies—quaking aspens and the creosote bush

Many plants that thrive in harsh conditions (heat, cold, low rain, etc.) have developed strategies that increase their chance for successful reproduction. Clonal reproduction is one such strategy. Clonal plants are species that, in addition to sexual (gamete) reproduction, can also sprout new offspring from the roots. The roots are often protected from the harsh environment and continue to spread even though the top of the plant may be dying. Moreover, the roots of clonal plants sprout new plants during times when seed germination is not possible.

The root systems of clonal colonies, such as quaking aspens and the creosote bush **(Figure 6.9)**, can live for thousands of years, protected from the aboveground elements. A 100-acre grove of aspens in southern Utah (named Pando, Latin for "I spread") has a root system that has been estimated to be 80,000 years old and to weigh almost 7000 tons, easily making it the heaviest organism on Earth. However, these estimates are based on algorithms derived from several indirect measures, including a questionable observation that this grove has shown no significant flowering or seed production for the past 10,000 years. In addition, some research indicates that a portion of Pando's roots have been dead for thousands of years and may have been replaced with younger roots. More recent estimates place the age of the root system closer to 10,000 years—still extremely old for a living organism.

BOX 6.1 HOW OLD ARE THEY?

Figure 6.9 Clonal plants. The root systems of these plants are estimated to be more than 10,000 years of age. Quaking aspens (*Populus tremuloides*) (*left*) range from southern Utah to northern Canada and Alaska. The creosote bush (*Larrea tridentata*) (*right*) is found primarily in the deserts of the southwest United States and northern Mexico. (*Left*, courtesy of P. Kunasz/Shutterstock; *right*, courtesy of Bufo/Shutterstock.)

The age estimates for the creosote bush root systems are similar to those for the aspens—10,000–13,000 years old. However, clonal colonies of the creosote bush grow over a much smaller range than the aspens. The key to the extended survival of the creosote bush lies in the extreme efficiency of its roots in holding water. Creosote bushes are known to thrive during periods when yearly rainfall may total less than two inches. One interesting aspect of creosote bush colonies is that they sprout new bushes in all directions at almost exactly the same distance from other bushes.

as the primer for replication of the lagging strand (see Chapter 4). Thus, telomeres function as a molecular clock that can determine the number of replicative cycles. Apical meristem cells contain telomerase, and thus telomeres do not shorten with each proliferative cycle. Moreover, the telomeres in apical meristem cells grown in culture actually *lengthen* with each proliferative cycle. Again, the mechanism underlying proliferative senescence in apical meristem cells has yet to be discovered.

Post-mitotic plant senescence involves programmed and stochastic processes

The vast majority of plant cells are post-mitotic, and most research has focused on the mechanism underlying senescence in these cells. Post-mitotic senescence is a highly regulated, highly ordered process that involves both the dismantling of post-mitotic cells and the recycling/remobilization of nutrients. We refer to the age-induced, regulated dismantling of post-mitotic cells as **programmed senescence**. Programmed senescence does not mimic apoptosis. Rather, programmed senescence induces a regulated dismantling of the chloroplasts and other organelles.

This dismantling of organelles eventually reaches a point at which normal cellular functions can no longer be sustained. Vital repair mechanisms that maintain cellular viability decline, along with the replacement of damaged proteins. At this point, **stochastic senescence** begins. The nuclear membrane and tonoplast break down and release their contents into the cytosol; much of this released material is toxic to the remaining cytoplasmic structures. The lipid bilayer of the cell membrane develops "holes" that allow extracellular molecules to enter the cell. That is, stochastic processes lead to breakdown of the chemical gradients imposed by the cell membrane. Chemical reactions within the cell stop, and cell death follows.

Leaves of *Arabidopsis thaliana* are the model for plant senescence

As discussed in Chapter 5, a worm (*C. elegans*), a fly (*Drosophila*), and a rodent (*M. musculus*) are widely used as models in describing animal senescence. The plant world also has its model for senescence: *Arabidopsis thaliana*. This plant, commonly known as thale cress and mouse-ear cress, is a small weed that grows wild throughout Europe, Asia, and northwest Africa **(Figure 6.10)**. The short life span of *Arabidopsis* (6–8 weeks in the laboratory) and the rapid induction of leaf senescence (4–5 days after full expansion) provide a significant advantage to researchers. Moreover, *Arabidopsis* is relatively easy to manipulate genetically.

Like animals, plants contain several cell types that can senesce at different rates and different times. Also like animals, plants show great diversity in the processes of senescence. Nonetheless, the study of plant senescence has, for the most part, been limited to cells found in *Arabidopsis* leaves. Unless otherwise noted, the description of plant senescence presented here is based on research performed using the senescent leaves of *Arabidopsis*.

Leaf senescence is a three-step process

The process of leaf senescence can be described in three phases: initiation, degradation, and termination **(Figure 6.11)**. The initiation phase coincides with the final stages of reproductive activity, including fruit ripening and the storage of nutrients in seeds. The initiation of senescence begins when the leaf cells receive a signal that induces the expression of genes aimed at degrading chloroplasts or decreasing photosynthesis. (It remains to be seen whether a decrease in photosynthesis initiates chloroplast degradation or chloroplast degradation leads to reduced photosynthesis.) This signal can be either (1) **abiotic**, a nonliving chemical or physical factor in the environment, or (2) **biotic**, a living organism. After the cell receives the signal, chloroplast-derived enzymes such as proteases and lipases catalyze reactions that break down the thylakoid membranes, releasing macromolecules into the cytosol. At this point, the leaf enters the degradation phase of senescence. The catabolic by-products of chloroplast breakdown are thought to induce the expression of nuclear genes that encode enzymes involved in macromolecular degradation. Although degradation of the chloroplast reduces the amount of sugar exported from the cell, the leaf cell remains a metabolic source by becoming a supplier of nutrients to the reproductive organs.

The leaf cell eventually reaches a point at which there are no more chloroplasts to degrade. At this point, the cell transitions from a source organ, supplying nutrients to other plant organs, to a sink organ,

Figure 6.10 *Arabidopsis thaliana*. This annual plant, native to Europe, Asia, and northwestern Africa, grows to 20–25 cm. Most of the green leaves are found in the rosette at the base of the plant, with a few, smaller leaves growing on the flowering stem. The small flowers (2–5 mm in diameter) are normally white to pale yellow. (Courtesy of V. Koval/123RF.)

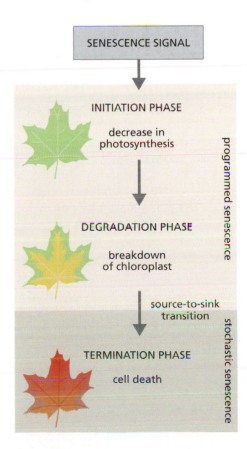

Figure 6.11 Leaf senescence. During the initiation phase, environmental or endogenous cues initiate a decrease in photosynthesis, which attenuates sugar production. This leads to breakdown of the chloroplasts and chlorophyll, resulting in leaf yellowing, and induces the expression of senescence-associated genes that code for proteins important to the catabolism of macromolecules and liberation of nutrients (degradation phase). In turn, the leaf cell begins its transformation from a nutrient source into a nutrient sink. During the termination phase, the cell membrane breaks down, leading to cell death and abscission.

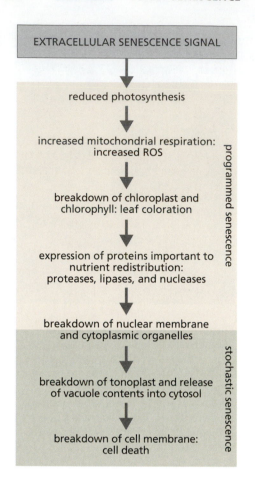

Figure 6.12 Leaf senescence at the cellular level. A reduction in photosynthesis marks the beginning of the programmed senescence stage. Programmed senescence transitions into stochastic senescence with the breakdown of the mitochondrial, nuclear, and vacuolar membranes.

receiving and holding nutrients and minerals. The transition from a source to a sink organ marks the beginning of the termination phase. The leaf cell dies, abscission occurs, and the nutrients of the dead leaf compost back into the surrounding soil, further supporting the nutritional needs of the plant.

The cellular events in plant senescence also follow a regulated pattern during the source-to-sink transition, although the exact order of events has yet to be established precisely **(Figure 6.12)**. Declining photosynthesis increase the demands on the mitochondria to supply ATP for dismantling the cell, thus increasing cellular respiration. As photosynthesis comes to an end and **monosaccharides** accumulate in the cytosol, the nucleus responds by expressing numerous proteases and lipases needed for the liberation of nutrients from macromolecules released from the chloroplasts. The cellular organelles—Golgi apparatus, mitochondria, microtubules, and so on—begin to degrade, releasing their contents into the cytosol **(Figure 6.13)**. As the organelles disintegrate, the regulated functions of the cell cease and stochastic senescence begins. The tonoplast ruptures, releasing the vacuolar contents, much of which can be toxic, into the cytosol. The cell membrane degrades, leading to the loss of cell integrity and, eventually, cell death.

Monosaccharides have an important role in leaf senescence

The senescence-related decline in photosynthesis correlates with an increase in cytosolic glucose concentration. The accumulating glucose does not combine with fructose to form sucrose, as normally occurs during pre-senescent metabolism. Rather, the increased concentration of glucose serves two important functions unique to programmed senescence: (1) glucose, or some factor associated with its metabolism, seems to be a signal that perpetuates programmed senescence; and (2) glucose supplies the substrate for the increased respiration by the mitochondria.

Recall from Chapter 4 that for glucose to be oxidized to produce ATP, it must first be phosphorylated. The plant cell uses the enzyme **hexokinase** to catalyze this reaction. Expression of hexokinase increases with the concentration of glucose. Interestingly, the concentration of hexokinase is greater than is needed for the phosphorylation of glucose. This "extra" hexokinase seems to bind to activator sites on genes that code for proteases. The proteases are responsible for the breakdown of proteins, such as those in the chloroplasts. Thus, the accumulation of glucose and/or hexokinase may be the signal that perpetuates programmed senescence.

While much of our current knowledge about the effect of glucose and hexokinase on programmed senescence remains circumstantial, two recent observations support their critical role. First, genetically engineered *Arabidopsis* that lacks the gene for hexokinase shows a delay in senescence, despite having an accumulation of intracellular glucose. On the other hand, overexpression of hexokinase in *Arabidopsis* leaves accelerates the start of programmed senescence. Second, the artificial accumulation of hexoses in young plants grown on a high-glucose medium initiates the expression of genes seen only during natural senescence.

The energy needs of the leaf cell increase during programmed senescence. Prior to the start of programmed senescence, the leaf mitochondria need only supply enough ATP for basic cell maintenance—protein

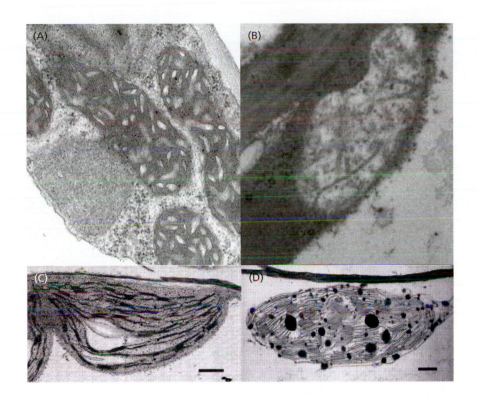

Figure 6.13 Breakdown of the mitochondria and chloroplasts in senescing _Arabidopsis_ cells. These electron micrographs show the difference between (A) mature and (B) senescent mitochondria and between (C) mature and (D) senescent rosette-leaf chloroplasts. (A, B, courtesy of E. Olmos; C, D, from M.T. Kaup, C.D. Froese and J.E. Thompson, _Plant Physiol._ 129:1616–1626, 2002. With permission from American Society of Plant Biologists.)

synthesis, vacuole function, and so on. After the start of senescence, the mitochondria must meet the new energy demands for nutrient-liberating catabolism, in addition to the energy needs for cell maintenance.

The increased energy demand on mitochondria during macromolecular degradation presents a potential problem for the senescent cell. That is, given that senescent cells are no longer producing glucose from photosynthesis, where does the substrate for mitochondrial respiration come from? Free fatty acids do not appear to solve this problem, as this substrate is reserved for seed development. The senescent cell continues to use only carbohydrates in respiration. During the early phases of programmed senescence, the leaf cell turns to its intracellular reserve of **starch**, the storage form of carbohydrates in plants, to supply glucose **(Figure 6.14)**. The switch from photosynthesis-derived monosaccharides to starch-derived glucose presents only minor metabolic problems for the cell. The leaf cell is accustomed to breaking down starch at night, when photosynthesis does not occur. Thus, the accumulation of glucose during the early stages of programmed senescence most likely reflects two factors: (1) an almost total absence of sucrose production, and (2) the release of glucose from intracellular starch stores.

The starch reserves in leaf cells at the beginning of programmed senescence are rarely sufficient to supply glucose throughout this phase. Late in senescence, the cell must turn to external sources of glucose

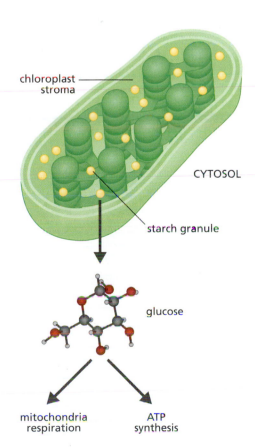

Figure 6.14 Conversion of starch to glucose during programmed senescence in leaves. As the leaf transitions into senescence, the chloroplast begins to degrade and can no longer supply the cell with sufficient glucose and fructose for the generation of ATP. Therefore, the leaf turns to its stores of starch to provide glucose for mitochondrial respiration and ATP synthesis.

to finish the cell-dismantling process. The source of glucose appears to be the phloem, which contains sucrose supplied by the leaf cell before the decline in photosynthesis. However, converting phloem-derived sucrose into glucose and fructose within the cytosol is difficult during programmed senescence, because intracellular levels of sucrose invertase, the enzyme that converts sucrose to glucose and fructose, are low when the cell is in source metabolism. The senescent cell has apparently overcome this problem by increasing sucrose invertase in the **apoplast**, the space between the cell wall and the cell membrane; this enzyme, known as apoplastic invertase, breaks down sucrose to fructose and glucose, which then enter the cytosol—a process known as **apoplastic uploading (Figure 6.15)**. In a mechanism that has yet to be elucidated, the uploading of glucose and fructose into the senescent cell induces the final transition from source metabolism to sink metabolism. While the mechanism inducing increased apoplastic uploading during late senescence remains obscure, some studies have shown that an increase in the phosphorylated form of free glucose, glucose 6-phosphate, in the cytosol during late programmed senescence induces expression of apoplastic invertase. Later we'll see that the plant hormone cytokinin may also affect levels of the apoplastic sucrose invertase.

The increase in glucose and fructose uptake by the senescent cell may also initiate the final steps in the leaf cell's transition from source to sink metabolism. Experimental evidence has shown that several enzymes involved in sink metabolism are up-regulated at approximately the same time as the senescence-induced apoplastic uploading of sucrose. Whether apoplastic uploading initiates sink metabolism or is the result of sink metabolism remains to be seen.

Breakdown of the chloroplast provides nitrogen and minerals for other plant organs

The chloroplast contains approximately 75% of the proteins found in photosynthetically active leaf cells. The breakdown of these proteins to free amino acids is an important source of nitrogen exported from the cell during programmed senescence. Liberation of free amino acids from their parent proteins occurs primarily in the cytosol. Dissociation of the nitrogen-containing amine group from the free amino acid takes place in the apoplast or phloem. Estimates place the amount of chloroplast **catabolites** (compounds produced by catabolism) released into the biosphere at more than 1 billion tons annually.

In addition to the liberation of nitrogen from the chloroplast proteins, chlorophyll must also be safely broken down. Chlorophyll and its catabolites are highly reactive and have the potential to cause excessive oxidative damage if not degraded and stored properly. The senescence-related degradation of chlorophyll has evolved as a mechanism to protect the cell from damage during programmed senescence. A metal-chelating protein removes the magnesium core from the **porphyrin** ring, and the Mg^{2+} diffuses into the cytosol and is transported into the phloem. Then, the **protoporphyrin** ring, porphyrin without its metal, proceeds through several steps to become the "fluorescent Chl catabolites" (FCCs), the final catabolites formed in the chloroplast. FCCs are transported out of the chloroplast and into the central vacuole, where they are converted (nonenzymatically) to nonfluorescent chlorophyll catabolites by the low pH of the vacuole. The nonfluorescent chlorophyll catabolites are released into the cytosol during stochastic senescence. The nonreactive phytol tail derived from chlorophyll is transported to a plastid called a plastoglobule, where it is stored and later used for synthesis of vitamin A.

Chlorophyll turnover is a normal process in developing and mature leaves and reflects a balance between synthesis and catabolism. Synthesis dominates during growth, resulting in a net accumulation

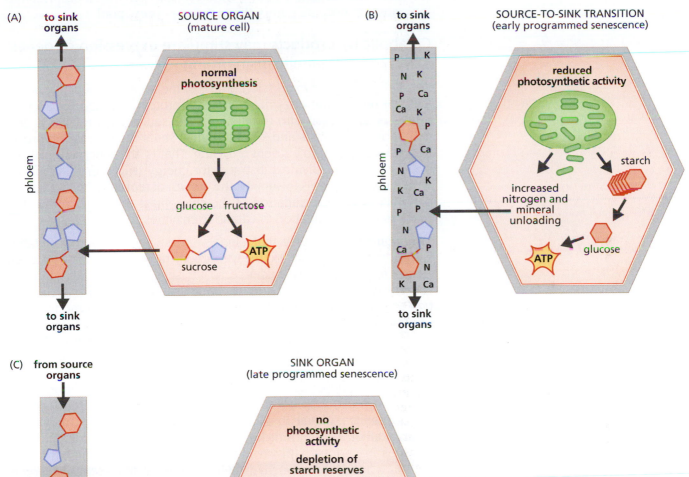

Figure 6.15 The transition from source to sink metabolism. (A) The mature, non-senescent cell supplies sucrose, through photosynthesis, to other organs of the growing plant and supplies glucose and fructose to the mitochondria for ATP production. (B) As photosynthesis wanes at the end of the growing season and the cell enters programmed senescence, the cell no longer exports sucrose but switches to a source organ for nitrogen and minerals. Nonetheless, the mitochondria must continue to supply energy, as ATP, for the dismantling process. The senescent cell derives the substrate for energy production from its intracellular stores of starch. (C) When the stores of starch are depleted, often before the end of programmed senescence, the cell imports glucose and fructose from sucrose circulating in the phloem. Sucrose is hydrolyzed to glucose and fructose in the apoplast in a reaction catalyzed by apoplastic invertase. The uploading of glucose and fructose into the senescent cell induces the final transition from source metabolism to sink metabolism.

of chlorophyll. Catabolic pathways during growth, development, and maturity are linked to synthetic pathways; catabolites from chlorophyll breakdown occurring when the plant is photosynthetically active are reused in chlorophyll synthesis. The senescence-related catabolic pathway has evolved separately from this pre-senescent process. End products of chlorophyll breakdown during senescence are not used for resynthesis of the pigment but are transported from the chloroplast into the cytosol to become part of the recycled nutrient pool.

Catabolic by-products may stimulate expression of genes involved in organelle dismantling

Recall that the gene expression profile of the photosynthetically active leaf cell reflects the enzymes and other proteins needed for sucrose synthesis and export into the phloem. The senescence-related degradation of chloroplasts and/or chlorophyll shifts the expression profile to genes important to nutrient recycling. That is, the senescent leaf cell switches from supplying substrates for energy (sucrose) to supplying nitrogen and minerals to support the growth of reproductive organs. As described above, degradation of the chloroplast and other organelles requires the up-regulation of catabolic enzymes.

The regulatory pathway that signals the switch in source metabolism from carbohydrate export to nutrient mobilization has not been clearly identified. Most evidence so far suggests that the signal for senescence-related reduction in photosynthesis and/or accumulation of glucose in the cytosol involves post-translational modification of proteins rather than expression of transcription factors. One such modification involves the interplay between **protein kinases** and phosphatases and most likely reflects a process similar to that described in *C. elegans* (see Figure 5.26). Some investigations have identified a group of Ca^{2+}-dependent and mitogen-activated kinases expressed in senescent *Arabidopsis* leaf cells, but it is far too early to present a model for a signaling pathway that induces or silences senescence-regulatory genes. Rather, our current knowledge is limited to the identification of genes in various pathways in senescent leaves. Recent investigations have identified more than 2000 genes in the senescent *Arabidopsis* leaf **transcriptome** including more than 100 transcription factors. The overwhelming majority of the mRNA expression in the senescent leaf codes for proteases, lipases, and nucleases.

Plant membranes degrade during leaf senescence

The rise of eukaryotic organisms correlated directly with their ability to separate one cell from another and to compartmentalize components within the cell through the development of lipid bilayer membranes. Biological membranes allow some molecules into a compartment while keeping others out, establishing selective permeability. Selective permeability primarily reflects the physical characteristics of the membrane, a lipid bilayer with **hydrophilic** (water-attracting) and **hydrophobic** (water-repelling) domains. The concentrations of other membrane constituents, such as proteins and glycoproteins, are such that precise electrical charges and fluidity are maintained within a very narrow range. Disruption of the membrane's spatial arrangement leads to a breakdown in selective permeability, loss of cellular compartmentalization, and cell death.

Recall that 75% of proteins broken down during programmed senescence originate in the chloroplast. It makes sense, therefore, that senescence-related membrane degradation would start with the chloroplast. The cell membrane and the vacuolar (tonoplast) and mitochondrial membranes must be maintained during chloroplast breakdown to provide support for degradation of the released macromolecules and export of nutrients from the leaf.

Degradation of the chloroplast membranes and release of the chloroplast contents into the cytosol initiate the expression of proteases and lipases targeted at organelle membranes. In the cytoplasm, there appears to be a hierarchy of degradation of the organelles. The precise order has yet to be described, but the membranes of the mitochondria and central vacuole are the last to be degraded. Rupture of the tonoplast and release of the toxic vacuolar contents into the cytosol hastens the degradation of the cell membrane, leading to cell death.

The process of senescence-related membrane degradation is identical in all membranes, regardless of the structure involved (**Figure 6.16**). Intracellular signals stimulate the expression of chloroplast and cytosolic lipases and proteases, in addition to cytosolic nucleases. Recent investigations have shown that as many as 130 genes coding for various kinds of lipases are expressed during leaf senescence.

While much of the lipolytic pathway of the membranes of leaf cells is not yet clearly established, we know that the free fatty acids resulting from catabolism of the phospholipid bilayer remain in the membrane for a period of time. These fatty acids are initially converted to sterol and wax esters in order to limit the destructive influence of highly charged lipids. Accumulation of sterol and wax esters occurs only in the early phases of membrane degradation, when organelle function must be maintained. During the later stages, free (de-esterified) fatty acids begin to accumulate and act as autocatalytic agents. That is, the detergent-like effects of free fatty acids are used to increase the speed of membrane disruption. De-esterified fatty acids have a tendency to "poke" holes in the membrane, causing instability in the bilayer. The instability increases permeability, leading to an accelerated rate of phospholipid breakdown and accumulation of de-esterified fatty acids. The free fatty acid-induced physical instability increases membrane leakiness, which, in turn, causes further instability in the membrane.

As the instability and leakiness increase, selective permeability is lost. Molecules and ions that were once kept outside the compartment now begin to pass freely through the membrane. The electrochemical gradient dissipates, and normal chemical reactions within the compartment cease. Without the electrochemical gradient and the barrier imposed by the lipid bilayer, stochastic processes result in rupture of the membrane and complete decompartmentalization.

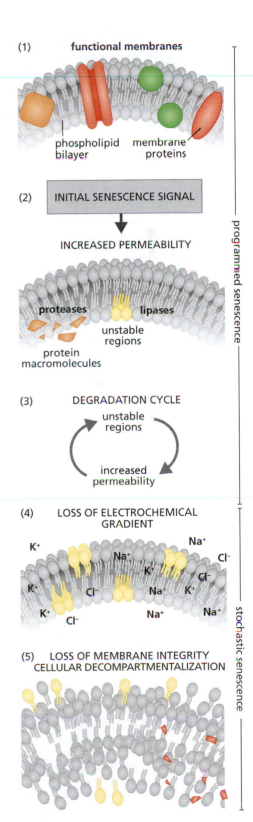

Figure 6.16 Membrane degradation during leaf senescence. (1) Many of the functional properties of membranes are coupled to their physical characteristics and the relationship between their lipid and protein phases. (2) On recognition of a senescence signal, lipases and proteases are expressed. The breakdown of phospholipids and proteins results in unstable regions of the membrane, which, in turn, increases its permeability to many compounds to which it was previously impermeable (membrane leakiness). (3) Increasing membrane instability leads to increased permeability that leads to further instability. (4) Eventually, the degree of permeability becomes so great that the differences in solute concentrations between the cellular compartments disappear and the electrochemical gradient no longer exists. (5) Without the electrochemical gradient, membrane integrity cannot be maintained and cellular decompartmentalization occurs.

INITIATING PLANT SENESCENCE

Most plants are autotrophic, and environmental conditions affect their senescence-related biology to a much greater degree than is observed in animals. With this in mind, we can expect that the signals initiating plant senescence most likely arise from environmental cues. These cues include light intensity, shade, temperature variations, soil nutrients, water content, toxins in the soil and atmosphere, predation (insects), and pathogens. With the exception of light intensity, environmental cues that can initiate senescence reflect stress conditions. Most plants in the wild experience some type of stress during their life span, and many enter senescence as a result.

Here we explore environmental factors that influence plant senescence. We approach this topic with the assumption that declining photosynthesis is the primary event of senescence. We consider first the mechanism by which light and light intensity influence the decline in photosynthesis observed at the beginning of senescence. We then examine the role of cytokinins in senescence and how stress-induced changes in other phytohormones stimulate the senescence-related loss of photosynthesis.

Light intensity affects the initiation of plant senescence

We have all observed how sunlight affects the growth of plants. If you take two plants of the same species and place one in full sun and the other in partial shade, both will grow, but the growth of the plant placed in the sun may far exceed that of the plant in partial shade. The failure of the shaded plant to achieve its maximum growth potential reflects the attenuated rate of photosynthesis. Plants needing full sunlight and grown in partially shaded areas bend toward the light in order to increase their photosynthetic activity, a phenomenon known as **phototropism**. The cells on the shaded side of the stem grow faster and longer than those on the sunny side, causing the plant to bend.

The mechanism underlying phototropism may also provide clues to the initiation of plant senescence. Mature plants, those that have reached their full growth potential, are faced with a problem when trying to maximize light exposure and, thus, photosynthetic activity for all of their leaves. Since the ability to synthesize more cells or to elongate existing cells in the stem is limited or nonexistent in mature plants, they cannot direct their leaves toward light. Only leaves in the top portion of the plant will receive maximal exposure to the sun, so the upper leaves are much more photosynthetically active than leaves at the bottom of the plant. This differential exposure to light leads to a reduction in overall photosynthesis and establishes conditions that may initiate senescence.

Chlorophyll absorbs light in the blue and red regions of the visible light spectrum. Evidence now suggests that the different absorption properties may have different functions in the plant. The recent discovery of **phototropins**, the photoreceptors responsible for phototropism, has helped to suggest a mechanism for initiation of plant senescence. We know that blue light may be most important to germination and plant growth. Blue light induces release of the phytohormone auxin, which preferentially acts on cells on the shaded side of the stem. Red light may have a greater effect on photosynthesis. Spectrum analysis also shows that photosynthetic activity induced by red light decreases as the light intensity decreases **(Figure 6.17)**. That is, red light (wavelengths ~620–700 nm) induces greater photosynthetic activity than does far-red

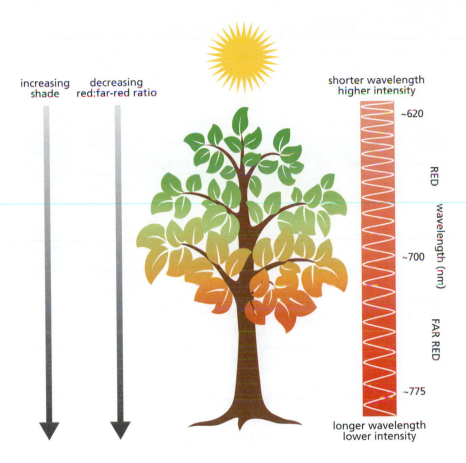

Figure 6.17 The ratio of red to far-red light and its effect on leaf senescence. Chlorophyll absorbs light in the red (~620–700 nm) and far-red (~700–775 nm) regions of the spectrum. Red light has the most powerful influence on photosynthesis. Far-red light has only a minor influence on photosynthesis, due to significantly fewer photoreceptors for this light. When a plant has reached maturity, the upper leaves absorb a greater amount of red light than the lower leaves, decreasing the red:far-red ratio. The resulting decline in photosynthesis may initiate the senescence program.

light (~700–775 nm). This effect appears to be a result of leaf cells having significantly more photoreceptors for red light than for far-red light.

The upper leaves of a mature plant absorb a greater amount of the high-intensity red light, and photosynthetic activity in these leaves remains vigorous. As the light filters through the plant, the ratio of red to far-red light decreases, resulting in attenuated photosynthesis in the lower leaves. That is, the decrease in the red:far-red ratio in the lower leaves may be the signal that initiates senescence in the upper leaves, although the mechanism for this effect has not been clearly defined.

Cytokinins delay senescence

The **cytokinins** function primarily as growth regulators in young plants. The cytokinin signaling pathway affects stem cells, meristems, seed development, chloroplast biosynthesis, and several other growth and development processes. Biosynthesis of cytokinins occurs in all plant tissues and seems to be regulated by nitrogen concentration. That is, as the levels of nitrogen (derived from the root) rise during plant growth, cytokinin concentration also increases, and vice versa. Expression of cytokinin-dependent response proteins is regulated through a phospho-relay system. Briefly, cytokinin attaches to a class of histidine kinase receptors that transfer the phosphate group to an activator protein **(Figure 6.18)**. The phosphate group then translocates to a transcription factor that induces expression of the cytokinin-dependent response proteins. These proteins stimulate various other metabolic pathways, leading to growth of the cell.

Cytokinins have long been recognized as senescence-delaying factors. Leaves provided with exogenous sources of cytokinin are prevented from entering the early stages of programmed senescence, resulting

Figure 6.18 General scheme of cytokinin-induced cell growth. In young plants, nitrogen in the phloem induces cellular biosynthesis of cytokinin. The binding of cytokinin to a kinase receptor causes phosphorylation of an activator protein, which crosses the nuclear membrane. Then, either the activator protein itself or the phosphorylation of another transcription factor initiates gene expression of cytokinin-induced proteins that stimulate growth or cell division. (Adapted from A. Santner, L.I.A. Calderon-Villalobos and M. Estelle, *Nat. Chem. Biol.* 5:301–307, 2009. With permission from Nature Publishing Group.)

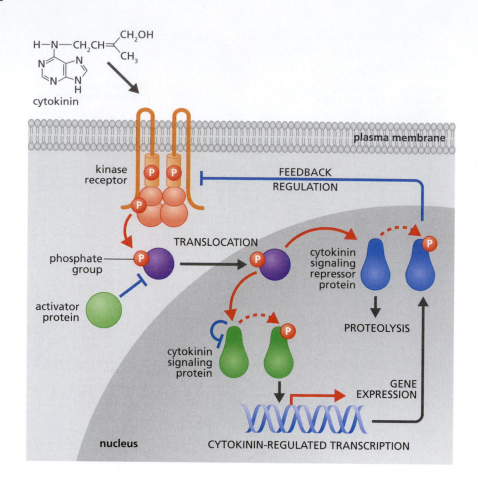

in increased longevity of the leaf. Conversely, inactivation of cytokinins in growing plants induces early senescence. As with much of plant senescence, the mechanisms underlying the cytokinin-dependent delay in senescence remain a mystery. Much of the uncertainty surrounding the role of cytokinins in senescence most likely reflects the stage of senescence the leaf is in during a particular observation. For example, cytokinin concentrations in leaves in the early stages of programmed senescence—that is, degradation of the chloroplasts—are similar to the concentrations found in pre-senescent cells. As programmed senescence enters its later stages, cytokinin concentration drops and the rate of organelle degradation increases.

The time-associated concentration differences of cytokinins during plant senescence suggest a model that is consistent with this phytohormone's function in growth and development. Recall that the process of leaf senescence has evolved to provide sufficient nutrients to the reproductive organs. This requires that cellular organelles and gene expression remain viable while the cell's chloroplasts degrade. Also recall that cytokinin biosynthesis responds, at least in part, to the levels of extracellular and intracellular nitrogen. Thus, it should not be surprising that the cytokinin concentration increases or remains stable as the levels of intracellular nitrogen or free amino acids increase during the early phase of programmed senescence. Moreover, there is evidence that the increase in the senescence-related concentration of nitrogen in the phloem stimulates the biosynthesis of cytokinin in tissues other than leaf cells. These cytokinins may be returned to the leaf cells through the xylem.

The next few steps in this model of cytokinin's effects on senescence are less clear. We are left with the question, How does a hormone usually involved in growth and development support a dying cell? The answer may come through the known effect of cytokinins on gene expression. Although cytokinin-dependent expression of specific genes important to senescence has yet to be identified, experiments show that a mutant *Arabidopsis* that overexpresses histidine kinase 3, the cytokinin receptor, demonstrates delayed senescence. Moreover, *Arabidopsis* mutants that lack a gene for histidine kinase 3 do not respond to exogenous cytokinin. These observations suggest that cytokinin-dependent gene expression occurs during the programmed senescence phase.

Finally, cytokinins may be important to sugar partitioning and the regulation of source-to-sink metabolism. Cytokinin concentration seems to vary with expression of the apoplastic invertase (the enzyme that hydrolyzes sucrose to fructose and glucose). That is, as the concentration of cytokinin increases during leaf senescence, so does that of apoplastic invertase, or vice versa. In our model, however, the timing of these two events would seem to be inconsistent with leaf senescence. The majority of apoplastic uploading occurs late in programmed senescence, a time when the cytokinin concentration should be declining. As with so much of our understanding of plant senescence, elucidation of the precise relationship between cytokinins and glucose accumulation requires significantly more research.

Other plant hormones induce senescence

The concentrations of abscisic acid, ethylene, jasmonic acid, and salicylic acid have all been shown to increase during various stages of senescence. The actions of these hormones in response to stress have, for the most part, been thoroughly worked out, but their role in senescence is less clear. The specific roles and possible interactions of these four phytohormones in non-stress senescence remain virtually unknown. The brief discussion here focuses only on how their functions during growth, development, or stress might relate to senescence.

Abscisic acid has several functions during growth, development, and stress that correlate well with senescence. This hormone induces closure of the stomata in response to drought conditions. Stomatal closure reduces water loss through transpiration but also reduces carbon dioxide intake, thus decreasing photosynthesis. The decreased photosynthesis may lead to initiation of the senescence program. Abscisic acid also responds to various stressors by inhibiting cell growth, especially in buds and leaves. Inhibition of cell division or cell elongation could lead to expression of proteins that are important to senescence. Finally, abscisic acid enhances the effect of ethylene on leaf abscission.

Ethylene has long been known to stimulate fruit ripening and leaf abscission **(Figure 6.19)**. Recent investigations have also shown a significant increase in ethylene concentration in senescent leaves. In addition, molecular analysis indicates that many genes associated with ethylene biosynthesis are up-regulated during senescence. The time during programmed senescence—early or late—when increased ethylene expression occurs has not been ascertained. As with the other three phytohormones discussed in this section, we do not yet know whether the presence of ethylene is the cause or the result of senescence.

Jasmonic acid (methyl jasmonate) and **salicylic acid** are noted for their response to wounds and pathogens, respectively. Jasmonic acid's correlation with leaf senescence arises from its effect on the chloroplast. In leaves brushed with jasmonic acid, chloroplast degradation occurred significantly faster than in control leaves. Plants genetically

Figure 6.19 Control of fruit ripening by ethylene. Ethylene triggers the fruit-ripening process by turning off anti-ripening genes. (Courtesy of L. Whitaker/Getty.)

altered to be insensitive to methyl jasmonate show delayed senescence. DNA microarray studies have shown that the salicylic acid-induced senescence transcriptome has several genes in common with the age-induced senescence transcriptome.

ESSENTIAL CONCEPTS

- Plant cells can be distinguished from animal cells by three unique structures: the cell wall, plastids, and the central vacuole.

- Photosynthesis, the synthesis of glucose from a reaction involving CO_2, water, and light, takes place in the chloroplast.

- Plant hormones, also known as phytohormones, are primarily involved in plant growth and development.

- Plant senescence is a developmental process evolutionarily selected for optimal reproductive success.

- Plant senescence occurs in phases. Programmed senescence consists of a deliberate and genetically regulated dismantling of the chloroplasts and other cellular organelles while simultaneously maintaining mitochondrial and nuclear functions. Stochastic senescence is an unregulated, stochastic degradation of the structures that normally maintain cellular integrity—the cell wall, cell membrane, and vacuole.

- The vast majority of information on plant senescence has come from the study of leaves. Leaf senescence can be described in three phases: initiation, degradation, and termination.

- Glucose and fructose accumulate in the leaf cell cytosol during the degradation phase of senescence, while synthesis of sucrose diminishes. The increased cytosolic glucose stimulates the expression of hexokinase, which, in turn, stimulates the expression of proteases.

- The accumulation of glucose in the cytosol during chloroplast degradation arises from two sources: intracellular starch reserves and the uploading of sucrose (as glucose and fructose) from the phloem.

- The degradation of chlorophyll is an orderly, genetically controlled process. It also marks a switch in the type of nutrients exported from the leaf to other plant parts.

- The by-products of chlorophyll degradation stimulate the expression of several enzymes that further dismantle the components of the leaf cell. The loss of integrity of cell membranes brings to a close programmed senescence and marks the beginning of stochastic senescence.

- The environmental cues that initiate leaf senescence are not fully understood, but it seems that both the amount of light and the type of light are important to the initiation of senescence in leaves.

- Stress-induced changes, such as drought, extreme temperature, and insect infestations, stimulate the synthesis of phytohormones and induce the senescence-related decline in photosynthesis.

DISCUSSION QUESTIONS

Q6.1 Explain why plant senescence is regarded as the final stage of plant development.

Q6.2 Briefly describe the cellular differences between programmed plant senescence and stochastic plant senescence.

Q6.3 List the order of events in programmed senescence, occurring at the cellular level.

Q6.4 Programmed leaf senescence results in a decrease in sucrose concentration in phloem. Does this observation reflect a decrease in the synthesis of intracellular glucose and fructose because of the decline in photosynthesis? Explain.

Q6.5 Describe why the plant cell uses more energy during programmed senescence than during non-senescent, normal cellular function.

Q6.6 The highly evolved genetic program for senescence in plants is evidenced by differences in the pre-senescent and senescent chlorophyll-degradation pathways. Explain.

Q6.7 Explain how the fatty acids found in normally functioning cellular membranes also play an important role in the degradation of membranes during leaf senescence.

Q6.8 The transition from programmed senescence to stochastic senescence can be characterized by the transition from source to sink metabolism. What is meant by the terms source and sink metabolism? What is the role of sink metabolism in the reproductive characteristics of the plant?

Q6.9 Discuss the differences between animal and plant mitotic cell senescence.

Q6.10 Explain how the lack of phototropism in an adult plant contributes to the initiation of leaf senescence.

FURTHER READING

BASIC PLANT BIOLOGY

Alberts B, Bray D, Hopkin K et al. (2010) Essential Cell Biology. New York: Garland Science, pp 453–494.

Sadava D, Heller HC, Orians GH et al (2007) Life: The Science of Biology, 8th ed. Sunderland, MA: Sinauer.

Stern KR (2006) Introductory Plant Biology, 10th ed. Boston: McGraw-Hill.

Whitmarsh J & Govindjee (1999) Concepts in Photobiology: Photosynthesis and Photomorphogenesis (Singhal GS, Renger G, Sopory SK et al., eds). New Delhi: Narosa, pp 11–51.

THE BIOLOGY OF PLANT SENESCENCE

Bleecker AB (1998) The evolutionary basis of leaf senescence: method to the madness? *Curr Opin Plant Biol* 1:73–78.

Gan S (2003) Mitotic and postmitotic senescence in plants. *Sci Aging Knowl Environ* 38:RE7.

Guo Y, Cai Z & Gan S (2004) Transcriptome of *Arabidopisis* leaf senescence. *Plant Cell Environ* 27:521–549.

Leopold AC (1975) Aging, senescence and turnover in plants. *Bioscience* 25:659–662.

Lim PO, Kim HJ & Nam HG (2007) Leaf senescence. *Annu Rev Plant Biol* 58:115–136.

Matile P, Hortensteiner S & Thomas H (1999) Chlorophyll degradation. *Annu Rev Plant Physiol Plant Mol Biol* 50:67–95.

Noodén LD (ed) (2004) Plant Cell Death Processes. San Diego: Academic Press.

Oparka KJ (1990) What is phloem unloading? *Plant Physiol* 94:393–396.

INITIATING PLANT SENESCENCE

Kim HJ, Ryu H, Hong SH et al. (2006) Cytokinin-mediated control of leaf longevity by AHK3 through phosphorylation of ARR2 in *Arabidopsis*. *Proc Natl Acad Sci USA* 103:814–819.

Müller B & Sheen J (2007) *Arabidopsis* cytokinin signaling pathway. *Sci STKE* 2007:cm5.

Niinemets U (2007) Photosynthesis and resource distribution through plant canopies. *Plant Cell Environ* 30:1052–1071.

HUMAN LONGEVITY

7

"OLD AGE ISN'T SO BAD WHEN YOU CONSIDER THE ALTERNATIVE."

-MAURICE CHEVALIER, ACTOR (1888-1972)

Thus far, our examination of aging and longevity has focused on basic biological mechanisms observed in simple eukaryotic organisms. We have been able to construct, with a fair degree of consistency, a hypothesis predicting that longevity evolved in animals and plants, whereas aging reflects a random, or stochastic, phenomenon. Longevity evolved through genes selected for survival to reproductive age. However, the declining force of natural selection with age discourages the fixing of genes that convey an "aging" genotype and phenotype.

As we turn our attention to our own species, we must address a fundamental question: With respect to aging and longevity, have humans followed a pattern of evolutionary development similar to that observed in other, nonhuman populations? It is possible that the environment, the primary driving force behind Darwin's theory of natural selection, has not affected adaptation in humans to the same degree as in other species. *Homo sapiens* is the only species, with the possible exception of a few nonhuman primates and marine mammals, that has a brain large enough to confer the ability to alter the environment to the species' advantage. Altering the environment can have a huge impact on survival. The construction of houses and fences or walls provides shelter and keeps out enemies; the making and wearing of clothing provides protection from harsh weather; and the development and use of agriculture prevents starvation. Thus, the possibility exists that because of our manipulation of the environment, environmental factors have influenced the origins of our longevity differently than for other species.

This chapter focuses exclusively on human longevity. We begin with an exploration of the origins of human longevity and discuss the theoretical basis for the life span that *Homo sapiens* has today. Note the use of the phrase "origins of human longevity" rather than "evolution of human longevity." This is done intentionally, to distinguish *Homo sapiens* from other species in the archaic *Homo* genus, such as *Homo heidelbergensis* and *Homo neanderthalensis*. Although the genes resulting from evolutionary selection for longevity were carried forward into *H. sapiens*, our superior intelligence most likely established life-span and mortality trajectories that differ significantly from those of our

Homo predecessors. These trajectories continue to evolve. The second section of the chapter presents some historical and recent observations that test the theoretical basis for the origins of human longevity and life span.

ORIGINS OF HUMAN LONGEVITY

Recall from Chapter 3 that results from laboratory experiments using artificial selection in *Drosophila* were consistent with the evolutionary theory of longevity proposed by Peter Medawar and with the mathematics of W. D. Hamilton. Similar laboratory experiments cannot be done in humans. Therefore, we must look to other, noninvasive methods for confirmation of the evolutionary theory of longevity. The emerging field of **biodemography**—the science that integrates biology and demography—provides the methods to investigate the origins of longevity that are uniquely human. Biodemography is primarily a mathematical and theoretical science that constructs models to predict the origins of human longevity. These mathematical models integrate the principles of mortality analysis with experimental observations from scientific fields such as archeology, physical anthropology, genetics, and evolution.

In this section, then, we explore the origins of human longevity through the science of biodemography. We start with a few general biodemographic principles that guide the development of models predicting the origins of human longevity, and these principles lead us to an emerging theory suggesting that the origins of longevity in humans may be unique among species, reflecting our superior intelligence and our ability to manipulate the environment.

Human mortality rates are facultative

Molecular and cellular observations made in simple eukaryotes and rodents have allowed us to separate, with a fair amount of precision, the differences among aging, longevity, and age-related disease. Results from highly controlled laboratory experimentation in animals have led to a definition of longevity as a by-product of genes selected for survival to reproductive age (see Chapter 3). Using similar laboratory techniques, researchers have observed that another measure of the length of life, the mean life span, seems to be more closely related to random, or stochastic, effects occurring as a matter of chance during development. As a result, we have been able to define longevity as the maximum age potential of a particular species, and life span as the length of life of an individual within the species.

The study of human mortality in the discipline of biodemography cannot easily separate the genetic, intrinsic rate of mortality from the environmentally dependent, extrinsic rate of mortality (see Chapter 2 for a discussion of "intrinsic" and "extrinsic" rates in the context of aging). Nor should environmental factors that affect human longevity be isolated. Human deaths due to age-related disease or environmental factors are accounted for mathematically in biodemographic predictions concerning the origins of human longevity. That is, biodemographers view environmental factors as critical variables in their mathematical models. Humans respond to environmental changes through cognitive reasoning, rather than simply reacting with pure instinct. Humans can change the environment to fit their needs, and this manipulation has resulted in mortality and longevity characteristics that are unique to *H. sapiens*.

The human ability to manipulate the environment reflects one of the most important principles of biodemography: mortality rates are facultative. The term **facultative**, as used in biodemography, means that environmental influences cause the mortality rate, or the trajectories of the mortality rate, to have significant plasticity. That is, the rates are not fixed, like those predicted by Gompertz analysis (see Chapter 2). This plasticity arises from the fact that the differences in environmental conditions that influence the mortality of individuals in human populations are infinite and are continually changing. Although the overall population mortality rate may take on a particular Gompertz mortality pattern, the rates for subpopulations will differ significantly. Many biodemographers believe that these subpopulations have had a significant influence on the origins of human longevity.

We can use the example of the seasonal variation in births in a historical population to demonstrate the facultative nature of mortality and its effect on longevity **(Figure 7.1)**. Babies born during periods in which the quantity and quality of nutrition are greatest—that is, the months during and immediately after the harvest (in the Northern Hemisphere, September–December)—have longer lives than those born in the winter months, when food shortages are most likely to occur (January and February). This simple correlation suggests that environmental factors during early life have established multiple subpopulations with distinctive rates of mortality. Because infant mortality would be lower for babies born during the harvest months, these subpopulations would have greater fitness, with a greater impact on life span and/or longevity.

Genetic factors cause significant plasticity in human mortality rates

Human mortality is highly variable, and this variability remains even after controlling mathematically or statistically for the confounding factors of age-related disease and environmental influences. Moreover, biodemographic research has shown that included within the overall human mortality rate are discrete subpopulations of individuals

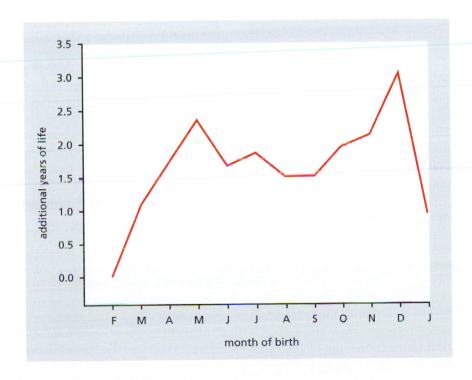

Figure 7.1 A biodemographic prediction of additional years added to the human female life span as a function of birth month. The research suggests that differences in longevity may reflect differences in nutritional factors at the time of birth. These results were generated using regression analysis of the length of life of 6908 women over the age of 30 born in European countries between 1800 and 1880. (From L.A. Gavrilov and N.S. Gavrilov, in Modulating Aging and Longevity [S.L.S. Rattan, ed.], Dordrecht, Netherlands: Kluwer Academic Publishers, 2003. With permission from Springer Science.)

sharing a specific Gompertz mortality rate that differs from that of the population as a whole. Together, the non-environmental variability and the existence of discrete subpopulations with their own mortality rates strongly suggest that genetic factors influencing human longevity also have plasticity and are not fixed. (Experimental evidence for the genetic plasticity of mortality rates was discussed in Chapter 2.) Studies in the Mediterranean fruit fly demonstrate that late-life mortality departs from the Gompertz rate by leveling off and then decelerating (see Figure 2.19). This late-life difference in mortality rates is generally accepted to mean that different genotypes for longevity exist within populations. Biodemographers suggest that these different genotypes could have existed throughout evolutionary history and may have played a role in the selection of genes that determine human longevity.

The effect of genetic plasticity on mortality rates provides biodemographers with an evolutionary basis for predicting the origins of longevity in humans. Imagine a population of early hominids having enough genetic plasticity to produce different mortality rates. This population would contain a subpopulation of individuals that died before reaching the end of their reproductive life span and, at the other extreme, a subpopulation that lived well beyond the end of their reproductive life span. The longer reproductive life span of the long-lived subpopulation would endow greater fitness, albeit only slightly greater than for the shorter-lived subpopulation. Over evolutionary time, genes producing the genotype of the longer-lived, longer-reproductive-period subpopulation would be selected over genes producing the genotype of the shorter-lived population. (See the discussion of genetic drift and mutation accumulation in Chapter 3.) The human genome would have drifted, and would still be drifting, toward increased longevity.

Mortality rates differ in long-lived humans

In the past, testing the possibility that genetic factors account for differences in the mortality rates of the longest-lived humans has been difficult, simply because not enough individuals reached an advanced age. Evidence for differences in mortality rates was generally limited to observations of families with several individuals who reached an advanced age. Today, however, several scientific studies are being conducted, throughout the world, that explore the relationship between genetics and life span in cohorts of long-lived humans, normally defined as people over the age of 100—**centenarians**. (In the United States, the number of people over the age of 100 is somewhere between 100,00 and 200,000.) These investigations have only just begun, and the analysis is limited, for the most part, to demographic studies. While some studies of centenarians are beginning to identify specific alleles associated with extreme longevity, these results are far too preliminary for discussion here.

One approach to investigating the genetic contribution to human longevity is to evaluate the mortality rates of centenarians' siblings or twins. Siblings or twins of centenarians are highly likely to have a similar genetic profile, providing demographic evidence that longevity may be related to genetics. Several studies have taken this approach and have shown similar results: the survival probabilities and mortality rates of the siblings or twins of centenarians differ significantly from the mortality rate in the general population. For example, the New England Centenarian Study found that female and male siblings of centenarians were, respectively, 8 and 17 times more likely than the general population to reach the age of 100. This same study showed that, at all ages

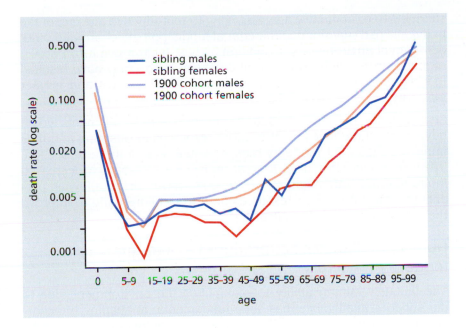

Figure 7.2 Age-specific death rates by gender in a 1900 birth cohort and in siblings of centenarians. These data show that age-specific death rates are significantly lower at all ages for the siblings of centenarians than for the general US cohort born in 1900. (See Figure 2.15 for an explanation of the change in slope of age-specific mortality.) (From T.T. Perls et al., *Proc. Natl Acad. Sci. U.S.A.* 99:8442–8447, 2002. With permission from National Academy of Sciences.)

throughout the life span, centenarians' siblings had mortality rates that were roughly half of those observed for the United States as a whole **(Figure 7.2)**. While much research is still needed to confirm the genetic component of human longevity, it appears that genes have a significant role in determining longevity.

Human intelligence has altered mortality rates

Darwin's principle of reproductive fitness closely links fitness to a species' adaptation to environmental conditions. For the majority of species on Earth, adaptation occurs purely as a chance event, when an allele arises in an individual that increases the probability of its survival to reproductive age. That is, for most species, mortality rates reflect an adaptation to the environment (see Chapter 1). The origin of longevity in *Homo sapiens* may have resulted from our advanced intelligence, which has allowed us to adapt the environment to our genes and thus change mortality characteristics. For example, the use of tools, such as sharpened stones and bones, enhanced our ability to hunt and increased the variety of foods in our diet. Early *H. sapiens* would have had significantly better nutrition, which contributed to our larger size compared with other primates—a significant survival advantage. Invention of the needle and thread led to better clothing, which allowed us to inhabit geographical locations different from those in which our physiology had evolved. We were no longer dependent on the local environment to provide food; we could move to where the food was and thus gain a significant survival advantage.

The intelligence that allowed *H. sapiens* to manipulate the environment for a survival advantage would probably have had a significant impact on the survival of mothers and infants. The increase in the quality and quantity of food and nutrition, along with advances in simple technology (tools), would have reduced maternal deaths at birth—which, most likely, would have had two outcomes that decreased mortality rates and increased longevity.

First, mothers surviving birth would live to have more children. Having more children increased the chance of at least one child, if not more, surviving to reproductive age. Surviving to reproductive age meant the individual had superior genes, better at fending off infections and other detrimental environmental conditions. Thus, what arose as a product of intelligence—reduced maternal deaths at birth—may have resulted in children with genes imparting lower mortality rates and greater longevity.

Second, better nutrition and better protection from the elements would have allowed the investment of more resources in offspring and thus improved the infant mortality rate. As you'll see in the next section, infant mortality rates have a great impact on life span. A decreased infant mortality rate might also have led to fewer births, allowing parents to concentrate their resources on fewer children and to improve the overall quality of their offspring.

Human intelligence has produced a unique longevity trajectory

The uniquely human ability to manipulate the environment, due to our superior intelligence, has been fundamental in shaping the characteristics of human longevity. However, researchers cannot directly measure intelligence in our phylogenetic predecessors—one cannot give intelligence tests to fossils. Instead, biodemographers often turn to the indirect measures of intelligence used in physical anthropology and evolution/ecology to support their mathematical predictions on the origins of longevity in *H. sapiens*. To this end, brain size, measured directly in living animals or deduced from the size of the brain cavity of fossilized skulls, has been used to approximate intelligence.

The positive correlation between brain size, body weight, and longevity in mammalian species has been known for several decades **(Figure 7.3)** (see also Chapter 1). In general, the larger the body and brain weight, the longer the life span. These early correlations also describe a distinctive phylogenetic separation between primates and non-primates. Note in Figure 7.3 that the data points for primates and humans fall above the regression line, whereas the data points for rodents and ungulates fall below the regression line. This means that the body/brain weight–life span correlation in primates (human and nonhuman) is separate from that in other phylogenetic groups. Moreover, the body/brain

Figure 7.3 Correlation of body weight and brain weight with life span in various mammalian groups. Note that the human life span is considerably greater than that of ungulates (hoofed animals, in several orders of mammals) of similar body weight. A significantly larger brain underlies the difference in life span between humans and the other three groups of mammals. (From G.A. Sacher, in Lifespan of Animals [G.E.W. Wolstenholme and M. O'Connor, eds.], London: J.A. Churchill, pp. 115–141, 1959. Little Brown and Co. With permission from Elsevier.)

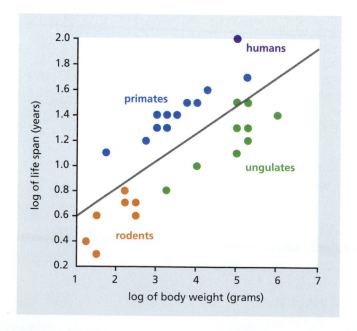

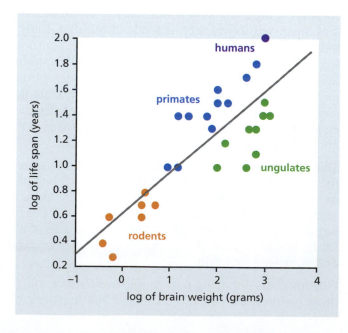

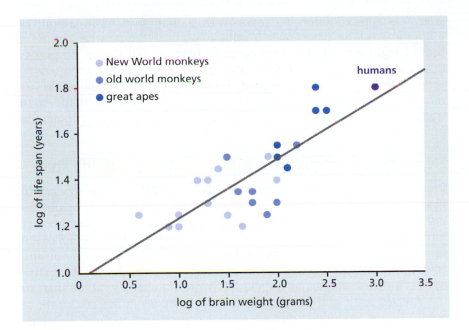

Figure 7.4 Correlation between brain weight and life span in humans and their immediate phylogenetic relatives. These data indicate that brain weight, even among morphologically similar species, has been a critical factor in the origins of human longevity. (From J.R. Carey, *Longevity: The Biology and Demography of Lifespan*, Princeton, NJ: Princeton University Press, 2003. With permission from Princeton University Press.)

weight–life span correlation in humans is significantly different from that in other primates, suggesting yet another phylogenetic separation for intelligence among primates. For these reasons, biodemographers generally limit their analysis of the origins of human longevity to the primate order.

New World and Old World monkeys have significantly smaller brains and shorter life spans than humans **(Figure 7.4)**. Moreover, the great apes with body weights similar to that of humans have smaller brains and a shorter life span. If we narrow the analysis even further by considering only hominids, the strong relation between brain size and life span remains **(Figure 7.5)**. There is also a tremendous shift in life span between our immediate phylogenetic relatives and modern humans. The life span difference between *Australopithecus afarensis* and *Homo*

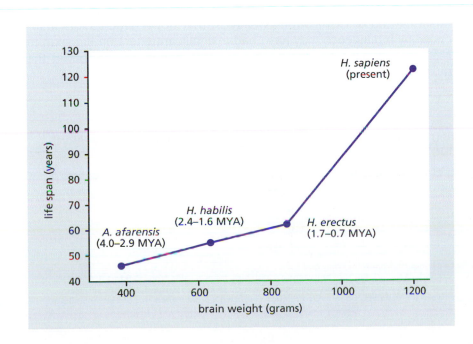

Figure 7.5 Correlation between brain weight and life span in humans and earlier hominids. Life spans of *Australopithecus afarensis*, *Homo habilis*, and *Homo erectus* were estimated from fossils on the basis of dental and bone development. Brain weights in these hominids were estimated from brain-cavity volume. MYA = million years ago. (From H.M. McHenry, *J. Hum. Evol.* 27:77–87, 1994. With permission from Elsevier.)

erectus is approximately 15 years, with a change in brain weight of 450 grams taking place over 3 million years. *H. sapiens* developed a brain weight almost 400 grams greater than that of *H. erectus* in less than 700,000 years, which translated into a gain in life span of 60–70 years. Thus, the anthropological data indicate that intelligence, as measured by cranial volume, set *H. sapiens* on a mortality and life-span trajectory distinctly different from that of our immediate phylogenetic predecessors.

Heredity has only a minor influence on human life span

We have stressed that intelligence allowed early *Homo* species to manipulate their environment, thus increasing survival and establishing conditions that could have resulted in a genetic drift toward greater longevity. In other words, human longevity seems to have followed the prediction of the evolutionary theories proposed by Medawar and Hamilton that you learned about in Chapter 3. The term "life span," as you'll recall, applies to individuals within a species rather than to the species itself and is determined largely by an individual's rate of aging. That is, an individual's life span is determined by random events occurring over time. So, if the evolutionary theories are valid, we would expect heredity to have a minor effect on life span.

We can test this possibility by evaluating the correlation between the age of death in parents and their children and in twins. Weak correlations would suggest that the nongenetic components of life span are more influential than genetic factors in *H. sapiens*.

Results for investigations, dating back to 1903, that compare age of death in a parent with age of death in offspring have consistently shown very weak correlations. That is, contrary to popular belief, our life span is not determined by our parents. **TABLE 7.1** shows the contribution of heredity to life span in a French Canadian population born during the nineteenth century. Although the contribution varies with age of death of the offspring and with gender, heredity accounts for no more than about 10–16% of the total influence on life span. Similar results have been found when evaluating the age of death in monozygotic (identical) and dizygotic (fraternal) twins: the contribution of heredity to life span was found to be no more that 25% in 2800 Danish twin-pairs. Together, comparisons of the age of death of parents and their offspring and of twins strongly suggest that nongenetic factors influenced by intelligence may have had a significant effect on life span in *H. sapiens*. You'll learn the nature of these nongenetic factors in the next section.

TABLE 7.1

CORRELATION COEFFICIENTS FOR THE CONTRIBUTION OF HEREDITY TO LONGEVITY, AS DETERMINED BY COMPARING THE AGE OF DEATH IN PARENTS AND OFFSPRING

Age of child at death	Father/son	Mother/daughter	Mother/son	Father/daughter
<20 years	0.043	0.241	0.014	0
>20 years	0.129	0.106	0	0.190
>50 years	0.101	0.112	0	0.067
All ages	0.101	0.161	0.052	0.072

From P. Philippe, *Am. J. Med. Genet.* 2:121–129, 1978. With permission from John Wiley and Sons.
Note: A correlation coefficient gives an indication of the strength of the relationship between two variables. It has a value from –1.0 to 1.0. Correlation coefficients greater than 0 indicate a positive relationship; less than 0, a negative relationship; equal to 0, no relationship. The closer the coefficient is to 1.0 or –1.0, the greater the strength of the relationship between the variables. Generally, correlation coefficients greater than –0.3 but less than 0.3 are considered weak.

THE RISE OF EXTENDED HUMAN LIFE SPAN IN THE TWENTIETH CENTURY

The science of biodemography has provided the theoretical framework suggesting that intelligence is the underlying reason for humans' unique trajectory in mortality rates and life span. This unique trajectory, however, did not arise "instantaneously," 1–1.2 million years ago, with the increase in brain size that accompanied the speciation of *H. sapiens* from *H. erectus*. Rather, humans had to develop socially and culturally before their advanced intelligence could operate to affect life span and mortality rates. For example, the sharing of information through oral and archaic written forms was essential to the spread of agricultural techniques that reduced starvation in hunter-gatherer societies. The transfer of this information from one group of people to another required the establishment of routes linking geographical regions, a process that necessitated the use of simple engineering skills that could be developed only with human intelligence. (See Chapter 1 for a discussion of how insects are being used to study the impact of social groups on longevity and mortality.)

The pace of knowledge accumulation that led to extended life span was rather slow throughout most of human history. The pace of discovery in biology and medicine quickened during the seventeenth century as the Age of Enlightenment progressed. The shift from largely agrarian societies to industrial economies required a more educated workforce. As education for the masses expanded, so did technology. Medical technology, such as the microscope, led to the discovery of bacteria and, eventually, to public health policies that reduced the spread of infections and increased life span.

At the beginning of the twentieth century, the technology created by human intelligence profoundly reduced the incidence of lethal diseases and increased life span. In a single 30-year period, from 1920 to 1950, the average human life span increased by more than 75%. Many demographers and biodemographers believe that this great leap forward in longevity placed our species on a new evolutionary track from that experienced just 100 years earlier. In this section, we examine the reasons for the unprecedented rise in human life span during the twentieth century. We begin with a brief look at life spans in historical societies, then discuss the rise of modern biological inquiry and how this led to the mortality rates and extended life spans observed in the developing countries of today.

For most of human history, the average human life span was less than 45 years

Determining the average human life span in historical populations can be difficult. The construction of life tables and determination of age-adjusted life expectancy did not begin until 1750–1800. Estimations of mean and maximum life span prior to that time are compiled from indirect data such as skeletal remains, tombstone inscriptions, and records of religious rituals. The accurate determination of age in skeletal remains depends, to a large degree, on the state of bone ossification. Individuals with poor ossification, such as the very young and the very old, tend to have faster rates of bone decomposition after death, which makes age determination difficult. Life-span data generated from skeletal remains tend to overestimate mean life span and underestimate maximum life span. Age of death determined from tombstone inscriptions may be unrepresentative of the entire population, as only some of the population may have been wealthy enough to afford a tombstone. Gravesite data also exclude cultures in which cremation was generally preferred to burial.

Despite the limitations in determining age prior to the recording of **vital statistics**, we know from archeological evidence that from the Neolithic Period (~5000 B.C.E.) to the Enlightenment (beginning ~1600 C.E.), the average human life span remained fairly stable at 30–40 years **(Figure 7.6)**. Age estimations made from suture closing in skulls recovered from a Greek cemetery, dated between 3500 B.C.E. and 1300 C.E., show a mean age at death of 32–38 years for males and 28–33 years for females. While no skeletons over the age of 60 were found in this gravesite, tombstone inscriptions from ancient Roman cemeteries indicate that a few individuals lived into their eighties and nineties. Thus, while mean life span was only half of that seen in economically developed countries today, the Roman cemetery inscriptions suggest that maximum life span potential may have been similar.

The fall of the Roman Empire in ~500 C.E. and the ensuing breakdown of an organized governing structure resulted in the loss of the census data needed to construct reliable life tables for more than 1200 years. Accurate census recordings of the general population did not begin again in Europe until around 1850. However, the Church of England began keeping Christian baptismal and death records of their members in 1541, and these records have been used to determine mean life span prior to 1850. These data show that mean life span had increased only moderately from the gravesite data in Greece and Rome, to 30–45 years **(Figure 7.7)**. Thus, based on data collected between 3000 B.C.E. and 1850 C.E., mean life span was less than 45 years of age throughout most of human history. Although biodemographic deductions have shown that early humans had a brain size similar to that observed today, that human intelligence did not contribute significantly to extended life span, at least until 1850 C.E.

Control of infectious diseases increased mean life span

A change in the pattern of mean life span in England seems to have begun somewhere around the beginning of the nineteenth century. Mean life span in the United States started to increase steadily around 1840 **(Figure 7.8)**, an increase directly related to invention of the

Figure 7.6 Mean age at death of males and females as estimated from skeletal remains recovered from a Greek cemetery. Individuals in this sample died between 3500 B.C.E and 1300 C.E. The data show that mean life span hovered around 30–35 years for most of human history. (Data from G.Y. Acsadi and J. Nemeskeri, History of Human Life Span and Mortality, Budapest: Akademiai Kiado, 1970, p. 346.)

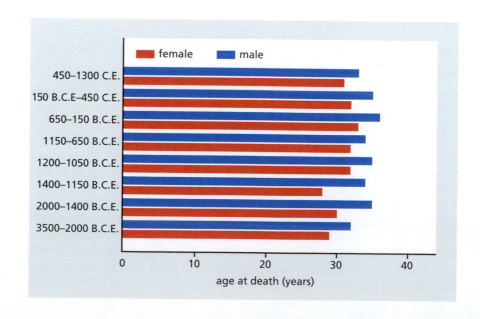

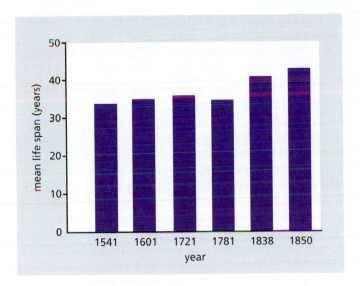

Figure 7.7 Mean life span in England and Wales as determined from records of the Church of England. (From W.E. Smith, Human Longevity, New York: Oxford University Press, 1993. With permission from Oxford University Press.)

microscope and the discovery of bacteria by Anton van Leeuwenhoek in 1676, a time when infectious diseases were the number one cause of death throughout the world. The discovery of bacteria and the development of microbiology in the 1700s gave rise to the understanding that bacteria were the cause of infection. As governments began to establish public health policies, the incidence of infectious diseases began to decline slowly and mean life span began to increase.

Edward Jenner introduced the first successful vaccination for smallpox in 1811, but widespread inoculation programs did not begin until Louis Pasteur provided the methods for producing large quantities of vaccines in 1880. (Until the discoveries of Pasteur, control of infectious diseases was mainly limited to isolating humans from the sources of bacteria.) In addition, milk pasteurization, first adopted in Chicago in 1908, contributed to the control of milk-borne diseases (e.g., gastrointestinal infections) from contaminated milk supplies. Eventually, procedures were developed for ensuring safe water and the treatment

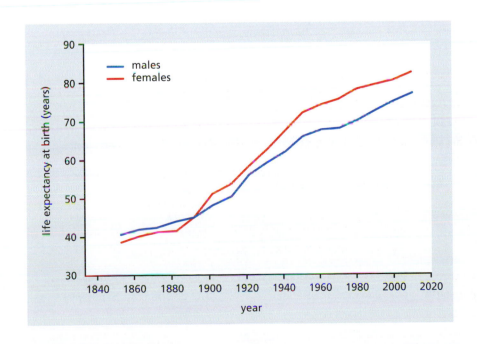

Figure 7.8 Life expectancy at birth for males and females in the United States between 1850 and 2010. Notice that life expectancy at birth was similar in males and females until about 1930; this has been attributed to the high level of maternal death during childbirth. When the majority of women began to deliver their children in a hospital, survival increased, as did life expectancy at birth. (From Department of Health and Human Services, National Center for Health Statistics, *Natl Vital Stat. Rep.* 54:19, 2006, www.dhhs.gov.)

of sewage. Life expectancy at birth began to increase dramatically only after the advent of widespread vaccinations and general acceptance of the need for proper sanitation. Note, in Figure 7.8, the rapid increase in mean life expectancy at birth beginning somewhere between 1890 and 1900.

Infectious diseases can affect all age groups equally (though is higher in infants, as described below) resulting in high mortality rates at all ages. For example, the leading cause of death in the United States in 1900 was pneumonia and influenza, followed closely by tuberculosis **(TABLE 7.2)**. The top three causes of death in 1900 were diseases that are now preventable (pneumonia and influenza, tuberculosis, diarrhea and food-borne illness), an observation supporting the contention that infectious diseases and inadequate sanitation led to early death and low life expectancy at birth. Reduction in preventable diseases that have high mortality rates, such as diarrhea and tuberculosis, resulted from government-enforced sanitation and vaccination programs during the latter half of the 1920s. Only pneumonia and influenza remained in the top 10 causes of death in 2006, and their ranking continues to decline as vaccination against influenza becomes more common. The containment of infectious diseases during the 1920s and 1930s led to a significant statistical shift: diseases with a strong genetic component—such as heart disease and cancer—replaced infectious diseases as the leading causes of death. Because diseases with a strong genetic component are more frequent in older age groups, mortality rates at younger ages are reduced.

Decreases in infant mortality increased life expectancy

Mean life expectancy is the statistical probability of achieving a certain age in the future given the current age distribution of the population (see the discussion of life tables in Chapter 2 and Appendix 1). Recall that mean life expectancy at birth is calculated from the ratio of the number of years lived in all age categories and the number of deaths occurring in a particular age category (in this case, 0–1 years). Thus, if the infant mortality rate (0–1 year of age) is high, the number of years

TABLE 7.2
THE TOP TEN CAUSES OF DEATH IN THE UNITED STATES FOR SELECTED YEARS

1900	1920	1940	1960	1980	2000	2006
Pneumonia and influenza	Pneumonia and influenza	Heart disease	Heart disease	Heart disease	Heart disease	Heart disease
Tuberculosis	Heart disease	Cancer	Cancer	Cancer	Cancer	Cancer
Diarrhea and food-borne illness	Tuberculosis	Stroke	Stroke	Stroke	Stroke	Stroke
Heart disease	Stroke	Kidney disease	Accidents	Accidents	Accidents	COPD[1]
Stroke	Kidney disease	Pneumonia and influenza	Infant mortality	COPD	COPD	Accidents
Kidney disease	Cancer	Accidents	Pneumonia and influenza	Pneumonia and influenza	Diabetes	Diabetes
Accidents	Accidents	Tuberculosis	Diabetes	Diabetes	Pneumonia and influenza	Pneumonia and influenza
Cancer	Diarrhea and food-borne illness	Car accidents	Atherosclerosis	Liver disease	Suicide	Alzheimer's disease
Senility	Premature birth	Diabetes	Diabetes	Atherosclerosis	Kidney disease	Kidney disease
Diphtheria	Puerperal fever	Premature birth	Liver disease	Suicide	Liver disease	Septicemia

[1] COPD = chronic obstructive pulmonary disease.

lived by all age categories (the numerator of the equation) declines, the death rate for infants (the denominator of the equation) increases, and the mean life expectancy at birth is low. Conversely, a decrease in infant mortality rate increases the number of years lived by all age categories and decreases the death rate for infants, thereby increasing mean life expectancy at birth.

As we've seen, infectious diseases were the major cause of death throughout much of human history. Infants are particularly susceptible to infectious disease, because the human immune system does not fully develop until many years after birth. Infectious diseases were the primary cause of death in infants until the 1920s. Moreover, before 1900, most births took place in the home and were attended primarily by midwives or other individuals who did not have specific medical training. Any complication threatening the life of the infant was, for the most part, left untreated. The high rate of infectious diseases and complications at birth resulted in a high infant mortality rate prior to 1900 **(Figure 7.9)**. The combination of public health programs, widespread inoculations, and hospital births decreased the infant mortality rate and increased life expectancy at birth. Note, in Figure 7.9, the opposing slopes in the infant mortality rate and life expectancy at birth that began around 1910.

Economically developed countries throughout the world have experienced a pattern of increasing life expectancy accompanied by decreasing infant mortality rate similar to that for the United States. However, high rates of infant mortality continue to suppress life expectancy at birth in less-developed countries **(TABLE 7.3** and **TABLE 7.4)**. For example, Angola has the world's highest infant mortality rate and lowest life expectancy at birth. Conversely, Singapore has the world's lowest infant mortality rate and highest life expectancy at birth.

Improved medical treatments account for the continuing increase in life expectancy

Although the rate of infant mortality in the United States began to flatten somewhere between 1970 and 1980, life expectancy at birth has continued to increase at nearly the same rate as that observed in

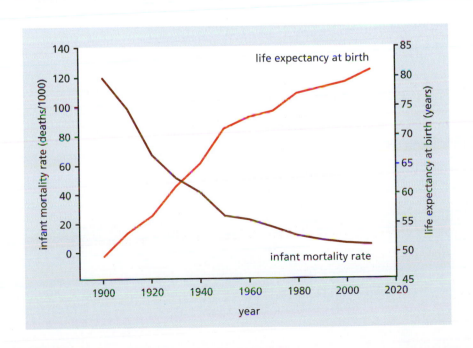

Figure 7.9 Life expectancy at birth and infant mortality rate for females in the United States. Decreasing infant mortality was the primary reason for the increase in the aged population until about 1970–1980. (From F.C. Bell and M.L. Miller, Life tables for the United States Social Security area 1900–2100, Washington, DC: Social Security Administration, 2005, p. 194.)

TABLE 7.3
INFANT MORTALITY RATE (IMR) AND LIFE EXPECTANCY AT BIRTH (LEB) IN SELECTED ECONOMICALLY LESS-DEVELOPED COUNTRIES

Country	IMR (per 1000)	World rank	LEB (years)	World rank
Angola	184.5	1	37.6	221
Liberia	149.7	4	40.6	216
Somalia	113.0	6	48.8	204
Chad	102.1	10	47.2	206
Zambia	100.7	12	38.4	220

Note: The rankings are based on data from 221 countries.

TABLE 7.4
INFANT MORTALITY RATE (IMR) AND LIFE EXPECTANCY AT BIRTH (LEB) IN SELECTED ECONOMICALLY DEVELOPED COUNTRIES

Country	IMR (per 1000)	World rank	LEB (years)	World rank
Singapore	2.3	221	81.8	1
Japan	2.8	219	82.0	3
France	3.4	216	80.5	10
Switzerland	4.2	209	80.6	9
Canada	4.6	199	80.3	13

Note: The rankings are based on data from 221 countries.

1910 (see Figure 7.9). Together, the relatively stable infant mortality rate and the increasing life expectancy at birth indicate that changes in infant mortality are no longer the cause of the continuing increase in life expectancy. Rather, the continuing increase since 1970 reflects improvements in care and treatment for noninfectious, life-threatening diseases in older age groups.

For example, in 1930, the number of years a person could expect to live after age 65 was a little over 12 years **(Figure 7.10)**. In 2010, those over

Figure 7.10 Age-specific life expectancy in the United States at 65 and 85 years of age. (From F.C. Bell and M.L. Miller, Life tables for the United States Social Security area 1900–2100, Washington, DC: Social Security Administration, 2005, p. 194.)

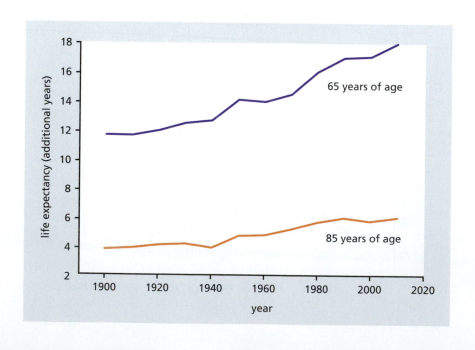

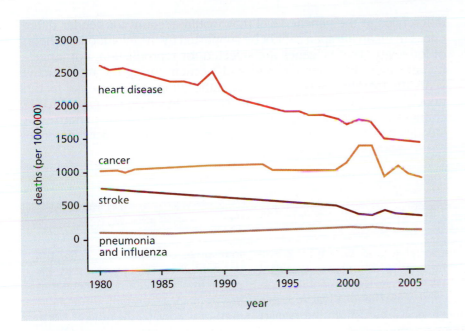

Figure 7.11 Death rates for the four leading causes of death in individuals over the age of 65. Deaths due to heart disease, the perennial leading cause of death, have declined by slightly more than 40% since 1980. This decline in heart disease has resulted in the significant gains in life expectancy at 65+ seen in Figure 7.10. (From National Center for Health Statistics.)

the age of 65 could expect to live an additional 18.1 years. The increase in life span after age 65 reflects, for the most part, the ability to treat noninfectious, life-threatening diseases such as heart disease and cancer. For instance, since its peak in 1980, mortality from coronary heart disease in the over-65 age group has fallen by almost 40%, from a high of 2500/100,000 to 1400/100,000 **(Figure 7.11)**. Factors accounting for this decrease include improved diagnostic procedures (early detection of coronary heart disease), prophylactic measures such as angioplasty and arterial stents, and replacement of coronary arteries (bypass surgery). That is, individuals who, 40 years ago, would have died due to lack of effective treatment are now able to enjoy a long, healthy life span **(BOX 7.1)**.

Women have a longer mean life expectancy than men

For most of the twentieth century, females born in economically developed countries outlived males, creating a "gender gap" in mean life span **(Figure 7.13)**. Reasons given for the gender gap are many, including genetic factors that favor females, but none of the explanations have been shown to be a direct cause. Plausible explanations are those suggesting that the gender gap is, for the most part, a uniquely human trait and reflects modifiable factors—that is, things we do to ourselves. Human males tend to participate more than females in behaviors that lead to greater rates of certain diseases. For example, from 1940 to 1985, the number of men who smoked cigarettes was almost double the number of women who smoked. This caused significantly greater early-age (40–60 years) mortality from heart disease and lung cancer in men than in women.

Homo sapiens appears to be one of only a handful of species that have a significant post-reproductive life span and show mean life-span differences between males and females. Several, but not all, studies have shown that no other mammals with long post-reproductive life spans show a gender difference.

While genetic factors accounting for the human gender gap cannot be ruled out, it seems unlikely that biological variation alone could explain the difference in longevity in males and females. At least two facts argue strongly against a purely biological explanation. First, biological differences at the genetic level that would account for gender differences in mean life span cannot be supported by the evolutionary theory of longevity. That is, genes are selected for reproduction, not longevity, and mortality rates for men and women during the reproductive years do not differ. Second, if genes rather than environmental factors underlie the difference in mean life spans, we would not expect to see changes in the gender gap. In fact, this gender gap has been narrowing steadily since 1950 (and perhaps longer; records of gender differences

BOX 7.1 BIOLOGY MEETS SOCIOLOGY: FUNDING SOCIAL SECURITY

Before getting your first formal job, one in which you were paid by payroll check rather than with cash, you most likely had to provide your employer with your Social Security number and to begin paying into the Social Security system, so that someday you'll be eligible for retirement benefits. Many individuals in the baby boomer generation, those born between 1940 and 1960, take the Social Security system for granted and will, most likely, collect all the retirement benefits due to them when they reach retirement age. Subsequent generations may not be as lucky, as questions have been raised concerning the solvency of the Social Security Trust Fund. Some even call the Social Security funding dilemma a "crisis." To understand why the retirement benefits of Social Security may be in trouble, we only have to look at the changes in mean life span during the twentieth century.

The Social Security system came into being in 1935 as a part of President Roosevelt's New Deal economic recovery program. The idea was to lessen the financial burden on the family's primary wage earner for care of elderly family members. This was to be accomplished by establishing a system in which the federal government would tax workers to pay for their retirement. The underlying theory was that, because the ratio of workers to retirees was so great, the current workers (the young) could pay for the retired generation. Each successive generation would be expected to do the same. Moreover, the tax structure for Social Security was set up to include additional money to fund a trust that would defray expenses during times of economic downturns and low tax revenues.

In 1935 the theory made sense, because of three important demographic facts: (1) the proportion of individuals over the age of 65, the designated retirement age, was 4% of the total population—a value that had remained constant throughout most of history; (2) the ratio of workers to retirees was 45:1; and (3) the age-specific life expectancy at 65 was 12 years **(Figure 7.12)**. That is, at any one time, very few people (as a proportion of the population) were eligible for receiving Social Security benefits, there were plenty of workers to pay for those who did retire, and the payout period was relatively short. Today, however, 15% of the population is eligible to apply for Social Security, the worker-to-retiree ratio is approaching equality, and retirees will remain eligible for benefits, on average, for almost 20 years (see Figure 7.12). Thus, we are asking a smaller group of workers to pay for an increasing number of retirees who may live a significant portion of their total life span in retirement.

The math tells us everything: the change in demographics, with the increase in the aging population, has placed significant pressure on our ability to adequately ensure a solvent Social Security system for individuals born today. Finding solutions to the coming financial difficulties in Social Security will not be easy and will require significant action by our political representatives. In this battle there is only one certainty: the percentage of individuals living long, healthy lives will continue to increase.

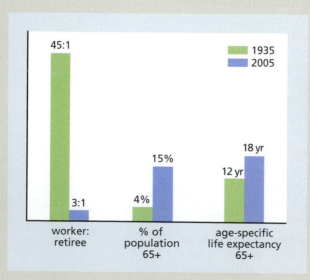

Figure 7.12 Variables affecting the Social Security Trust Fund.

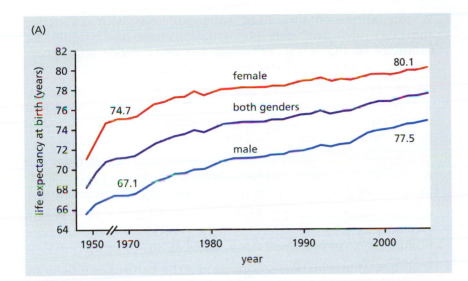

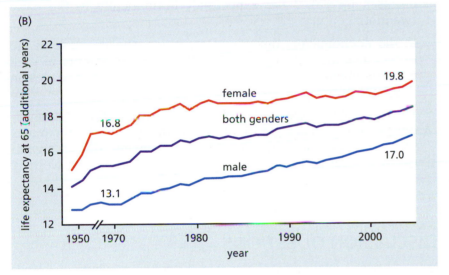

Figure 7.13 The "gender gap" in life expectancy at birth and at age 65. The difference in mean life span between men and women has been decreasing steadily since accurate recordkeeping began (~1950). (A) The difference in life expectancy at birth between men and women in 1970 was 7.6 years. By 2003, the difference had declined to 2.6 years. (B) A similar pattern of decrease in the gender gap can be seen in life expectancy after the age of 65.

in longevity are unreliable). For example, as shown in Figure 7.13, a male born in 1950 could expect to live 65 years, whereas a female had a mean life expectancy at birth of 71 years, a gap of 6 years. By 2003, the gap was only 2.7 years. The narrowing of the gender gap was caused by a 13.4% gain in mean life span for men, compared with a modest 6.7% gain for women. Genetic differences would persist, for the most part, regardless of environmental stimuli.

ESSENTIAL CONCEPTS

- Biodemography combines the classical techniques of demography—mortality analysis—with the biological principles of longevity. Biodemography is primarily a mathematical and theoretical science that constructs models predicting the origins of human longevity.

- Humans respond to environmental changes on the basis of cognitive reasoning and thus can change the environment to fit their needs, rather than instinctively allowing the environment to control their longevity.

- A population contains many subpopulations with distinctive mortality rates, and such subpopulations have had a significant influence on the origins of human longevity. Biodemographers refer to human mortality rates as facultative.

- Biodemographers suggest that the different genotypes resulting in different mortality rates could have existed throughout evolutionary history and may have influenced the selection of genes that determine human longevity.

- The origin of longevity in *Homo sapiens* may result from our advanced intelligence, which allows us to adapt the environment to our genes and thus change mortality characteristics.

- Archeological evidence indicates that from the Neolithic Period (~5000 B.C.E.) to the Enlightenment (~1600 C.E.), the average human life span remained fairly stable at 30–40 years.

- Widespread inoculation programs and the control of infectious diseases increased life expectancy dramatically in the early twentieth century.

- In the economically developed countries, the increase in life expectancy at birth after 1970 reflects improvements in care and treatment for noninfectious, life-threatening diseases in older age groups.

- Throughout the twentieth century, females born in the developed countries had a greater life expectancy at birth than males The gender gap in longevity is now narrowing, due to significantly greater gains in mean life span for men than for women.

DISCUSSION QUESTIONS

Q7.1 Briefly explain what is meant by the statement "human mortality rates are facultative."

Q7.2 Describe the evidence suggesting that genetic factors influencing longevity have plasticity.

Q7.3 Describe how the intelligence of early *Homo sapiens* allowed parents to give more resources to fewer children. What effect would this have on human longevity?

Q7.4 What evidence suggests that environmental factors have had a greater influence than genetic factors on human life span?

Q7.5 What evidence suggests that *Homo sapiens* is on distinctly different mortality and longevity trajectories than our immediate phylogenetic predecessors?

Q7.6 Explain why the invention of the microscope in the mid-seventeenth century would lead to increased longevity.

Q7.7 Consider the following graph **(Figure 7.14)**. Describe (1) why life expectancy at birth began to increase around 1900–1910 and (2) why life expectancy at birth for females started to diverge from that of males around 1920.

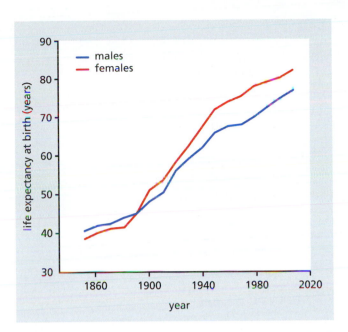

Figure 7.14 (From Department of Health and Human Services, National Center for Health Statistics, *Natl Vital Stat. Rep.* 54:19, 2006, www.dhhs.gov.)

Q7.8 Explain why, after 1970, increases in life expectancy at birth in the United States were due to medical treatments for age-related diseases, not a continuing decrease in infant mortality rate. Give an example to support your explanation.

Q7.9 List three factors that have placed the US Social Security system in financial crisis.

Q7.10 Has the gender gap in life expectancy at birth been increasing or decreasing since 1950? Give some likely reasons for the increase or decrease.

FURTHER READING

ORIGINS OF HUMAN LONGEVITY

Carey JR (2003) Longevity: The Biology and Demography of Lifespan. Princeton, NJ: Princeton University Press.

Carey JR & Judge DS (2001) Principles of biodemography with special reference to human longevity. *Population* 13:9–40.

Gavrilov LA & Gavrilov NS (2003) Modulating Aging and Longevity (Rattan SLS ed). Dordrecht, Netherlands: Kluwer Academic Publishers, pp 27–50.

Herskind AM, McGue M, Holm NV et al. (1996) The heritability of human longevity: a population-based study of 2872 Danish twin pairs born 1870–1900. *Hum Genet* 97:319–323.

Perls TT, Wilmoth J, Levenson R et al. (2002) Life-long sustained mortality advantage of siblings of centenarians. *Proc Natl Acad Sci USA* 99:8442–8447.

Philippe P (1978) Familial correlations of longevity: an isolate-based study. *Am J Med Genet* 2:121–129.

Vaupel JW (2010) Biodemography of human ageing. *Nature* 464:536–542.

THE RISE OF EXTENDED HUMAN LIFE SPAN IN THE TWENTIETH CENTURY

Acsadi GY & Nemeskeri J (1970) History of Human Lifespan and Mortality. Budapest: Akademiai Kiado, p 346.

Kinsella K & He W (2009) An Aging World: 2008. Washington, DC: US Government Printing Office. www.census.gov/prod/2009pubs/p95-09-1.pdf.

Smith WE (1993) Human Longevity. New York: Oxford University Press.

THE PHYSIOLOGY OF HUMAN AGING

8

"MIDDLE AGE IS WHEN YOUR AGE STARTS TO SHOW AROUND YOUR MIDDLE."

-BOB HOPE, ACTOR (1903–2003)

We all know what human aging looks like. A 20-year-old looks different from a 40-year-old, who looks different from a 60-year-old. Looks can be deceiving, however, when it comes to the internal physiological changes that accompany the external signs of aging. Although research on populations has shown that every physiological system declines with age, the amount of decline, the systems affected, and the age at which the decline begins are highly variable and specific to each individual. This variability in age-related physiological loss in humans occurs because aging is a random, or stochastic, process caused by a loss in molecular fidelity. Because of the random nature of aging, it is very difficult to predict the amount of age-related loss in any specific physiological system in any specific individual.

The inability to precisely define the amount and timing of human age-related physiological loss means that we must use generalities to describe functional decline. The amount of age-related loss in a specific physiological system described in this chapter is based on averages for populations, using data from both cross-sectional and longitudinal studies. Assigning a discrete age at which a particular physiological decline begins is impossible. Although we will refer to time-dependent changes in physiology using the term "age-related," keep in mind the following important and, as yet, nonmeasurable variables that affect the start and rate of functional decline in humans: (1) the trajectory or rate of aging may be significantly influenced by events that took place during early growth and development; (2) environment and lifestyle choices introduce significant variation in the start and rate of aging; and (3) the loss of or decline in reproductive ability seems to accelerate age-related functional loss.

In this chapter and the next, we are concerned with age-related functional decline in various human physiological systems. This chapter focuses on physiological decline that does not, in general, increase the risk of disease or mortality. Physiological systems that are more likely to develop age-related diseases leading to an increase in mortality rate or **morbidity**—the circulatory system, nervous system, and skeletal system—are discussed in the next chapter. Not all systems and organs are covered in this text. For example, we exclude the lungs (respiratory system) and the liver (including the gallbladder), which show extremely minor changes with aging, changes that are limited, for the most part, to alterations in their circulatory components that lead to age-related disease.

CHANGES IN BODY COMPOSITION AND ENERGY METABOLISM

The shape and composition of the human body reflect, for the most part, the relationship among the amounts of its four primary constituents: water, protein, lipid, and bone. Although lifestyle and personal choices can have a major impact on body composition, the amounts of protein, lipid, and bone are also influenced by the physiological stage of life (the percentage of water in the body remains remarkably consistent throughout life).

From infancy through puberty, the body focuses on the growth and development of its internal organs, bones, and cells of the immune system, and the rate and amount of protein deposition are higher during this period than at any other time in the life span. Prior to puberty, the shape and composition of the human body do not differ significantly between the sexes, but the onset of puberty initiates a greater focus on the growth of sex-linked characteristics. Muscle development in males outpaces that in females, while females generate greater fat deposition. Both sexes experience significant growth in bone mass, proportional to their final body height. The shape and composition of the human body during growth and development have a strong genetic component, but as the current epidemic of childhood obesity demonstrates, environmental influences are also important.

Once growth and development have ended—somewhere between 25 and 30 years of age—the size, shape, and weight of the human body reflect, for the most part, factors over which we have direct control. While the normal, active, and developmentally mature adult experiences a slight age-related decrease in the mass of muscle, internal organs, and bone, fat stores remain dynamic throughout life, increasing or decreasing depending on nutrition and exercise. Near the end of the life span, decreases in all components of the body accelerate.

In this section, we examine the normal age-related changes in the composition and size of the human body, along with the mechanisms that underlie these changes. We begin with a brief summary of how energy intake and expenditure affect the amount of fat we store. Then we look more closely at specific age-related alterations in body composition and weight.

Energy balance is the difference between intake and expenditure

Humans, like all organisms, need energy to drive the basic chemical reactions of life. Our energy comes from the food we eat in the form of fat and carbohydrate. (Humans do not, to any great degree, use protein as an energy source, except during starvation.) The biochemical mechanisms by which fat and carbohydrate give rise to ATP, the cellular form of energy, were detailed in Chapter 4. Here we look at ways to evaluate whether the combination of food intake (energy intake) and physical activity (energy expenditure) leads to maintaining, losing, or gaining weight. We begin by calculating **energy balance**, using Equation 8.1:

TABLE 8.1
THE ENERGY CONTENT OF THE FOUR MACRONUTRIENTS IN FOOD

Macronutrient	Energy content	
	kJ/g	kcal/g
Fat	38.9	9.3
Protein	16.7	4.0
Carbohydrate	17.5	4.2
Alcohol	31.4	7.5

Energy balance = energy intake − energy expenditure   (8.1)

Our energy intake is easily determined by calculating the amount of energy in the fat, protein, and carbohydrate of the food we eat. This energy is measured in joules. A **joule (J)** is defined as the energy expended in applying a force of 1 **newton** through a distance of 1 meter. Because scientists are concerned with large amounts of energy, food energy is generally expressed in **kilojoules (kJ)**: $1 \text{ kJ} = 10^3 \text{ J}$. You might also see food energy expressed in **kilocalories (kcal)**, an older convention that is still often used. A **calorie** is the amount of energy required to heat 1 gram of water 1°C ($1 \text{ kcal} = 10^3 \text{ cal}$). The conversion factors for kilojoules and kilocalories are $1 \text{ kJ} = 0.239 \text{ kcal}$, and $1 \text{ kcal} = 4.184 \text{ kJ}$. **TABLE 8.1** shows the energy content of the four macronutrients in food types: fat, protein, carbohydrate, and alcohol. By summing the total amount of energy of the macronutrients in food we consume in a day, we can determine our energy intake for the day. Although protein and alcohol are not generally used as starting substrates for producing ATP, they do have the potential to be converted to fat and carbohydrate and stored in the body's energy reserves, and thus they must be included in the calculation of energy balance.

Determining energy expenditure is a bit more complicated and is normally accomplished through **indirect calorimetry**, a calculation of energy expenditure based on the measurement of respiratory gases **(Figure 8.1)**. Total energy expenditure (TEE) includes three components: resting energy expenditure, physical activity, and diet-induced

Figure 8.1 Indirect calorimetry. Oxygen consumption by the whole body can be determined through an analysis of respiratory gases. (A) The subject breathes through a one-way valve, which forces air to move in one direction (*solid arrows*). (B) The expired air travels through a hose to a device that measures its volume and temperature. (C) A sample of the expired air goes to the oxygen analyzer (D); the most common type of oxygen analyzer heats the air to a temperature that causes ionization of the oxygen molecules. (E) An ionized-oxygen detector inside the analyzer sends a digital signal (*dotted lines*) with information on the raw oxygen amount to a computer. The computer calculates the actual oxygen amount by adjusting the raw amount based on the temperature of the expired air. Subtracting the amount of expired oxygen from the amount of inspired oxygen gives the oxygen consumption. Multiplying the oxygen consumption (in liters) by 20.92 kJ gives a good estimate of energy expenditure. (Photo courtesy of Jennifer Ruhe.)

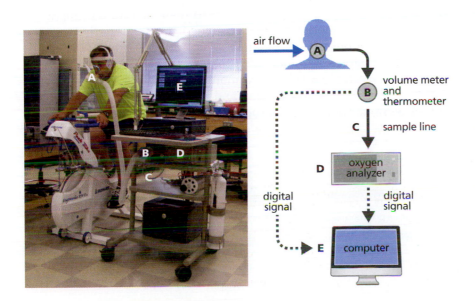

air flow

A

volume meter and thermometer — B

C   sample line

oxygen analyzer — D

digital signal

digital signal

E   computer

thermogenesis **(Equation 8.2)**. TEE is also measured in kilojoules or kilocalories.

Total energy expenditure (TEE) = resting energy expenditure (REE) +
physical activity + diet-induced thermogenesis (DIT)   (8.2)

Resting energy expenditure (REE) is the amount of energy required to maintain the essential functions of life—heart rate, body temperature, brain function, and so on—and makes up 60–70% of TEE. Physical activity includes all movement of skeletal muscles, no matter how slight, and, on average, contributes about 20% to TEE in the adult human. However, the contribution of physical activity to TEE varies significantly among individuals. For example, a highly trained endurance athlete may expend as much 30% of his or her TEE on exercise, whereas a truly sedentary individual may expend only 5% of TEE on physical activity. **Diet-induced thermogenesis (DIT)** is the amount of energy required for digestion, absorption, and nutrient storage and accounts for the remaining 10–20% of TEE. The contribution of DIT to TEE does not vary much among individuals or with age, is difficult to measure, and is normally included as a constant in the TEE equation.

Accumulation of fat occurs throughout maturity

The typical mature human body contains almost 40 days' worth of energy in the form of fat, whereas we store only about 1 day's worth of energy in the form of glucose. Thus, when we measure the energy state of the mature human body—the energy balance—we are actually measuring the dynamics of fat storage. Positive energy balance—greater energy intake than expenditure—results in an increase in fat storage and in body weight. Negative energy balance—greater expenditure than energy intake—results in decreased fat storage and body weight. A person is said to be in energy balance when intake and expenditure are equal. In this case, body weight does not change.

So, once we have attained full growth and development—that is, maturity—changes in our body weight reflect alterations in the amount of fat we store. Data collected from the ongoing National Health and Nutrition Examination Survey III (NHANES III) show that body weight increases, on average, by approximately 15% between the ages of 20 and 60–69 **(Figure 8.2)**. The increase in weight occurs slowly, as small changes in energy balance over long periods have large effects on body weight. We can use a simple example to demonstrate the impact of small changes in energy balance sustained over long periods on the storage of fat. Consider a woman weighing 50 kg (about 100 pounds) with 10 kg of stored fat, equivalent to 389,000 kJ of energy (38.9 kJ/g × 10,000 g; see Table 8.1). This woman is currently in energy balance, with a daily food intake of 8370 kJ and a constant and stable REE and DIT of 5857 kJ and 837 kJ, respectively. Using Equation 8.2, we can calculate the amount of energy she uses per day in physical activity **(Equation 8.3)**.

Total energy expenditure (TEE)   = 8370 kJ

Resting energy expenditure (REE) = 5857 kJ

Diet-induced thermogenesis (DIT) = 837 kJ

Thus,

Physical activity = TEE – REE – DIT

$$= 8370 \text{ kJ} - 5857 \text{ kJ} - 837 \text{ kJ} = 1676 \text{ kJ} \qquad (8.3)$$

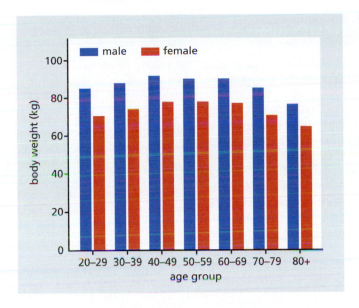

Figure 8.2 Average weights of men and women in the United States in 2004. These data are from the most recent year of a large-scale (19,000 participants) study that includes individuals over the age of 70. Cross-sectional studies completed since 2004 have supported these data. (Data from M.A. McDowell et al., *Natl Health Stat. Rep.* 10:1–48, 2008.)

If the woman begins to eat a little more each day, say just 418 kJ (equal to about ¼ cup or 3 tablespoons of vanilla ice cream) while keeping her physical activity the same, her energy balance becomes positive and she begins to store fat. Thus, in approximately 100 days she will gain about 1 kg of body weight, and in 1 year she will gain almost 4 kg (1 kg of body fat is equivalent to about 38,900 kJ). Thus, even small changes in energy intake can have a significant impact on body weight over the long haul. Similarly, slight changes in energy expenditure would have the opposite effect. An increase in physical activity of 418 kJ daily, such as jogging at a moderate pace for 1 mile (or walking 1.5 miles), would bring the woman back into energy balance and prevent fat accumulation.

As indicated by this example, fat accumulation can result from excessive calorie intake, decreased physical activity, or a combination of the two. Aging affects both of these. (**BOX 8.1** describes a useful way of considering changes in body fat with age.) Calorie consumption tends to decrease with age **(Figure 8.3)**—although, as Figure 8.3

Figure 8.3 Average calorie consumption by men and women in various age groups in 1970 and 2000. Numbers above the bars are the percentage increase between 1970 and 2000. (Data from J.D. Wright et al., *Mortal. Morbid. Wkly Rep.* 53:80–82, 2004.)

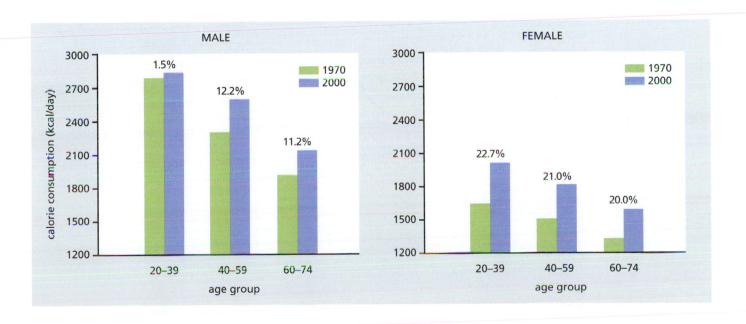

BOX 8.1 BODY MASS INDEX AND MORTALITY RISK

The **body mass index (BMI)**, the ratio between weight and height (Equation 8.4), can be another useful way to describe changes in fat content. Since genetics plays a large role in determining one's final stature, and height generally remains stable when growth and development are complete, BMI is a more sensitive and individualized measure of body fat than just weight alone. More importantly, BMI has been shown to be highly correlated with mortality due to health problems associated with being underweight, overweight, or obese **(Figure 8.4)**.

$$BMI = weight\ (kg)\ /\ height^2\ (m^2)$$
$$(8.4)$$

The calculation of BMI dates back to the early 1800s, when a Belgian social scientist, Adolphe Quetelet (1796–1874), found that body weight during growth varied as a function of height squared. The index, then called the Quetelet index, had limited use as a measurement relating obesity to health, primarily because obesity was virtually unknown in the general population at that time. It

was not until the end of World War II that statisticians working for life insurance companies noted a slight increase in mortality rates among individuals weighing more than the average body weight for the population. This led to the development of the weight-to-height tables and the concept of an ideal body weight for a specific height. It was soon found that the large variation among individuals in what constituted an ideal body weight rendered the weight-to-height tables too inaccurate for use by physicians as a measure of health.

In 1972, Ancel Keys (1904–2004) was the first to note that the Quetelet index was the best proxy for body fat and suggested its use as an easy method for physicians to discuss possible health problems associated with obesity. Keys popularized the name "body mass index." The results of several investigations conducted during the last two decades of the twentieth century showed a close relationship between BMI and the risk for health problems. These findings led the National Institutes

of Heath (NIH) to establish specific BMI values as guidelines for healthy weight (shown in Figure 8.4).

It has become clear that the NIH's BMI values need to be adjusted for the older population. Several longitudinal studies have found that BMIs in the range of 25–28 (a range considered overweight and with higher risk for health problems, according to NIH guidelines) do not increase health risk for those over the age of 75 **(Figure 8.5)**. However, BMIs of less than 18.9 (considered underweight) remain a **risk factor** for health problems in the elderly. Most clinical experts agree that in the older population, being underweight poses a greater risk for health problems than being obese. Many clinicians have stated that individuals older than 75 who are within the 25–30 BMI range should not consider a weight-reduction plan, except for vanity reasons, and should never drop below a BMI of 22.

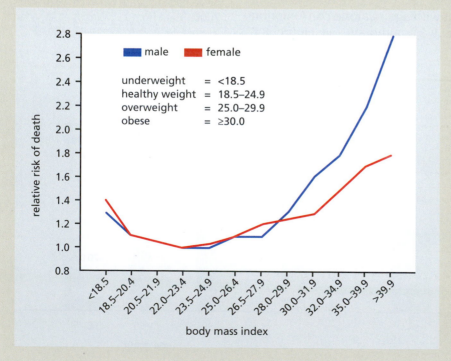

Figure 8.4 Relationship between body mass index (BMI) and relative risk of death. These graphs, based on data from 457,758 men and 588,369 women, collected between 1984 and 1996, clearly show that the risk of death is greater for those who are underweight or overweight. The BMI categories (*listed above the curves*) are commonly used as age-independent markers for possible health problems associated with fat content in otherwise healthy adults over the age of 20. (Adapted from E.E. Calle et al., *N. Engl. J. Med.* 341:1097–1105, 1999. With permission from New England Journal of Medicine.)

BOX 8.1 BODY MASS INDEX AND MORTALITY RISK

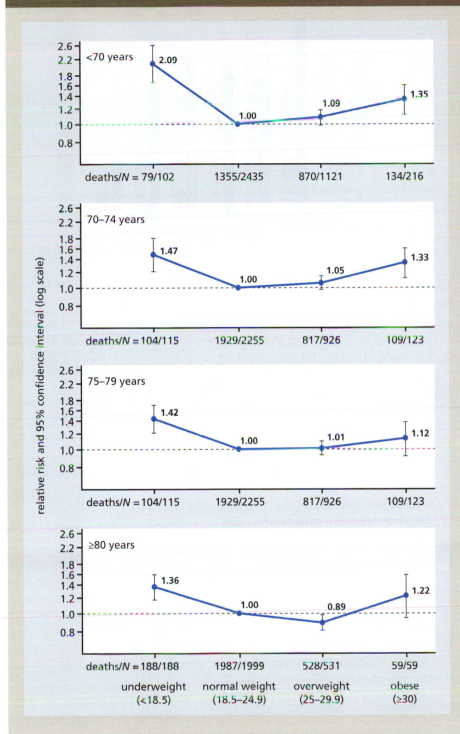

Figure 8.5 Relationship between BMI and relative risk of death in age groups between 45 and 101. Note the progressive decline in relative risk in the overweight category from ages <70 years to ≥80 years. Many experts believe that the older population should have a different set of BMI values from those used for the general population, to reflect these findings and to help prevent underweight. This and many other investigations find that underweight (BMI <18.9) and obesity (BMI ≥30) remain risk factors for early mortality in the older population. Deaths/N, on the horizontal axes, indicates number of deaths in the sample size N for each BMI category. (From M.M. Corrada et al., *Am. J. Epidemiol.* 163:938–949, 2006. With permission from Oxford University Press.)

shows, it increased, overall, by an average of 11% in men and 22% in women between 1970 and 2000, and data from recent cross-sectional investigations suggest that calorie consumption continues to increase. Despite this increase in calorie consumption in recent years, the age-related decrease remains. In addition, older individuals are less likely than younger people to engage in voluntary physical activity **(Figure 8.6)**. Since energy intake seems to decrease with age, the only logical

Figure 8.6 Percentage of individuals engaging in regular leisure-time physical activity sufficiently vigorous to affect the mass of body fat. Regular leisure-time physical activity is defined as engaging in vigorous physical activity for 30 minutes or more five or more times per week or engaging in light to moderate physical activity for 20 minutes or more three or more times per week. (Data from J.R. Pleis, B.W. Ward, and J.W. Lucas, *Vital Health Stat.* 10:217, 2010; C.A. Schoenborn, J.L. Vickerie, and E. Powell-Griner, Health Characteristics of Adults 55 years and Over: United States, 2000–2003, Hyattsville, MD: National Center for Health Statistics, 2006.)

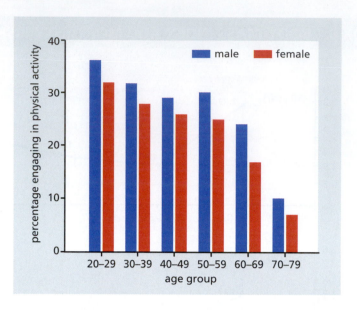

explanation for the age-related increase in body weight is that the decrease in energy expenditure over time must be greater than the decrease in energy intake, a conclusion supported by the data in Figure 8.6. The lack of physical activity can also lead to a loss of muscle tissue exceeding that expected for normal age-related skeletal muscle loss (a condition known as sarcopenia, discussed later in the chapter). The loss of muscle mass due to physical inactivity leads to a decrease in REE, exacerbating the age-related decrease in energy expenditure. After age 70, the body weight loss most likely reflects an association with decreased calorie intake near the end of life, as discussed later.

Excessive loss of body weight near the end of the life span increases mortality rate

Although body weight increases throughout much of the human life span, there comes a time when weight begins to decline. This can be seen in Figure 8.3, where the body weight of those 70 and older is less than that in the younger age groups. The decreased body weight in this age group reflects the inclusion in the data of individuals near the end of their life span. End-of-life weight loss has only recently been accepted as an age-related phenomenon, and the underlying mechanism has not yet been identified. We do know, however, that both fat and muscle mass seem to decline during this period.

Recent data suggest that a decrease in food intake parallels the end-of-life decline in body weight and can lead to a clinical condition known as the **anorexia of aging**. The four hallmarks of the anorexia of aging—excessive weight loss, decreased appetite, poor nutrition, and inactivity—are the primary predictive symptoms of a larger and more general syndrome known as **geriatric failure to thrive**. This multifactorial syndrome, often indicating the end of life, includes impaired physical function, malnutrition, depression, and cognitive impairment. An algorithm for diagnosing geriatric failure to thrive has been established and suggests when physicians might begin discussing end-of-life options with patients and families, to prevent needless interventions that may only prolong suffering **(Figure 8.7)**.

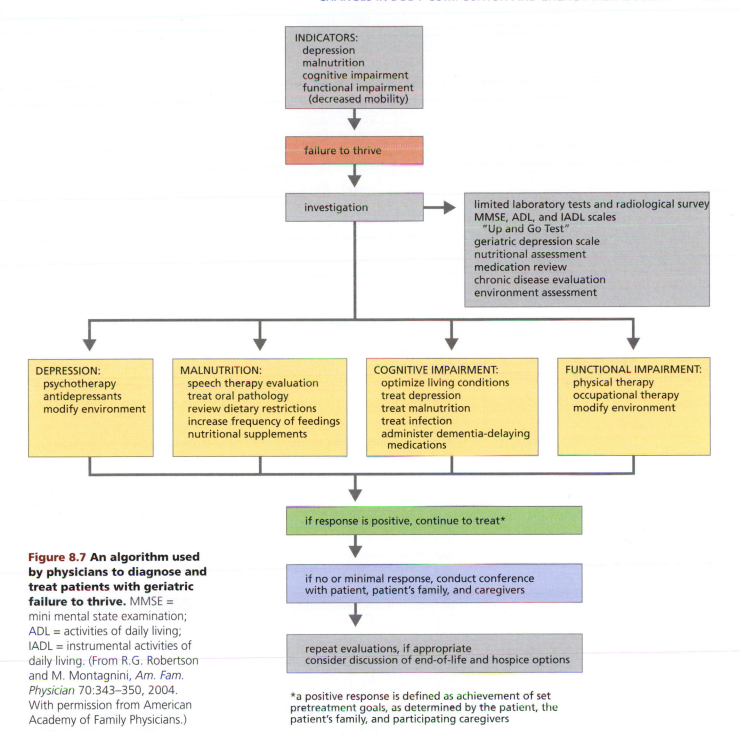

Figure 8.7 An algorithm used by physicians to diagnose and treat patients with geriatric failure to thrive. MMSE = mini mental state examination; ADL = activities of daily living; IADL = instrumental activities of daily living. (From R.G. Robertson and M. Montagnini, *Am. Fam. Physician* 70:343–350, 2004. With permission from American Academy of Family Physicians.)

Sarcopenia is the age-related decline in skeletal muscle mass

An age-related decrease in muscle mass, known as **sarcopenia**, occurs during the maturity phase of the human life span. The decrease happens regardless of factors known to increase muscle mass, such as increased physical activity. The loss in muscle tissue with age reflects a decrease in both muscle cell number and cell size, and it seems to be greater in men than in women. The loss in both cell number *and* size differentiates sarcopenia from loss of muscle mass due to disuse, in which only cell size decreases. A series of cross-sectional and longitudinal studies have shown that after the age of 40, muscle mass declines at a rate of about 2–3% per decade in men and 1–2% per decade in women.

The mechanism underlying sarcopenia is mostly unknown, although some changes are consistently observed in studies investigating the causes of sarcopenia. Fat deposits and connective tissue seem to replace the contractile components of muscle. Since only myocytes (muscle cells), or muscle fibers, can generate force, increased fat and connective tissue in the muscle will decrease strength.

The type of **muscle fiber** affected by sarcopenia appears to be selective. Muscle cells are classified based on their content of isoforms of the contractile protein **myosin**. Cells with high concentrations of the myosin isoform that results in slow contraction are known as **type I fibers** (sometimes called slow-twitch fibers). Muscle fibers with high concentrations of the myosin isoform that results in fast contractions are known as **type II fibers** (fast-twitch fibers). The sarcopenia-related muscle fiber loss seems to have a greater effect on type II fibers than type I. Thus, muscles that require a high speed of contraction, such as those in the hands and other areas of fine motor control, tend to be more affected by sarcopenia than the large muscle groups. Interestingly, recent investigations have discovered a third type of muscle fiber, **type IIX**, that appears only during aging. It has characteristics of both type I and type II fibers. The role of type IIX fibers in sarcopenia has yet to be investigated.

A recent observation made in rodents, as models for human aging, is that the number and function of neuromuscular junctions, or **motor end plates**, decline with age. The motor end plates regulate how many muscle fibers are used during a contraction: the greater the number of motor end plates recruited by the nerve innervating the muscle, the greater the strength of the contraction. Thus, the age-related decline in motor end plates may lead to fiber disuse and signal the start of **apoptosis**. Direct evidence for disuse and apoptosis in sarcopenia in humans has yet to be published, but several investigations on completely sedentary young adults (experimentally induced bed rest) and individuals returning from extended stays in space (zero gravity) show apoptotic events related to a decline in motor end plate activity.

The loss of muscle tissue can have significant health consequences. Muscles use considerable amounts of energy at rest to maintain their form and to stay prepared for contraction. The muscles use more than half of REE. Thus, a decrease in muscle mass reduces resting metabolic rate, and without an equivalent reduction in energy intake, this leads to fat accumulation. Men experience an approximately 2.5% decrease in resting metabolic rate per decade, whereas women seem to have a slower rate of decline, at 1.5% per decade **(Figure 8.8)**. These values are remarkably close to the age-related loss in muscle mass. Lack of physical activity will lead to a greater loss in muscle tissue and accelerate the age-related drop in REE.

The primary job of human skeletal muscles is to provide strength and stability to the body. If muscle mass declines, one's ability to react to situations requiring a certain level of strength and stability will also decline. Consider the simple act of walking. We don't really think much about it until we take a misstep off a curb or step onto an uneven surface. During young adulthood, an injury from this misstep is unlikely, because we have enough muscle mass to react accordingly and maintain balance. With the age-related decline in muscle mass, an individual

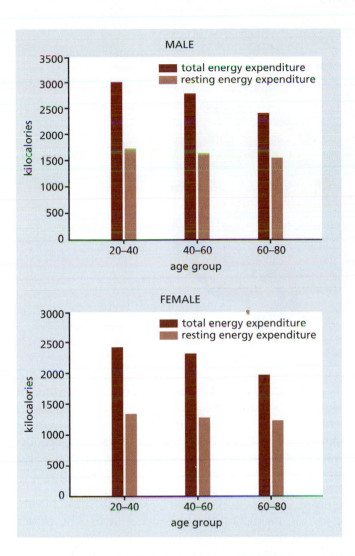

Figure 8.8 Total and resting energy expenditure at various ages in men and women. (Data from S.B. Roberts and G.E. Dallal, *Public Health Nutr.* 8:1028–1036, 2005.)

may have insufficient strength to keep upright when the misstep occurs. Indeed, falls resulting from the age-related decline in muscle mass are a major risk factor for morbidity and mortality.

CHANGES IN THE SKIN

Human skin is the largest organ of the body and serves as the boundary between our bodies and the outside world. Our skin not only contains our internal body components but protects them from the sometimes harsh and harmful external environment. For example, harmful solar radiation, such as ultraviolet (UV) light, does not penetrate below the level of human skin. Our skin also participates in the innate immune system (discussed later in this chapter) by establishing a barrier against toxic invasion. In addition to its protective role, skin serves as an organ of the thermoregulatory system and as a sensory organ for touch. In this section, we discuss the basic structure of skin, how aging affects it, and the role of environmental factors in age-related changes in the skin. The skin's role in the immune system is briefly described in the later section "Changes in the Immune System."

The skin consists of three layers

There are three layers of skin: **epidermis**, **dermis**, and **subcutaneous fat tissue (Figure 8.9)**. The epidermis is the outer layer of skin.

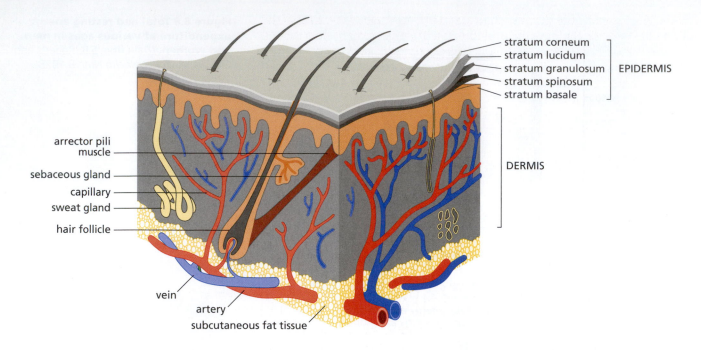

Labels on the diagram:
- stratum corneum
- stratum lucidum
- stratum granulosum — EPIDERMIS
- stratum spinosum
- stratum basale
- DERMIS
- arrector pili muscle
- sebaceous gland
- capillary
- sweat gland
- hair follicle
- vein
- artery
- subcutaneous fat tissue

Figure 8.9 Cross-sectional diagram of human skin. The three layers of the skin—epidermis, dermis, and subcutaneous fat tissue—protect the internal organs from environmental insult, provide our sense of touch, and help to thermoregulate the body through sweating (if hot) and vasoconstriction (if cold).

It consists of five layers: the stratum corneum, stratum lucidum, stratum granulosum, stratum spinosum, and stratum basale. Although each layer has a specific function, the overall purpose of the epidermis is to prepare the living skin cells, the **keratinocytes**, produced in the stratum basale, to become the flat, dead, hard cells of the stratum corneum. This dead skin sheds off about every two weeks and is replaced by new dead skin cells that have moved upward through the layers. Through this process, cells damaged by solar radiation or other insults are sloughed off, protecting the underlying layer. The epidermis also contains several specialized cells, including **melanocytes**, which produce pigments; **Langerhans cells**, which play a role in the immune system; and **Merkel cells**, which are a form of sensory receptor.

The dermis consists mostly of connective tissue and is much thicker than the epidermis. It contains the organelles of the skin: exocrine (sweat) glands, hair follicles, sebaceous (oil) glands, capillaries, and nerve endings. These structures are supported by a matrix of elastic proteins, **collagen**, and elastin fibers. (Collagen accounts for about 70% of the dermis's dry weight.) The dermis is responsible for the skin's pliability and helps transmit the sensations of touch and pressure. It is also involved in the regulation of body temperature. The sweat glands release water onto the surface of the skin, allowing evaporation and cooling when the internal body temperature increases. Conversely, during cold exposure, blood vessels in the skin constrict and shunt blood to the internal organs.

Subcutaneous fat tissue, making up the bottom layer of the skin, contains adipose (fat) tissue and blood vessels. The fat layer provides cushioning and insulation. It also helps to regulate the temperature of both the skin itself and the body.

Wrinkles are caused by a loss of skin elasticity and subcutaneous fat

Wrinkles are one of the most visual consequences of aging; they occur in 100% of the human population and in all parts of the body. The underlying cause of wrinkles is a decrease in the number of skin cells developed by the dermis, changes in the normal functioning of the skin cells, and declining subcutaneous fat. The age-related decrease in number of skin cells results in a thinning of the skin and occurs primarily in the epidermis. Cell loss appears to result from a slowdown of cell division caused by the shortening of telomeres (see Chapter 4). Some studies have also found that the number of mitotic cells in the stratum basale declines with age. As a result of the loss in skin cells, production of elastin and collagen decreases, and the skin loses significant elasticity. The skin also loses elasticity due to an increase in nonenzymatic cross-linking in collagen (**BOX 8.2**). With each new cross-link, the elastic properties of the protein decline and the dermis flattens. With the loss of elasticity, the epidermis no longer smoothly covers the irregularities in the underlying dermis and subcutaneous fat.

The most significant cause of the irregularities in skin shape that arise with age is the decline in stored lipids in the subcutaneous fat layer. The majority of body fat in the average adult human is in subcutaneous fat. Young adults (20- to 40-year-olds) tend to store fat equally throughout the body. As we age into the sixth decade and beyond, the fat stores are distributed more to the abdomen (men), hips (women), and buttocks (both sexes). Moreover, and for unknown reasons, subcutaneous fat in the face, arms, and legs decreases. As we age, the loss of subcutaneous fat in these regions causes irregularities that are covered by a relatively inelastic epidermis, thus creating wrinkles.

Ultraviolet light causes significant damage to the skin over time

Extrinsic factors such as pollution, tobacco smoke, and excessive alcohol consumption can cause significant damage to the skin. However, more than 90% of the time-dependent environmental damage to skin is caused by sun exposure. **Photoaging**—the long-term damage to skin caused by UV light—occurs mainly within the dermis and results from an accumulation of abnormal elastic tissue, a condition known clinically as **solar elastosis**. The accumulation of damaged elastic and collagen fibers results in a dysfunctional extracellular matrix and leads to loss of elasticity of the skin, wrinkle formation, and **telangiectasia** (appearance on the skin's surface of small dilated blood vessels). In addition, repeated bouts of overexposure to the sun's radiation (that is, sunburn) damage the melanocytes. Alterations in melanocyte function and structure can lead to a form of skin cancer called **melanoma** in genetically at-risk individuals. The importance of the ability of melanocytes to produce the protective skin pigment melanin is evident from the fact that dark skin photoages at a much slower rate than lighter skin.

The etiology of photoaging and solar elastosis has yet to be conclusively determined. Most research indicates that UV-generated role of reactive oxygen species (ROS; also called oxygen-centered free radicals) may have a central role. Recall from Chapter 4 that the unpaired electrons of ROS react quickly with other molecules and alter **protein structure** and function. Since collagen and elastin have very slow turnover rates, damaged elastic fibers tend to accumulate in the matrix of the dermis. Damage to the elastic fibers also initiates the immune response (see p. 246), which can cause further damage to the dermis. Many investigations have shown that repeated bouts of immune system response to ROS-damaged tissue accelerate the rate of skin aging.

Photoaging can be completely prevented—do not expose yourself to the harmful UV radiation of the sun or tanning booths. However, since

BOX 8.2 THE MAILLARD REACTION AND AGING

Much of the age-related decline in physiological function may be the result of a biochemical rearrangement of proteins that has many similarities to the process that causes foods to turn brown when heated. The biochemical pathway leading to the browning effect of food, known as the **Maillard reaction** (named after Louis Camille Maillard, who discovered the pathway), causes our arteries to stiffen, decreases joint movement, thickens the lens of the eye, and plays a part in many other non-life-threatening changes. Here we briefly discuss the biochemistry of the Maillard reaction and its effect on human aging.

The Maillard reaction occurs in three steps. First, glucose or fructose molecules react with an amino acid to form a **Schiff base**, an organic compound in which the nitrogen atom of an amino group is double-bonded to a carbon atom of glucose or fructose **(Figure 8.10)**. The enzymatic formation and degradation of Schiff bases serves an important synthetic and signaling function in normal biochemistry and is highly regulated.

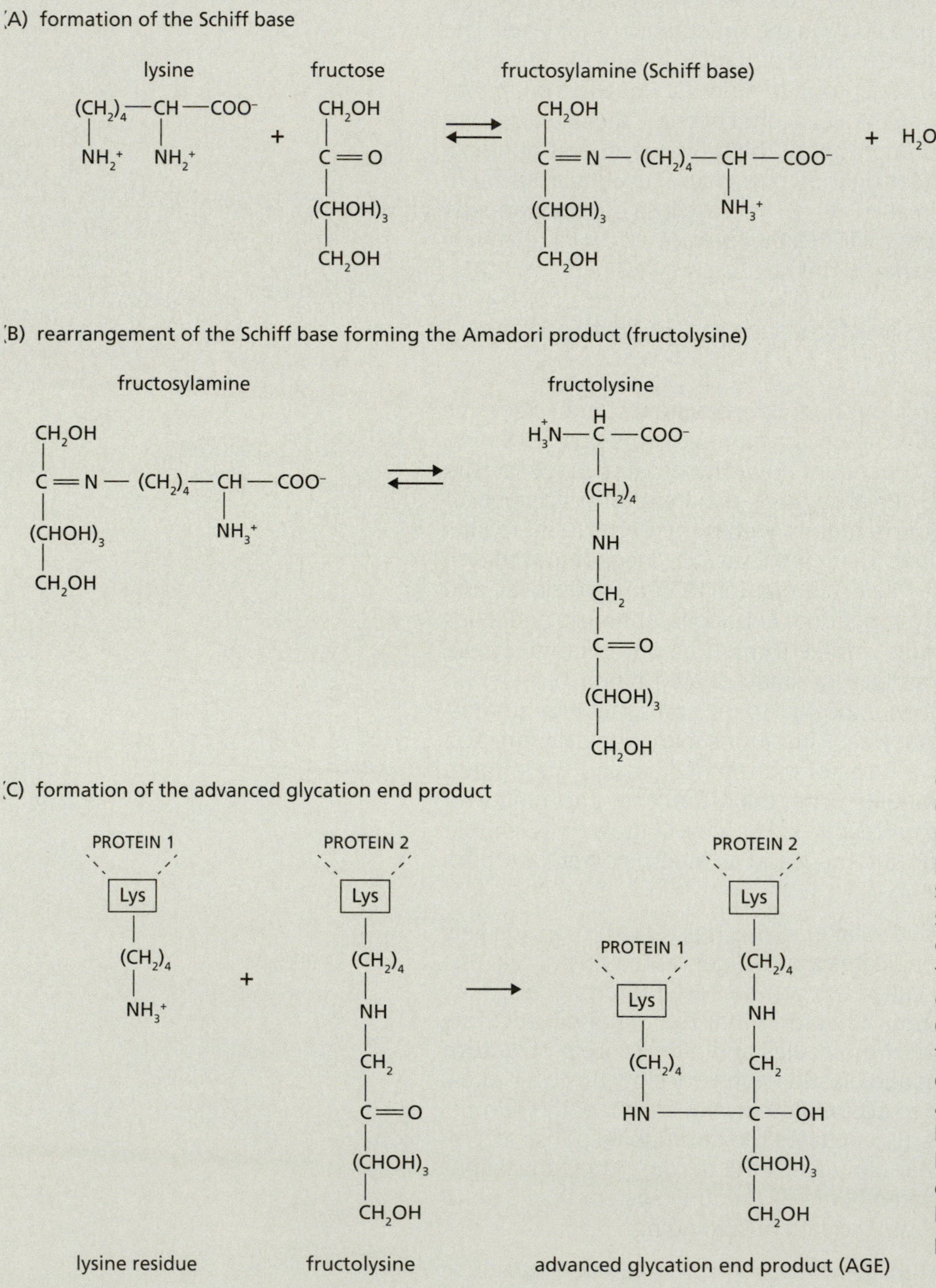

(A) formation of the Schiff base

(B) rearrangement of the Schiff base forming the Amadori product (fructolysine)

(C) formation of the advanced glycation end product

Figure 8.10 The Maillard reaction and the nonenzymatic formation of an advanced glycation end product (AGE). (A) An amino acid, such as lysine, forms a double bond with an aldose, such as fructose, to form a Schiff base (fructosylamine). (B) The Schiff base can undergo rearrangement to form the more stable ketone known as an Amadori product, represented here as fructolysine. Unlike Schiff bases, Amadori products can accumulate in the cell. (C) Over time, the Amadori product attached to one protein molecule (or protein subunit) bonds with a second protein molecule, creating a cross-link and forming an advanced glycation end product (AGE). (The linkage between the two proteins is shown only as a representation and is not intended to be chemically correct.) Formation of the AGE cannot be reversed and may lead to inactivation of the protein.

BOX 8.2 THE MAILLARD REACTION AND AGING

However, Schiff bases can also be formed through nonenzymatic processes, which are not regulated, and can lead to the accumulation of non-degradable cellular metabolites.

In the second stage of the Maillard reaction, the Schiff base undergoes nonenzymatic rearrangement to form an **Amadori product.** Amadori products are much more stable than Schiff bases and can accumulate within the cell. The third and final stage of the Maillard reaction results in the formation of an **advanced glycation end product (AGE),** consisting of two separate protein molecules linked together by the Amadori product. These linkages, called cross-links, change the protein's structure and thus alter its function. AGEs are highly insoluble and are not easily degraded. When these products are heated, they cause browning.

AGEs differ from normal, enzymatically cross-linked products and can result in age-related loss of physiological function. For example, collagen makes up more than 25% of the total protein in our bodies and represents the majority of protein in the extracellular matrix, the substance that holds cells together. The collagen fibril consists of three α-chains twisted into a triple helix and held together by hydrogen bonds **(Figure 8.11).** During translation, as collagen fibrils are synthesized, the enzyme lysyl oxidase catalyzes a reaction that joins a lysine residue in the protein to the aldehyde group of an aldose, usually glucose or fructose, creating a glycosylated lysine residue. The glycosylated lysine residues of several collagen fibrils bond spontaneously to form the collagen fiber. In turn, collagen fibers bind together to form a collagen bundle. The enzymatic bonding of one fibril to another provides collagen with its unique strength and elasticity.

The formation of collagen occurs with just enough cross-linking to provide sufficient strength while retaining elasticity. (If you pull your skin away from the underlying muscle, it will snap back into place because of the elasticity of collagen.) Collagen turnover is extremely slow, and it is not unusual to find organs in which the collagen laid down during development remains intact throughout the life span. This property of collagen leaves the protein highly susceptible to the random and unregulated process of nonenzymatic glycosylation and the formation of AGEs. Each nonenzymatic cross-link formed in the collagen fiber increases the collagen's strength (the collagen becomes stiffer) and decreases its elasticity (reducing its flexibility).

The importance of AGE formation in inactivating proteins cannot be overemphasized. Because AGEs have no known catabolic pathway, the accumulation of these products has been suggested to cause cellular and physiological dysfunction. As you'll see in this chapter and in Chapter 9, the formation of nonenzymatic cross-links and AGEs is suggested as the likely cause of age-related physiological decline.

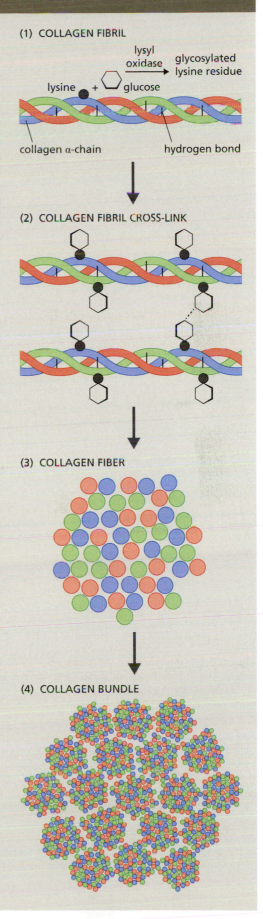

Figure 8.11 Enzymatic cross-linking of the collagen fiber. (1) The collagen fibril, the basic unit of the collagen fiber, contains three α-chains twisted into a helix held together by hydrogen bonds. Lysyl oxidase catalyzes a reaction between a lysine residue in an α-chain and an aldehyde, shown here as glucose, to form a glycosylated lysine residue. (2) The glycosylated lysine residue will bond spontaneously with other glycosylated lysine residues to form cross-links between fibrils. (3) Several cross-linked fibrils make up the collagen fiber. (4) Several collagen fibers bond together to form the final product—the collagen bundle.

many people are unable or unwilling to take this drastic step, limiting the amount of time spent in the sun and using sun block can significantly reduce photoaging and the risk of skin cancer.

CHANGES IN THE SENSES: HEARING, VISION, TASTE, AND SMELL

Hearing and vision are two of the few physiological systems that undergo age-related declines in all individuals. Neither type of decline, however, necessarily causes significant impairment to daily life, as the medical sciences have developed treatments that effectively negate the age-related losses: hearing aids increase volume and change the pitch; reading glasses improve close-up vision. Taste and smell are our two chemical senses. Contrary to popular belief, age-related changes in taste buds and olfactory centers are minimal. In this section, we examine how our sensory organs, and therefore our perceptions and interactions with the outside world, can change as we age.

The sense of hearing is based on the physics of sound

To understand the physiology of hearing and any age-related change that may occur, you must first understand the physics of sound. A sound wave begins when an object vibrates in a medium (gas, liquid, or solid). Remember that vibrations are oscillating in nature. That is, a vibration pushes outward beyond the resting state of the object, causing **compression** of the molecules in the surrounding medium and an increase in pressure. The outward motion of the vibration then retracts beyond the resting state of the object, causing a decrease in pressure in the medium, or **rarefaction**.

The height difference between a compression peak and a rarefaction peak defines the **amplitude** of the sound wave **(Figure 8.12)**. The loudness of a sound is directly proportional to the wave's amplitude. As the sound wave radiates outward from the object creating the vibrations, the amplitude becomes less and the sound softer. The loudness of a sound is measured in **decibels**. One decibel is the logarithmic increase in loudness above a sound that is barely audible to the human ear. The human ear can tolerate sound between 0 and 130 decibels. Sounds greater than 130 decibels cause pain. The **frequency** of the vibration, defined as the distance between consecutive peaks of either

Figure 8.12 Components of a sound wave. The outward push of a vibrating object decreases the distance between molecules in the surrounding medium and increases the pressure (compression). The inward contraction of a vibrating object increases the distance between molecules in the medium and decreases the pressure (rarefaction). The distance between the compression and rarefaction peaks defines the amplitude of the sound wave. Loudness of the sound is directly proportional to amplitude. The frequency of the wave, or how fast the vibrations occur, determines the pitch of the sound.

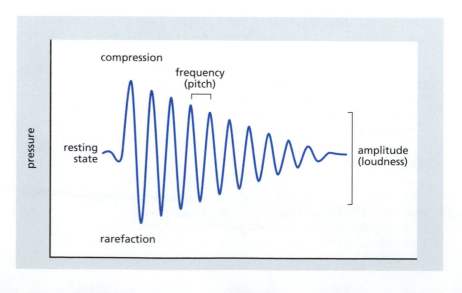

compression or rarefaction, determines the **pitch** of the sound: the higher the frequency, the higher the pitch; the lower the frequency, the lower the pitch. Although the human ear can hear sounds between 20 and 20,000 Hz (hertz, or cycles per second), it best detects sounds between 1000 and 4000 Hz.

Transmission of sound through the human ear occurs in three steps

To hear sounds, the human ear has to do three things: (1) direct the sound waves to the hearing part of the ear; (2) sense the fluctuations in air pressure caused by the sound vibrations; and (3) translate these pressure fluctuations into a signal that the brain can understand. Each of these tasks is performed by a different part of the ear.

The first step of sound transmission involves the outer ear, comprising the **pinna**, the external auditory canal, and the **tympanic membrane**, or eardrum **(Figure 8.13)**. The pinna serves to "collect" the sound waves and channel them through the external auditory canal to the tympanic membrane. Vibration of the tympanic membrane occurs at the same frequency as that of the sound wave, and the membrane depresses inward a distance proportional to the pressure exerted on it. Thus, the distance moved by the tympanic membrane determines the loudness of the sound we hear.

The tympanic membrane stretches across the external auditory canal and separates the outer ear from the middle ear. Vibrations of the tympanic membrane are transferred to the three bones of the middle ear, the **malleus, incus**, and **stapes**. These bones amplify the sound wave before it enters the inner ear. The inner ear contains fluids. The transmission of a sound wave through fluid is more difficult than through air, but the force per unit area on the oval window of the middle ear is considerably greater than that on the tympanic membrane, thus the sound is amplified. The middle ear also contains the opening to the

Figure 8.13 Anatomy of the human ear.

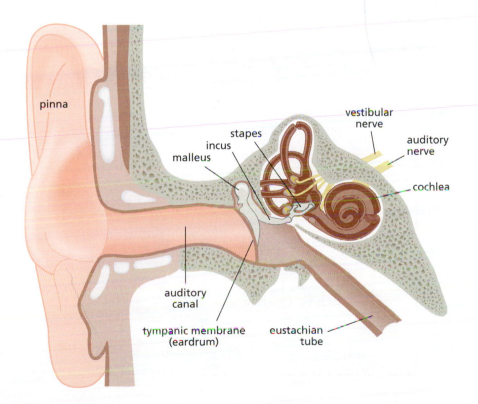

eustachian tube, which allows pressure equalization between the middle ear and the external, environmental pressure.

Conversion of a sound wave into a neural impulse occurs in the inner ear. As the stapes vibrates on the oval window, the vibrations are transmitted through the fluid-filled cochlear duct and into the **cochlea (Figure 8.14)**. The cochlea houses the **organ of Corti**, which contains highly specialized receptor cells with hairlike projections called **stereocilia**. The stereocilia move a distance directly proportional to the vibrations of the fluid in the cochlea. As the stereocilia move, the pressure of the movement opens voltage-gated calcium channels in the receptor cell membranes. In turn, the receptor cells release neurotransmitters that stimulate the cochlear nerve, and the signal is transmitted to the hearing centers of the brain—we hear the sound.

Loss of stereocilia contributes to age-related hearing loss

Most experts agree that the primary cause of the age-related decline in hearing, known as **presbycusis**, is alterations in the inner ear, although the reasons for these changes are unknown. Most notable

Figure 8.14 Components of the inner ear. (A) Fluid-filled cochlea. (B) The cochlea consists of three tubes that run the length of the organ. The vibration of a sound wave is transferred to the fluid of the cochlea at the oval window. These fluid vibrations change the pressure exerted against the walls of the cochlea, which, in turn, is detected by the organ of Corti in the cochlear duct. (C) Cells, called hair cells, of the organ of Corti have hairlike projections, the stereocilia, that move a distance directly proportional to the pressure exerted by the cochlear fluid. (D) Movement of the stereocilia causes voltage-regulated channels to open, and the influx of calcium into the cell induces the release of neurotransmitters—generating an electrical signal.

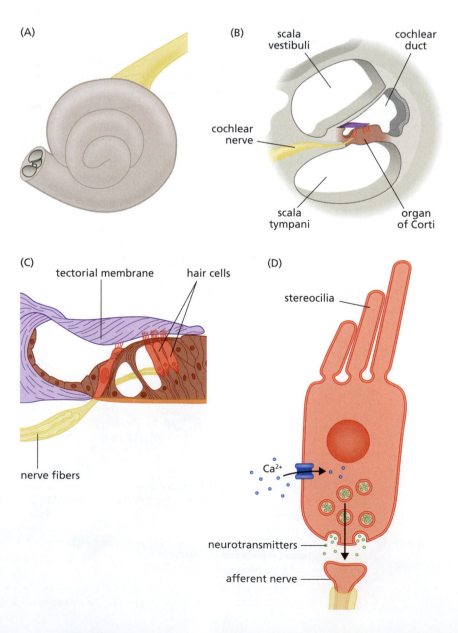

among the many changes in the inner ear are the loss of hair cells and/or of stereocilia in the organ of Corti. A decrease in the number of stereocilia on each cell reduces the rate of neurotransmitter release, resulting in a diminished ability to detect the loudness and higher pitch of sound. Other changes include loss of neurons throughout the auditory pathway and thickening of the capillary walls (slowing of blood flow). Presbycusis may also result from attenuation in the movement of the tympanic membrane and the bones of the inner ear. Recall that the sound wave becomes a physiological event through the transfer of sound vibrations to the tympanic membrane, causing the bones of the middle ear to move in direct proportion to the pressure. Any attenuation in the movement of these bones will cause a discrepancy between the actual wave and the sound we hear. The malleus, incus, and stapes move only as much as their ligaments and tendons allow. As we age, collagen—the primary protein of ligaments and tendons—becomes stiffer as a result of nonenzymatic cross-linking and the formation of AGEs (see Box 8.2). The cross-linking of collagen in these ligaments and tendons causes a reduction in both the distance and the speed of movement by the middle ear structures. Thus, the ability to detect the loudness (amplitude) and high pitch (frequency) of sound declines in the post-reproductive period.

The sense of sight is based on the physics of light

The human eye consists of two main components that work together to create our sense of vision. The outer portion of the eye, which is exposed to the environment, contains the optics for focusing on an object and includes the pupil, cornea, iris, lens, and ciliary body **(Figure 8.15)**. The inner portion of the eye contains the structures that transform light into a neural impulse and includes the retina, fovea centralis, optic disc, and optic nerve.

Light reflecting off objects travels in straight lines in all directions. The ability to capture light waves and develop an image of the object requires an optical system that can focus the many rays of reflected light onto a single point. The optical system of the human eye—the **cornea** and **lens**—focuses the light from external sources onto a single point on the **retina**, which transforms the light wave into electrical impulses for interpretation by the brain. To focus the light onto a single point in the area of the retina's **fovea centralis**, the cornea must bend

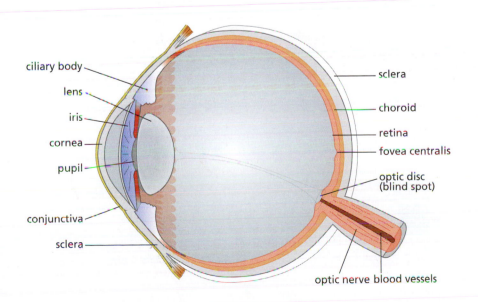

Figure 8.15 Anatomy of the human eye.

ciliary body

lens

iris

cornea

pupil

conjunctiva

sclera

sclera

choroid

retina

fovea centralis

optic disc (blind spot)

optic nerve blood vessels

(refract) the light rays. The amount of bending of light is referred to as **refractive power**; the greater the bending, the greater the refractive power.

The human lens is attached to a sphincter-shaped (circular) muscle, the **ciliary body**, by **zonular fibers**. Relaxation of the ciliary body (the muscle elongates) stretches the zonular fibers, causing elongation of the lens **(Figure 8.16)**. Contraction of the ciliary body (the muscle shortens) relaxes the zonular fibers, allowing the natural elastic elements of the lens to recoil and form a more spherical shape. The ability of the lens to change shape allows us to view both near and far objects. When viewing objects that are far away, the lens has a flat shape caused by relaxation of the ciliary body, and the cornea provides all the refraction. Light coming from near objects strikes the cornea at greater angles and requires a greater power of refraction. Contraction of the ciliary body causes the lens to become more spherical, increasing the refraction of light. Thus the lens assists the cornea with refraction, and this process of reshaping the lens is known as **accommodation**.

Presbyopia can be explained by age-related changes in the refractive power of the lens

All individuals over the age of 50 have undergone changes in the optic portion of their eyes that affect their ability to focus on close objects—a condition known as **presbyopia**. People over the age of 40 or 50 can often be seen holding reading material at a distance from the eyes in order to focus on the print. This behavior directly reflects the inability of the lens to recoil, form a spherical shape, and increase its refractive power sufficiently to focus on near objects. Although the precise cause of presbyopia has yet to be clearly established, several factors are known to contribute. First, the cells of the lens are not replaced once they are formed; that is, they are **terminally differentiated**. Terminal differentiation also results in the loss of organelles in the lens cells. The consequent inability to replace or repair damaged cells in the lens can result in loss of elasticity. Second, as described above, collagen, the protein that holds the cells together and provides the lens with its elasticity, becomes stiffer with age, and this prevents the lens from shortening into the spherical shape needed to focus on close objects.

Figure 8.16 Mechanism of refraction and accommodation. Focusing by the human eye requires that light rays converge on a single point on the retina, the fovea centralis. Light reflected off far objects (A) hits the cornea at lesser angles than light reflected off close objects (B) and requires less refraction—the lens does not assist, or accommodate, the cornea and is flat in shape. Note that the lens becomes more spherical when focusing on close objects. Close objects require significant accommodation by the lens for proper refraction.

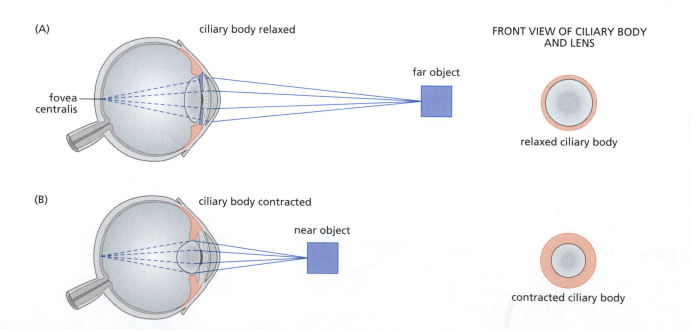

Finally, there is a slight decrease in the number of smooth muscle cells in the ciliary body; this decreases the strength of contraction and thus decreases refraction.

Terminal differentiation of lens cells leads to the formation of cataracts

Approximately 3.9% of individuals over the age of 60 have visual impairments that cause low vision or blindness **(Figure 8.17)**. Cataracts, the most common visual disease affecting the elderly, are at the interface between normal aging and disease. That is, while the clarity of the lens declines with aging, the formation of a cataract occurs in only 3.5% of individuals over the age of 60.

Cataracts can be defined as any opacity in the lens of the eye **(Figure 8.18)**. The cause of age-related cataracts remains to be discovered, but we know that years of environmental insults such as photo-oxidation, increased osmotic pressure, and other stresses contribute significantly to the development of opacity. In most tissues, insults to cells are prevented or repaired by processes in the cytoplasmic organelles; severe insult triggers apoptosis and cell death. Because lens cells are terminally differentiated, there is no apoptosis. In addition, maintaining the transparency of the lens requires functioning organelles—which, as noted above, decrease in number with age. Thus, environmental damage to the aging lens cannot be repaired.

How does environmental damage to the lens lead to age-related cataracts? There is, as yet, no definitive answer to this question, but recent evidence suggests that protein misfolding and the development of insoluble protein aggregates provide one explanation for the age-related increase in lens opacity. The human lens contains considerable amounts of the protein **crystallin**. The tertiary structure of crystallin supports the transparency of the lens. Like all proteins, crystallin undergoes denaturation (unfolding) from time to time. Other tissues repair the unfolding of proteins with the help of chaperone proteins. The human lens does not have the apparatus to synthesize chaperone proteins. Rather, it contains α-crystallin, a protein with a chaperone-like

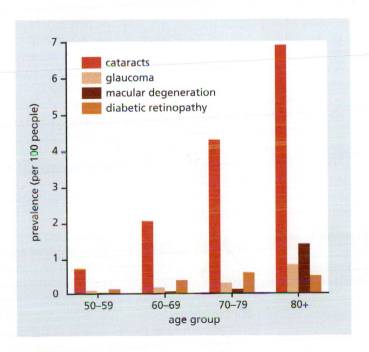

Figure 8.17 The prevalence of some age-related diseases of vision. Data for the 80+ age group are likely to be overestimates, as the number of people in the study population was small. (Adapted from Eye Disease and Prevention Group, *Arch. Ophthalmol.* 122:487–494, 552–563, 564–572, 2004. With permission of The Eye Disease and Prevention Group.)

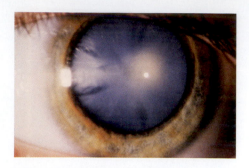

Figure 8.18 Opacity of the lens caused by the development of an age-related cataract. (Courtesy of Biophoto Associates/PR Science/Visual Photos.)

domain. In the lenses of young humans, α-crystallin helps itself to refold and regain its function, keeping the lens transparent. The amount of α-crystallin and/or of the chaperone domain decreases significantly with age. Denatured crystallin proteins are then unable to refold into their functional tertiary structure, and transparency is reduced.

This leads to a further question: How does the age-related increase in misfolded (or denatured) proteins lead to cataracts? Again, the underlying mechanism remains unknown. We do know, however, that unfolded proteins are highly susceptible to forming insoluble aggregates, and chemical analysis of cataracts reveals the presence of protein aggregates. When proteins denature, they expose amino acids that are not normally on the surface of the protein. These amino acids can then bond with amino acids in another denatured protein molecule—bonding that would not occur if a chaperone protein were present. That is, two proteins bond together and there is no mechanism for breaking them apart. After this process occurs many times over, a protein aggregate forms. (Protein aggregate formation is described in greater detail in Chapter 9.)

Cataracts can seriously impair vision, but they can be surgically removed. There are two types of cataract surgery: phacoemulsification and extracapsular cataract extraction. **Phacoemulsification**, the most common procedure, involves emulsifying (making soluble) the lens with ultrasonic waves. Once the lens becomes soluble, it can be removed by suction. Phacoemulsification does not remove the lens capsule. **Extracapsular cataract extraction** involves removing the lens while the elastic capsule that covers the lens is left partially intact. Leaving the elastic capsule allows for implantation of an artificial lens. In both surgeries, an artificial lens is placed into the cornea. The artificial lens cannot contract or lengthen and does not correct for the loss of accommodation, although some artificial lenses that will improve accommodation are now being developed. Cataract surgery has become extremely common in the United States. Of the 3.5% of individuals over the age of 60 who develop age-related cataracts, 95% have the cataracts removed.

The senses of taste and smell change only slightly with age

The mouth and nose provide us with our hedonic sense of taste, a vital sensory process that urges us to eat when we are hungry. (If you don't think this is true, think about how unpleasant or uninteresting eating becomes when you have a head cold.) The sense of taste arises from our two chemical senses, taste and smell. The sensory organs for taste, the **taste buds**, are found primarily on the tongue but are also located on the roof of the mouth. Flavor detection by the taste buds has been divided into five general categories: salty, sweet, bitter, sour, and **umami** (the taste associated with salts of glutamic acid and other amino acids). Taste buds respond to food and send signals to the brain about the type of taste by detecting changes in ionic concentration for salty and sour tastes or through the stimulation of receptors specific for sweet, bitter, and umami flavors.

Approximately 80% of the flavor we experience in food arises from our sense of smell. The **olfactory nerve** in the epithelium of the upper nasal cavity detects aromatic compounds in food, based on their chemical structure **(Figure 8.19)**. Humans have more than 1000 different olfactory receptors that are specific to groups of odoriferous molecules. Each olfactory neuron contains one to four receptors. When a scent molecule binds to the receptor, a neural signal is sent to the **olfactory**

Figure 8.19 Anatomical location of the structures of the olfactory system and pharynx.

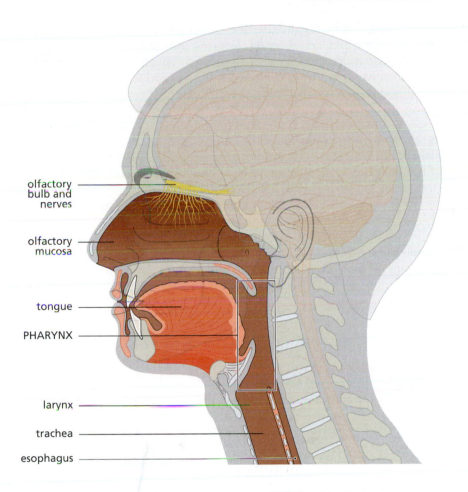

olfactory bulb and nerves

olfactory mucosa

tongue

PHARYNX

larynx

trachea

esophagus

bulb. The olfactory bulb decodes the signal, identifying which receptor was stimulated and how much stimulation occurred, then forwards that message via the olfactory nerve (first cranial nerve) to the olfactory center of the brain **(Figure 8.20)**. The signals from the olfactory center are relayed to the brain's **limbic system**, which determines whether the smell is pleasant or noxious. Areas in the limbic system also integrate the signal for taste with that for smell to give us the overall sense of taste.

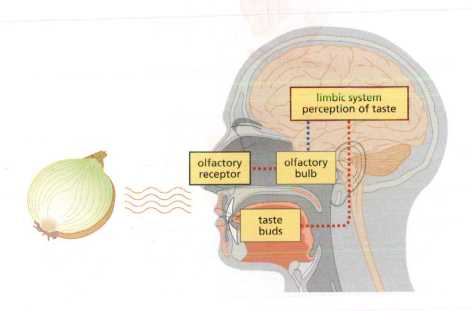

limbic system perception of taste

olfactory receptor

olfactory bulb

taste buds

Figure 8.20 The relationship between flavor and smell. The human sense of taste depends on both the olfactory (smell) and the taste-bud systems, with the sense of smell contributing about 80% to our overall sense of taste. When food is present, the aromatic compounds attach to one of more than 1000 specific receptors on olfactory neural cells in the nose. Binding of a chemical causes a signal to be sent to the olfactory bulb, which interprets the neural impulse. That message is then sent to the olfactory center in the limbic system of the brain. At the same time, the taste buds are stimulated by the chemical components of the food. The signals from the taste buds are relayed to the brain and integrated with the signals from the olfactory centers, and we perceive taste.

Early investigations evaluating the effect of aging on sense of taste often included individuals with diseases that affected the physiology of both taste buds and olfactory structures. This led to a general belief that the sense of taste declines with age. However, more recent and properly designed studies have, by and large, shown that age-related changes in taste buds and olfactory centers are minimal. To date, research has been unable to detect significant age-related changes in the number or neural activity of taste buds or olfactory bulbs or in the turnover of olfactory neurons. Some research suggests that the number of molecules needed to stimulate an olfactory receptor, known as the "threshold," increases with age. But the change in threshold appears to be small—it may be associated with specific disorders and may not be physiologically significant. Some individuals have a decreased ability to discriminate between odors, suggesting a disruption in the neural circuitry that connects the olfactory receptor to the olfactory centers in the brain. These effects are not usually seen in the general population and are more likely to be associated with a disease process than with age per se.

CHANGES IN THE DIGESTIVE SYSTEM

The human digestive system, or gastrointestinal (GI) system, runs as one long, continuous tube from mouth to anus **(Figure 8.21)**. The sole purpose of the digestive system is to extract the energy and nutrients from food that are needed for sustaining life and to eliminate solid substances that cannot be absorbed. All multicellular organisms have a digestive tract, with a basic structure that does not differ much from ours. The difference between the digestive tract of a worm and

Figure 8.21 Major organs of the human gastrointestinal tract.

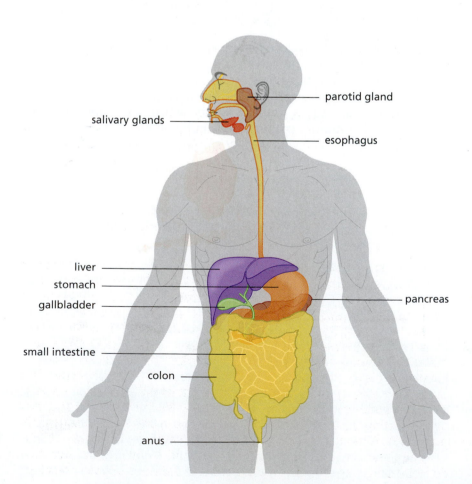

a human lies in the additional organs and systems—liver, pancreas, immune system, and neural control—that support the digestive process in humans. These additional organs allow us to eat a greater variety of foods than other species. Indeed, it could be argued that the ability to eat a wide variety of foods is one of the reasons that the human species has been so successful. We are the only species that can survive and even thrive in all parts of the world. The Inuit of the Arctic region survive and flourish as well on a diet consisting primarily of fish and marine mammals as does the largely vegetarian population in the interior of Southeast Asia.

In general, the digestive system as a whole shows little functional decline with age. Most of the age-related changes are associated either with disease processes, such as cancer or diabetes, or with nutritional "bad habits." This is not to say that aging does not affect the function of the digestive system. Common ailments such as gastritis, diarrhea, and constipation occur much more often in the elderly than in younger populations. Nonetheless, it remains unclear whether the age-related decrease in GI function is the result of physical changes in the system or the accumulated effects of years of poor eating habits. In this section, we examine some of the age-related changes that do occur, starting with the mouth and working through the small intestine. Age-related changes in the large intestine, rectum, and anus are associated with disease, not with aging per se.

Age-related changes in the mouth and esophagus do not impair digestion

Digestion begins in the mouth through the action of chewing and the initial preparation of food by the **salivary glands**. Chewing breaks down food into smaller pieces for ease of swallowing and creates a greater surface area for the action of the enzymes of the digestive system. The loss of teeth that can occur with age has the potential to significantly affect proper digestion, and it may limit the types of foods eaten and thus may lead to malnutrition. The reasons for age-related tooth loss remain controversial. While anatomical changes such as age-related bone loss (in the jaw) and loss in strength of the ligaments holding the teeth in position do occur, proper dental hygiene throughout life may offset these potential problems. Modern dentistry has eliminated the majority of age-related problems associated with poor dentition, thus age-related loss of teeth does not significantly affect the digestive process for most people living in developed countries. Lack of access to dental care and lack of education regarding dental health are a more likely cause of poor dentition in the elderly population than is biological aging.

The salivary glands perform several functions that prepare food for digestion by the stomach and intestines. **Saliva**, the product of the salivary glands, is composed of water, electrolytes, mucus, antibacterial compounds, and various enzymes. The saliva adds lubrication to food so that it can pass through the esophagus with relative ease. The water in saliva also helps to make dry food more soluble so that the aromatic compounds can be released (important for our sense of taste). Finally, saliva contains two very important enzymes, **lysozyme** and **α-amylase**. Lysozyme kills many types of bacteria and helps to prevent bacterial build-up in the mouth. α-Amylase initiates starch digestion by converting the long chains of glucose into **maltose** (a two-glucose molecule). Although there may be a slight decrease in saliva volume with age, the concentrations of lysozyme and α-amylase in the saliva do not change, and the decreased volume does not impair digestion.

However, some neurological disorders such as stroke, Parkinson's, and Alzheimer's disease can alter saliva volume to the point where digestion is impaired.

Once food has been chewed and saliva added, the tongue begins the swallowing process by lifting up and pushing the bolus of food into the **pharynx**. The pharynx connects the mouth to the **esophagus.** It contains the voice box, which separates the **trachea** from the esophagus. The presence of food in the pharynx causes the esophagus to begin its **peristaltic contractions**, the rhythmic contractions of smooth muscle that push food into the stomach. Age-related physical disorders of the tongue and esophagus that cause problems of swallowing are relatively uncommon (as they are in all age groups). Disruption of proper swallowing most often occurs in association with psychological disorders and, like salivary dysfunction, with neurological disease.

Decline in stomach function is most often associated with atrophic gastritis

The human stomach provides four basic functions in the digestive process: (1) it acts as a storage area that allows us to eat large meals that are then released slowly into the small intestine; (2) it liquefies food before it enters the intestine; (3) it continues the digestive process through the secretion of enzymes and other molecules; and (4) it secretes **exocrine hormones** that prepare other organs of the digestive tract for incoming food.

The stomach lies between the esophagus and the small intestine and is divided into four regions, each having a different function: (1) the contents of the esophagus empty into the cardia; (2) the fundus, formed by the organ's curvature, adds digestive juices to the food; (3) in the body or central region, food is mixed with the digestive juices; and (4) the pylorus facilitates further mixing of the stomach contents and passage into the small intestine **(Figure 8.22)**. The cells of the stomach wall secrete hydrochloric acid (HCl) and enzymes necessary for the digestive process **(Figure 8.23)**. The mucosa layer contains the glands that secrete HCl and other digestive enzymes; the submucosa, a matrix of connective tissue, provides structure for the blood vessels; muscles in the muscularis layer contract the stomach for mixing of the food; and the connective tissue of the serosa forms the barrier between the stomach and other organs of the digestive tract. The cells of the stomach wall are arranged into glands that form **gastric pits**, where food comes in contact with the HCl and digestive enzymes. Each region of the stomach has a slightly different arrangement of glands and different cells within the glands **(TABLE 8.2)**.

When food enters the stomach, the parietal cells secrete HCl. Hydrochloric acid helps to break down large particles of food into smaller pieces, which are then liquefied into a semi-solid fluid called **chyme** before entering the intestine. In addition, the acidic nature of the stomach causes the unfolding of large proteins (**denaturation**). This allows the enzyme pepsin, secreted by the stomach, to more efficiently break the bonds between amino acids (only peptides of two or three amino acids and single amino acids can be absorbed by the intestinal cells). The secretion of HCl also stimulates the secretion of other enzymes and hormones needed for proper digestion.

Evaluating possible age-related anatomical or physiological decline in the stomach can be challenging, due in large part to a condition known as **atrophic gastritis**. Atrophic gastritis causes inflammation of the stomach mucosa, leading to reduced numbers of parietal and chief cells.

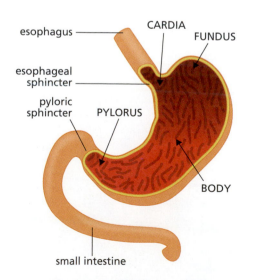

esophagus

CARDIA

FUNDUS

esophageal sphincter

pyloric sphincter   PYLORUS

BODY

small intestine

Figure 8.22 Anatomy of the human stomach.

TABLE 8.2
SUBSTANCES SECRETED BY CELLS OF THE STOMACH

Secreted substance	Secretory cell type	Primary function of secretion
Alkaline mucus	Mucosal cells	Protects stomach lining from the harmful effects of hydrochloric acid; aids in the liquefaction process
Hydrochloric acid (HCl)	Parietal cells	Breaks down large food particles to aid in the liquefaction process; especially important for the unfolding of large proteins
Rennin (chymosin)	Chief cells	Enzyme that coagulates milk, necessary to extract the protein in milk
Pepsin	Chief cells	Enzyme that breaks peptide bonds between the amino acids of proteins
Intrinsic factor	Parietal cells	Factor secreted in response to presence of protein in the stomach; needed for proper absorption of vitamin B_{12} (see text discussion of atrophic gastritis)
Gastrin	G-cells	Endocrine hormone that promotes secretion of HCl in response to protein in the stomach
Cholecystokinin	Mucosal cells	Hormone that stimulates secretion of bile and pancreatic digestive enzymes into the small intestine; may signal the brain when the stomach is full

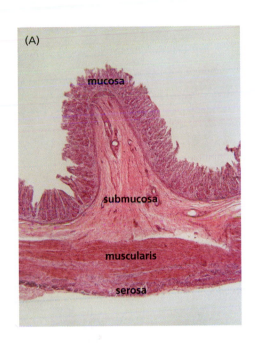

(A)

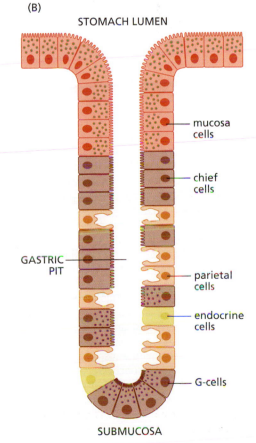

(B)

These cells are replaced by fibrous tissue. As a result, the stomach's secretion of essential substances such as HCl, pepsin, and intrinsic factor is impaired. Atrophic gastritis can lead to severe digestive problems and nutrient deficiencies. For example, decreased secretion of intrinsic factor by parietal cells can lead to vitamin B_{12} deficiency and the development of megaloblastic anemia (the production of underdeveloped red blood cells). While there are many causes of atrophic gastritis, about 90% of cases are the result of persistent infection by the bacterium *Helicobacter pylori (H. pylori)*. When treated with the appropriate antibiotic, atrophic gastritis can be cured and the problems associated with the disorder alleviated. The question remains, however, of why

Figure 8.23 The layers of the stomach wall. (A) Cross-sectional micrograph of the stomach wall in the region of the fundus, showing its four layers. (B) Arrangement of cells in the glands of the mucosa layer that form the gastric pits. Glands in different parts of the stomach may contain different proportions of the four types of stomach cells (see Table 8.2), but the anatomical arrangement is the same. Mucosa cells, which secrete protective alkaline mucus, are at the top, followed by chief cells and parietal cells, with G-cells at the bottom of the pit. (A, courtesy of J.Harshaw/Shutterstock.)

the aged population has a greater incidence of *H. pylori* infection. The answer awaits further research.

Changes in the small intestine can affect digestion and nutrient absorption

The digestion of food and absorption of nutrients occur primarily in the small intestine in what can best be described as a mass-action process. The chyme—the mixture of water, food, and digestive juices—moves through the intestinal lumen by peristaltic contraction. The peristaltic action also causes mixing of the chyme so that enzymes aiding in digestion come into closer contact with the nutrient molecules. When chyme enters the small intestine from the stomach, its presence stimulates the release of hormones that induce the pancreas to secrete several digestive enzymes and the gallbladder to release **bile salts (Figure 8.24)**. Fat digestion begins when the bile salts, secreted into the intestine from the liver via the gallbladder, emulsify lipids and thus increase their solubility. The greater surface area created by the bile salts increases the efficiency of the **lipases** secreted from the pancreas, enzymes that break apart large fat molecules into free fatty acids. Pancreatic enzymes catalyze reactions that break the bonds holding together the sugars in carbohydrates, the amino acids in proteins, and the fatty acids in lipids.

The intestinal wall has structures called **villi** that increase the surface area of the intestine and enhance absorption **(Figure 8.25)**. The epithelial cells also have protrusions on their membranes called microvilli, which form the **brush border**; this further increases surface area.

The human digestive process has evolved to break down food into its smallest basic nutrient components—fatty acids, monosaccharides and **disaccharides**, amino acids, vitamins, and minerals—so as to facilitate their absorption. Three types of absorption are used by the small intestine: passive diffusion, protein channels, and mediated transport. Recall from Chapter 4 that the cell membrane (plasma membrane) consists primarily of nonpolar (without a net charge) lipids. Small nonpolar molecules in the intestinal chyme—free fatty acids, some vitamins, and a few amino acids—are highly soluble in the nonpolar membrane and are absorbed by **passive diffusion**. That is, these molecules diffuse through the membrane into the intestinal cell without the help of membrane proteins **(Figure 8.26)**.

Figure 8.24 Digestion in the small intestine. As the chyme enters the small intestine from the stomach, it stimulates cells (*blue arrows*) at the interface between the stomach and small intestine to release hormones that, in turn, stimulate the release of bile salts from the gallbladder (*green arrow*) and digestive enzymes from the pancreas (*red arrows*). Bile salts increase the solubility of fats, and digestive enzymes break apart large molecules of lipids, proteins, and carbohydrates into their smallest possible components: free fatty acids, amino acids, and monosaccharides, respectively. Vitamins and minerals are normally bound to these large molecules and are released during the digestive process.

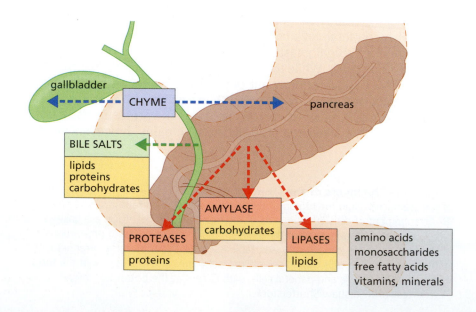

SMALL INTESTINE WITH VILLI (cross section)

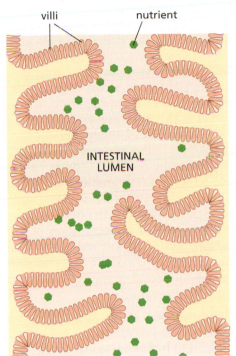

villi   nutrient

INTESTINAL LUMEN

INDIVIDUAL VILLUS WITH BRUSH BORDER

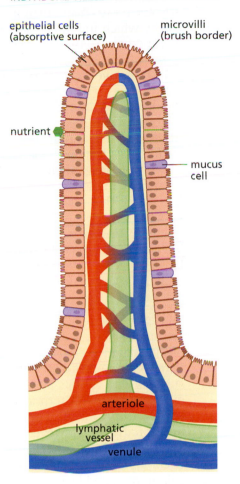

epithelial cells (absorptive surface)

microvilli (brush border)

nutrient

mucus cell

arteriole

lymphatic vessel

venule

Figure 8.25 Macroanatomy and microanatomy of the small intestine. The interior surface of the small intestine (*left*) contains protrusions into the lumen that are covered with villi. The villi increase the surface area of the small intestine and thus enhance absorption. The surface of each villus (*right*) is covered with epithelial cells that also contain protrusions that increase surface area: the microvilli, forming the brush border. Absorbed nutrients pass through the epithelial cells and into the general circulation via capillaries in the villi.

Ions are small and highly polar (having a net negative or positive charge) and cannot be absorbed by passive diffusion. Rather, ions are absorbed through channels in the membrane (Figure 8.26). The permeability of ion channels depends on a concentration gradient between the contents of the chyme in the intestinal lumen and the cytosol of the intestinal cell. When the concentration of the ion is higher in the chyme than in the intestinal cell, the channels open and ions flow through the membrane. When the intracellular ion concentration exceeds that in the chyme, the channels close.

Mediated transport makes use of specialized membrane proteins to transport large polar and nonpolar molecules across the membrane. There are two types: facilitated diffusion and active transport.

Figure 8.26 Passive diffusion and ion channels. Small nonpolar molecules can pass through the plasma membrane of the intestinal epithelial cell unobstructed, because the membrane is also nonpolar. Small polar ions are absorbed into the intestinal cell through ion channels. The channels open only when the concentration of the ion in the chyme (in the intestinal lumen) is greater than the intracellular concentration.

PASSIVE DIFFUSION

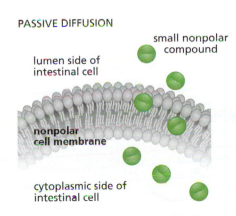

small nonpolar compound

lumen side of intestinal cell

nonpolar cell membrane

cytoplasmic side of intestinal cell

ION CHANNELS

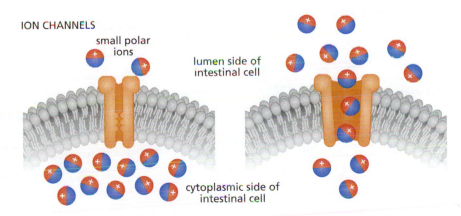

small polar ions

lumen side of intestinal cell

cytoplasmic side of intestinal cell

Facilitated diffusion occurs when the concentration of the molecule is greater in the chyme than in the intestinal cell. This is a simple process by which the molecule binds to a specific transport protein in the cell membrane that facilitates transport into the cell. The monosaccharide fructose is used as an example of facilitated diffusion in **Figure 8.27A**. **Active transport** also uses membrane proteins to transport large molecules across the membrane, but this is a two-step process (Figure 8.27B). In the first step, the molecule to be transported, with a lower concentration in the chyme than in the intestinal cell, binds to a transport protein in the cell membrane. Because of the higher concentration inside the cell, the molecule cannot overcome the unfavorable concentration gradient without help. That help usually comes in the form of a small ion with a favorable concentration gradient: higher in the chyme than in the cell. By using the ease of transport associated with the ion, the large molecule enters the cell against its concentration gradient. However, the increase in intracellular concentration of the ion would eventually inhibit further transport of the ion and, with it, the large molecule, if the ion is not removed from the cell. The second step of active transport uses energy to move the ion from the cell back into the intestinal lumen, against its concentration gradient.

The effects of aging on the form and function of the small intestine are somewhat controversial, because we do not know the full extent of the influence of disease and poor eating habits on normal function. Many early studies on the effect of aging on intestinal function included subjects with various diseases, and tissue samples were often collected during surgery. Not surprisingly, these initial studies suggested that age has a detrimental effect on the small intestine, including the malabsorption of nutrients. However, with the introduction of safe and somewhat pain-free biopsy methods, tissue samples have been collected from the small intestine of healthy individuals. These investigations show that villi number and height (a measure of surface area) do not change significantly with age. Moreover, the replicative capacity of epithelial cells, a property critical to maintaining absorption in the small intestine, does not change with advancing age.

Proper digestion and absorption rely on the ability of the intestinal tract to move the chyme efficiently through the lumen by peristaltic action. Attenuation of this rhythmic muscular action can lead to a decrease in flow and the development of constipation—a condition often associated with age. There does seem to be an age-related reduction in the force of contraction generated by the intestinal smooth muscle, resulting in a slight increase in transit time of the chyme. However, this slight increase does not result in increased constipation or malabsorption syndromes. Age-related constipation more likely reflects poor eating habits than physiological changes in the small intestine.

Figure 8.27 Facilitated diffusion and active transport. (A) Facilitated diffusion occurs when large polar or nonpolar molecules, such as fructose, attach to a specific transporter molecule and are transported into the cell. Facilitated diffusion predominates when the concentration of the molecule is higher outside (in the chyme) than inside the intestinal epithelial cell. (B) Active transport is a two-step process that uses a co-transporter. For example, glucose is bound to its transport protein along with sodium. The favorable concentration gradient of sodium (lower inside the cell) helps glucose enter the cell against its concentration gradient. Sodium is then "pumped" out of the cell using the energy of ATP and the exchange of potassium ions for sodium. This pump, called the sodium/potassium ATP pump, helps to maintain the electrical potential across the plasma membrane in many types of cells.

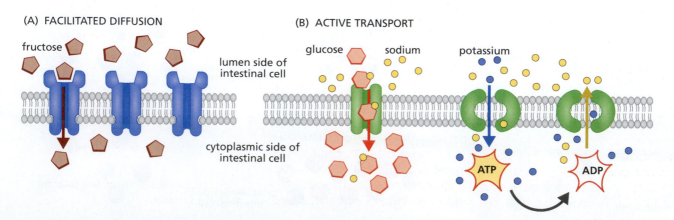

(A) FACILITATED DIFFUSION

fructose

(B) ACTIVE TRANSPORT

glucose   sodium   potassium

lumen side of intestinal cell

cytoplasmic side of intestinal cell

ATP   ADP

Controversy also exists concerning the effect of age on the absorption of nutrients. Traditionally, the absorption of **micronutrients**—essential vitamins and minerals needed in only very small quantities for proper growth and metabolism—has been suggested to decrease with age. This view has recently been challenged by the finding that specific proteins that transport micronutrients do not decrease with age. Nonetheless, calcium absorption may decline with age in women and may be associated with a decrease in the activity of vitamin D, the vitamin needed for proper absorption of calcium. The reasons for an age-related decrease in vitamin D activity are unclear, but it is most likely associated with the normal postmenopausal loss in bone mineral content (see Chapter 9). The finding that calcium absorption does not seem to decline with age in men suggests that malabsorption of this mineral reflects normal hormonal shifts during the post-reproductive period in women. That is, the age-related decline in calcium absorption may be closely associated with gender rather than a universal effect of aging.

CHANGES IN THE URINARY SYSTEM

The urinary system consists of two kidneys, two **ureters**, the **bladder**, and the **urethra (Figure 8.28)**. With the exception of the kidneys, there are no significant age-related alterations in the system that impede the elimination of urine. Age-related **urinary incontinence**, the inability to control urination, is the result of abnormal muscle weakness, disease processes that affect the nerves controlling bladder function, and/or excessive urine production by the kidneys. Therefore, the focus here is on age-related changes in the kidney. We begin by examining normal renal function.

The kidneys remove metabolic waste products from the blood

All human cells excrete metabolic waste product into the blood. These products must be removed, otherwise the delicate balance of blood pH—keeping it in the 7.2–7.4 range needed to carry out normal metabolic functions—would be affected; also, some waste products would prove toxic if allowed to accumulate. Maintenance of normal cell function also requires a stable blood concentration of certain minerals, collectively known as **electrolytes**, such as sodium, potassium, calcium, magnesium, and phosphorus. Thus our kidneys are responsible for both the elimination of metabolic waste products from the blood and the maintenance of blood electrolyte and pH levels.

The kidney receives blood through the renal artery; this blood then passes through a structure of the kidney called a **nephron (Figure 8.29)**. Filtration of the blood by the nephron occurs in three different processes in different locations: (1) **glomerular filtration**, in which fluid and low-molecular-weight molecules in the blood are filtered across the capillaries of the **glomerulus** and into the **Bowman's capsule**; (2) **tubular reabsorption**, the movement of substances from the **renal tubule** into the **peritubular capillaries**; and (3) **tubular secretion**, the movement of substances from the peritubular capillaries into the renal tubule.

Glomerular filtration results from a pressure gradient between the glomerular capillaries and Bowman's capsule. Blood in the glomerular capillaries has a pressure of 60 mm Hg, whereas Bowman's capsule has a pressure of about 44 mm Hg. The greater pressure in the capillaries forces fluid and low-molecular-weight molecules out of the capillaries and into Bowman's capsule. The fluid in Bowman's capsule, called

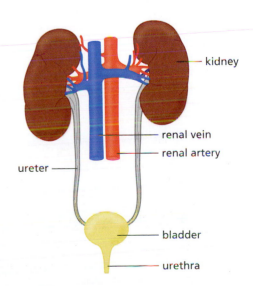

kidney

renal vein

renal artery

ureter

bladder

urethra

Figure 8.28 Anatomy of the human urinary system.

(A) KIDNEY

(B) NEPHRON (kidney tubule)

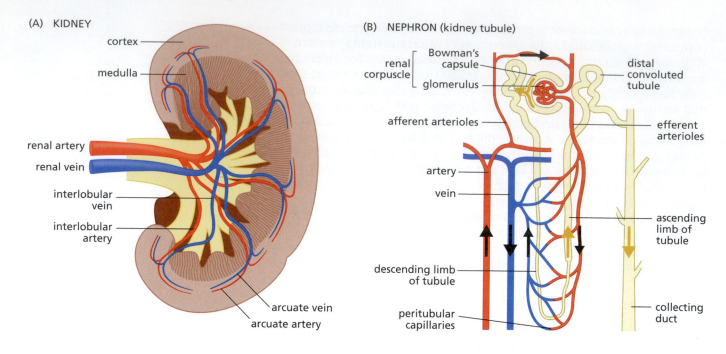

Figure 8.29 Anatomy of the kidney and nephron. (A) The gross structure of the kidney. (B) An individual nephron, showing the filtration system. *Black arrows* indicate the direction of blood flow through the arterioles, glomerulus, and peritubular capillaries. *Yellow arrows* indicate the direction of filtrate flow from Bowman's capsule to the collecting duct.

the **glomerular filtrate**, has the same concentration of substances as blood **plasma**.

Tubular reabsorption and secretion can occur by passive diffusion or mediated transport. Which process the nephron uses depends on the concentration of the substance in the renal tubule and in the peritubular capillaries. For example, the concentration of **urea**, a nitrogenous product of protein metabolism, in the glomerular filtrate is the same as that in the peritubular capillaries. As the glomerular filtrate moves through the tubule, water is reabsorbed and the concentration of urea in the filtrate relative to that in the peritubular capillaries increases. As a result of this concentration gradient, urea passively diffuses from the glomerular filtrate into the peritubular capillaries. Nitrogen in the urea is thus returned to the circulation and is used to synthesize amino acids.

Mediated transport is used when a substance in the renal tubule or the peritubular capillaries must be transported against its concentration gradient, from lower to higher concentration. An example of mediated transport is the reabsorption of sodium through a membrane channel called the **sodium/potassium ATP pump**. Maintenance of proper cellular function requires keeping the intracellular concentration of sodium lower than that in the extracellular fluid. Conversely, the intracellular concentration of potassium must remain higher than the extracellular concentration. Thus, sodium passively diffuses from the lumen of the tubule into the epithelial cells lining the tubule. However, in tubular reabsorption, sodium passes from the tubular epithelial cells into the peritubular capillaries, against its concentration gradient. The sodium/potassium ATP pump allows sodium to move against its concentration gradient by exchanging sodium for potassium, using ATP as energy source.

The kidneys help regulate blood pressure

Another important job of the kidney is to help the body maintain blood pressure by adjusting the amount of water in the blood. Moving water from the blood into the urine by tubular secretion decreases blood pressure. Tubular reabsorption maintains blood pressure by transferring water from the glomerular filtrate into the peritubular capillaries.

The movement of water in the nephron is coupled to sodium reabsorption and is regulated by several hormones. Recall that sodium passively diffuses from the tubular lumen into the tubular epithelial cells. The removal of sodium lowers the osmolarity (i.e., raises the water concentration) of the glomerular filtrate, and water passively diffuses into the tubular epithelial cells—that is, water follows sodium. If the osmolarity of the peritubular capillaries is *lower* than that of the glomerular filtrate, water moves from the peritubular capillaries into the tubule. The coupling of water and sodium reabsorption explains the different colors of urine. If you do not drink enough water, the kidney reabsorbs water and the urine is yellow due to a high concentration of solutes. If you drink more water than your body needs, the kidney produces clear urine, because the concentration of solutes declines with the tubular secretion of water into the urine.

The amount of water reabsorbed by the kidney also depends on the permeability of the renal tubule. Tubular permeability is under physiological regulation by the hormone **vasopressin**, or **antidiuretic hormone (ADH)**, secreted by the pituitary gland. When the hypothalamus (a region of the brain) senses a drop in blood volume, it sends a signal to the pituitary gland, which secretes ADH into the blood. ADH causes the opening of water channels (aquaporins) in the tubular epithelial cells, and the reabsorption of water increases. Conversely, an increase in blood volume inhibits the secretion of ADH, and more water remains in the tubule, resulting in a higher concentration of water in the urine.

Renal blood flow and kidney function decline with aging

Renal blood flow decreases with aging in terms of both total amount and percentage of cardiac output (the rate at which blood is pumped through the heart). The decrease in renal blood flow reflects a loss in the overall number of blood vessels in the kidney and the narrowing of the vessels. This loss of vessels occurs at all levels of the kidney, but the glomeruli seem to be the most affected. Beginning in a person's early thirties, the number of vessels in the glomerulus declines by about 10% per decade. Moreover, the remaining vessels are irregular and twisted, further limiting blood flow in the glomerulus.

Many of the arterial changes observed in the aging kidney are similar to those seen in other organs and include arteriolosclerosis and hypertrophy of the inner wall of the artery (see Chapter 9 for details on these conditions). One age-related change in the vasculature that seems to be more common in the kidney than in other organs is **fibrointimal hyperplasia**, abnormal growth in the intimal layer of the arterial wall. This condition is caused by the accumulation of plasma proteins in the vessel walls, a condition known as **insudation**, which causes vessel narrowing and reduced blood flow.

The age-related decline in the number and function of glomerular capillaries can also lead to a lower rate of fluid filtration from the glomerulus into the Bowman's capsule. The rate, or volume per unit of time, at which this fluid is filtered is known as the **glomerular filtration rate (GFR)**. A young 60- to 70-kg person has a GFR of about 180 L/day. Given that total blood volume, on average, is about 3 L, the body's entire blood supply is filtered 60 times per day. Such a high volume of filtering allows the kidney to respond to and rapidly remove large quantities of metabolic waste products. The decrease in filtration due to aging is highly variable among individuals, ranging from 0% to about 20% of normal GFR. However, any change in the GFR will affect blood and body homeostasis.

The underlying cause of a reduced GFR is **glomerulosclerosis**, the degenerative scarring of the glomeruli. Some estimates have suggested that by age 80, 30–40% of a person's glomeruli have stopped functioning due to this condition. Controversy exists as to whether glomerulosclerosis is a normal age-related event or is the result of high blood pressure. Age-related glomerulosclerosis cannot be distinguished from hypertension-related glomerulosclerosis in younger individuals. Furthermore, the condition is almost always found in older individuals with hypertension. Thus it is not clear whether age-related glomerulosclerosis precedes or follows hypertension.

CHANGES IN THE IMMUNE SYSTEM

Roy Walford (1924–2004), who conducted groundbreaking research in biogerontology, proposed 40 years ago that declining immune function might cause normal aging. This proposal, known as the immunological theory of aging, has yet to be completely tested, but we do know that mortality rate due to infectious diseases can be three to four times greater in people over the age of 65 than in younger age groups. Moreover, the percentage of people over 65 who are protected by vaccination against novel influenza viruses is only about half that found in the younger population. Clearly, advanced age compromises the effectiveness of the human immune system.

In this section, we explore the physiology of the immune system and discuss how aging affects the body's ability to repel attacks from potentially harmful invaders. Human immunity is an extremely complex function involving several redundant systems that work together to prevent infection. This complexity precludes a thorough discussion of the human immune system in this text. Instead, we focus on some functions that have been shown to decline with age.

Innate immunity provides the first line of defense against infection

Humans have two separate but equally important immune systems: **innate immunity** and **acquired (adaptive) immunity**. We are born with a fully intact innate, or natural, system of immunity that either prevents infection or reacts quickly to nonself agents. Innate immunity prevents infection by erecting barriers such as the skin, the mucosal linings of body cavities, and digestive acids and proteases. If a foreign substance gets through these protective barriers, the innate system stimulates **phagocytosis**, the process of engulfing and ingesting foreign particles; this is carried out by specialized blood- and tissue-borne cells such as **neutrophils** and **macrophages** **(TABLE 8.3)**.

Neutrophils circulate in the blood and migrate into tissue where foreign matter has entered the body. Both neutrophils and macrophages can carry out phagocytosis, but tissue macrophages are much larger than neutrophils and phagocytose larger particles. Small fragments of the macrophage-digested invader are carried to the lymph tissue by specialized cells called **dendritic cells**, which activate the acquired immune system.

When tissue injury occurs, the affected site releases compounds, such as histamines and prostaglandins, that increase blood flow, capillary permeability, and migration of neutrophils to the area. Collectively, these responses characterize inflammation and activate the phagocytotic process of the innate immune system **(Figure 8.30)**. If bacteria or other foreign agents enter a wound, tissue macrophages surround

TABLE 8.3
CELLS OF THE IMMUNE SYSTEM

Cell type	Where and/or how produced	Functions
Neutrophils	Bone marrow	Phagocytosis Release of chemicals involved in inflammation (vasodilators, and an assortment of protein degrading agents)
Macrophages	Bone marrow; in tissues, differentiate from monocytes (a type of white blood cell)	Phagocytosis Present antigens to helper T cells
Naive T cells	Synthesized in bone marrow; undergo further development in thymus gland	Stored in lymph tissue and await first exposure to a specific antigen
Helper T cells	Formed from naive T cells	Secrete cytokines that activate B cells and other types of T cells
Natural killer (NK) cells	Formed from naive T cells	Bind to virus-infected cells and cancer cells, inject toxins, and kill the cells
Cytotoxic T cells	Formed from naive T cells	Bind to antigens on the plasma membrane of virus-infected cells and cancer cells and kill them
B cells	Bone marrow; stored in lymph tissue	Initiate antibody-mediated immune response
Plasma cells	Formed from B cells; terminally differentiated cells	Secrete antibodies

the invaders within minutes and begin to phagocytose the organisms. Macrophages also induce the secretion of **cytokines** and **chemokines**, proteins that attract white blood cells, such as neutrophils, to the wound by **chemotaxis**. Because the tissue concentration of macrophages is low, neutrophils perform most of the phagocytosis at the inflammation site over the first few hours.

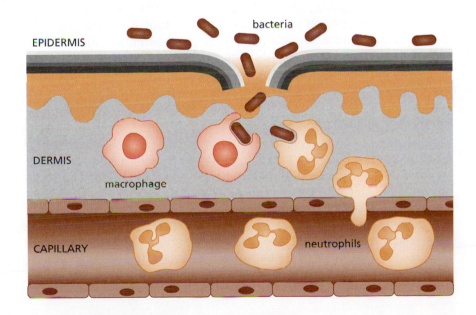

Figure 8.30 Inflammation and innate immunity. If bacteria breach the protective barrier of the innate immune system (represented here by a cut in the skin epidermis), the area around the invasion site becomes inflamed. The tissue macrophages are the first line of defense against the invading bacteria and begin the phagocytotic process within minutes. Inflammation also increases capillary permeability, allowing passage of neutrophils into the tissue. Infiltration by neutrophils, occurring by chemotaxis, and their subsequent phagocytotic activity start within an hour of injury and may last for 2–3 days.

Acquired immunity relies on lymphocytes reacting to antigens

The same processes that maintain life—breathing and eating—also expose human beings to millions of harmful bacteria, viruses, and toxins. While innate immunity provides an effective barrier against invasion by these agents, the sheer number of harmful agents in the environment can easily overwhelm this system, allowing many invaders to get through the innate immune system defenses. Once this happens, the responsibility for protecting the body from infection falls on the **lymphocytes—T cells** and **B cells**—of the acquired immune system (see Table 8.3).

T cells are small white blood cells (lymphocytes) that are made in the bone marrow and mature in the **thymus gland** before being stored in the lymph tissue. At this stage, they are **naive T cells**, not yet transformed into immune cells that attach to and kill invaders. T cells are activated when dendritic cells, carrying fragments of macrophage-digested foreign particles, enter the lymph tissue. The binding of these fragments to the surface receptors of a specific type of T cell stimulates the production and release of three different T-cell clones from the lymph tissue. These T-cell clones are responsible for destroying the invading **antigen**, any substance capable of inducing a specific immune response.

The vast majority (75–80%) of activated T-cell clones released from the lymph tissue are **helper T cells**. As the name implies, helper T cells "help" the immune system respond adequately to the invader. Helper T cells do not attack the antigens directly; rather, they secrete compounds known as **lymphokines** (interleukins and interferon) that stimulate the proliferation and differentiation of other cells that do attack the antigen (Figure 8.30). For example, lymphokines secreted by helper T cells induce B cells to synthesize and release antibodies specific to the antigen. The other two important types of T cells are **cytotoxic T cells** and **natural killer (NK) cells** produced for very specific purposes.

In brief, antigens entering lymph tissue are destroyed by macrophages, which then present antigen fragments to naive T cells **(Figure 8.31)**. Binding of antigen fragments to the surface of the naive T cells induces differentiation and proliferation of helper T cells and NK cells or cytotoxic T cells. Helper T cells secrete lymphokines that stimulate the production of antibodies in activated B cells. NK cells attach to antigens on the membranes of virus-infected cells and inject toxic proteins that destroy the cells. Naive B cells engulf macrophage-produced antigen fragments, and this induces the B cells to differentiate into **plasma cells** that are capable of synthesizing antibodies. The antibodies form extremely strong bonds with the antigen causing a change in the antigen configuration. As a result, the antigen can no longer bind to body cells and cause damage. The antibodies also mark the antigen for degradation by other immune cells.

The first time an antigen is encountered, it takes several days for the body to produce sufficient T cells and B cells to neutralize it. After the antigen has been eliminated and no longer poses a threat to the body, the majority of T and B cells are destroyed. But some T and B cells remain after this first encounter with an antigen; these are called **memory cells**. Memory cells contain surface receptors that are specific to the antigen that was eliminated, and they are able to produce clones quickly when encountering that antigen again. In fact, enough clones can be produced in a single day to effectively eliminate the antigen at the next exposure.

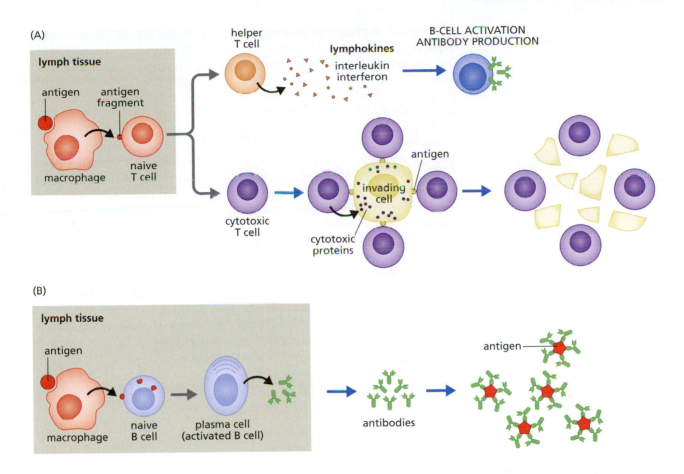

The phagocytotic function of neutrophils and macrophages declines with age

The effectiveness of the innate immune system relies, in large part, on the ability of neutrophils and macrophages to infiltrate the affected area. The number of neutrophils and macrophages available to the innate immune system does not decline with age, and the synthesis of these cells from hematopoietic stem cells in the bone marrow appears to be well conserved throughout life. Therefore, age-related functional loss in the innate immune system probably reflects a loss of the phagocytotic function of the neutrophils and macrophages **(TABLE 8.4)**.

Figure 8.31 Mechanism of T-cell and B-cell immunity. (A) Macrophages from the affected tissue present partially digested antigen fragments to naive T cells. Naive T cells are transformed into helper T cells, cytotoxic T cells (as shown here), or NK cells. (B) Helper T cells activate B cells, transforming them into plasma cells—immune cells that secrete antibodies. The antigen is immobilized by the strong bonding of antibodies.

TABLE 8.4
AGE-RELATED CHANGES IN THE INNATE IMMUNE SYSTEM

Cell type	Age-related change
Neutrophils	No change in number circulating or in bone marrow Slower release from bone marrow stores Slight decrease in phagocytosis Slight decrease in chemotaxis
Macrophages	No change in number Decrease in phagocytosis Decrease in cytokine and chemokine production
Natural killer (NK) cells	Increased number No change in binding or cytotoxicity No change in chemokine production

The decline in phagocytotic function of the innate immune system may explain why wound healing takes significantly longer in post-reproductive individuals than in younger populations. The healing delay may also result from a decline in cytokine and chemokine production by macrophages. Recall that macrophages stimulate the release of neutrophils from bone marrow stores through cytokine production and direct the neutrophils to the injured site through chemokine release. Attenuation of cytokine and chemokine release would reduce the number of neutrophils arriving, or delay arrival, at the wound site. In addition, two other age-related changes contribute to a delay in wound healing. First, the density of vessels and capillaries in the skin decreases with age. Because the healing of skin wounds relies on the ability of neutrophils to diffuse from capillaries into the damaged tissue, a lower blood vessel and capillary density would decrease the total number of neutrophils arriving at the wound site. Second, repair of skin wounds relies on the ability of fibroblasts to secrete the necessary connective tissues that bind replacement cells together. As we saw earlier in the chapter, the mitotic activity of aged skin cells declines, so there are fewer fibroblasts to provide the proteins for new connective tissue. Together, the decline in vessel density and in fibroblast production results in an increase in healing time and an increased risk of infection.

The production of naive T cells, number of B cells, and effectiveness of antibodies all decline with age

Recall that the thymus gland, a small gland just behind the breastbone, is the site where T cells develop. Aging results in a significant atrophy of the thymus gland, which causes an age-related reduction in the production of naive T cells. Thymic involution reflects the shrinkage of the epithelial space—the location of T-cell development—and replacement of the population of maturing T cells with connective tissue. It has been estimated that the epithelial space in the thymus declines by 90% by the age of 70. Interestingly, the epithelial cells that remain within the involuted thymus do not seem to have attenuated T-cell production. That is, the age-related decline in T-cell production reflects solely the loss of thymic epithelial cells. With the decline in production of naive T cells, the ratio of memory T cells to naive T cells increases. Moreover, the capacity for the clonal expansion of memory T cells into helper T cells, a vital process in mounting an appropriate immune response to an antigen, may also decline with age.

Together, the age-related decline in peripheral naive T cells and the decreased clonal expansion of memory T cells into helper T cells explain the overall age-related decline in immune function. Recall that the acquired immune system uses memory T cells to shorten the time required to respond to an invader. Helper T cells also stimulate the release of antibodies by B cells. If the memory T cell does not function at peak performance, the immune response is blunted and the person is at greater risk for developing complications as a result of the invading antigen. The age-related decline in naive T cells reduces the individual's ability to respond to a novel antigen. This may help explain why, as noted above for influenza, vaccinations are less effective in the elderly than in the younger population.

The number of mature B cells also falls with age, reflecting decreased B-cell synthesis in the bone marrow; this leads to an increase in the ratio of memory B cells to mature B cells, although the total peripheral B-cell component does not decline with age. As occurs for memory T cells, memory B cells have a lower capacity for clonal expansion in older people. With the age-related decline in peripheral mature B cells, the ability to mount an immune response to novel pathogens is blunted.

Antibody function also declines with age, reflecting the diminution in number and function of helper T cells. That is, the decreased clonal expansion of memory T cells into helper T cells reduces the amount of cytokine release, which, in turn, reduces the release of antibodies from B cells. In addition, the number of high-affinity antibodies (antibodies with multiple antigen-binding sites) also falls with age. This results in a weakening of the immune response, because the strength of the antigen–antibody bond determines, for the most part, antibodies' effectiveness in neutralizing invaders.

CHANGES IN THE REPRODUCTIVE SYSTEM

Human reproductive aging has, historically, been associated with female aging, and it reflects three important changes. First, for women, fertility comes to a rather abrupt end with the cessation of menses and the onset of menopause, usually at 50–60 years of age. Second, ovarian production of sex-linked hormones, such as estrogen and progesterone, drops dramatically at the time of menopause. And third, an age-related decline in the genetic quality of eggs leads to an increased occurrence of genetically damaged embryos, birth defects, and spontaneous abortions. As you learned in previous chapters and will learn more about in Chapter 9, the end of a woman's reproductive life span also correlates with other age-related physiological dysfunctions.

Men do not experience an abrupt end to fertility. This observation has often been misinterpreted to mean that men do not experience a decline in their ability to conceive. In the past couple of decades, researchers have found that men undergo many of the same age-related reproductive changes that are observed in women. Although men do not experience an end to their reproductive life span, male fertility and sex hormone production do decrease with age. Moreover, the genetic quality of sperm appears to decline with advanced age, resulting in an increased risk for genetic problems in offspring. In this section, we briefly examine how declining hormone production affects the reproductive systems of women and men.

Menopause is caused by declining secretion of sex hormones by the gonads

The female reproductive system produces a single egg on a 28-day schedule called the **menstrual cycle (Figure 8.32)**. The cycle begins when **luteinizing hormone (LH)** and **follicle-stimulating hormone (FSH)**, secreted by the pituitary gland, stimulate an **ovarian follicle**— the structure containing a maturing **ovum**—to release its mature ovum and to secrete **estrogen** and **progesterone** into the blood. Estrogen and progesterone serve two functions: (1) they are necessary for the maintenance of female sexual characteristics and sex organs, and (2) they have a feedback effect on the ovary, stimulating development of the ovum in the follicle. Increasing blood concentrations of LH, FSH, and estrogen induce a developing follicle to rupture and release the ovum, a process known as **ovulation**. The ruptured follicle becomes a structure called the **corpus luteum**, which increases the secretion of progesterone. Increased blood concentrations of estrogen and progesterone also have a feedback effect on the pituitary gland, inhibiting the secretion of LH and FSH and ensuring that only one follicle develops a mature egg. Estrogen and progesterone prepare the uterus for possible implantation of a fertilized egg by stimulating cell growth in and blood flow to the endometrial wall. If fertilization does not occur, estrogen and progesterone levels fall and the thickened layer of the endometrium is sloughed off in the menstrual flow. The fall in estrogen

Figure 8.32 The menstrual cycle. These graphs and diagrams show the relationships among body temperature, hormone levels, uterine changes, and ovarian morphology over the course of the menstrual cycle. Antral follicles (also called Graafian follicles) are follicles at a late stage of follicle maturation.

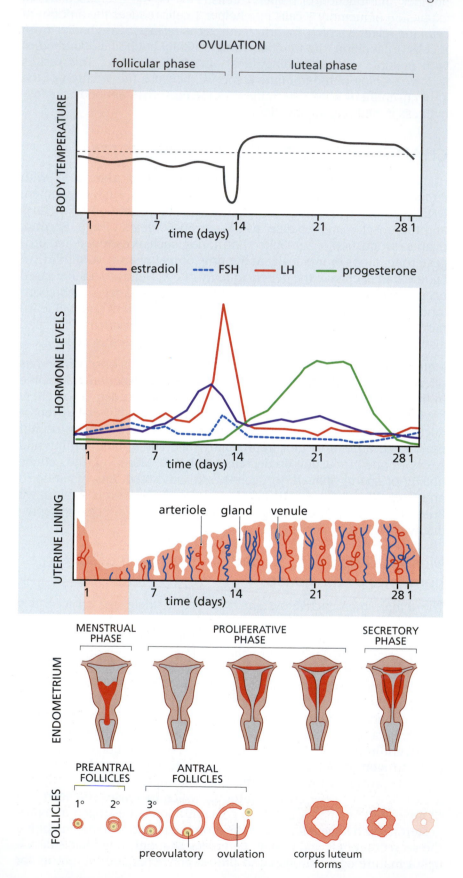

and progesterone also causes the pituitary to begin secreting LH and FSH, starting the cycle again.

At birth, the ovary contains a female's complete supply of **oocytes**, undeveloped eggs—about 750,000. The number of oocytes declines with age so that by the time a woman reaches menopause, at 50–60 years of age, fewer than 5000 oocytes remain. The number of ovarian follicles falls to zero at the time of menopause. These two age-related ovarian changes ultimately lead to a cessation of estrogen and progesterone secretion by the gonads. It is the termination of the ovaries' production of these two hormones that marks the start of **menopause**.

The cessation of estrogen and progesterone production at menopause causes significant age-related changes, in addition to the loss of oocyte development. Without the monthly cycle of ovarian hormones, the uterus begins to shrink and can decrease in size as much as 70% within 15–20 years after menopause. The vagina becomes smaller and less elastic, and its endothelial wall thins. With the loss of endothelium, the vagina becoming less effective as a barrier against abrasion, resulting in an increase in pain and risk of injury during intercourse. The reduction in endothelial tissue also results in a reduced amount of glycogen secreted into the vagina. The pH of the vagina rises, resulting in an increased risk of infections.

Estrogen also has effects on many other, nonsexual organs and tissues. As you'll learn in Chapter 9, normal bone growth in women can be influenced by estrogen. The loss of estrogen during menopause results in significant bone mineral loss. The hot flashes experienced by many menopausal and postmenopausal women appear to be related to loss of estrogen's inhibition of LH secretion. Numerous studies have shown that the hypothalamic neurons that regulate the pituitary secretion of LH may also be involved in thermoregulation. Without estrogen, these neurons are stimulated and this may affect thermoregulation. Finally, the loss of ovarian production of estrogen and progesterone has been related to an increasing risk for heart disease and cancer. Although the mechanism for this increased risk has yet to be elucidated, it is thought that estrogen acts as a "protective" agent against cell damage during the childbearing years.

Male fertility declines slightly with age

The issue of age-related alterations in men's ability to conceive at advanced ages has only recently become a topic of intensive research. It is now generally accepted that male fertility declines slightly with age. Anatomical changes cannot, in general, account for these declines. **Impotence**, or **erectile dysfunction**, is the inability to achieve or maintain an erection and most often reflects secondary symptoms arising from other medical problems. Age-related alterations in blood flow to the penis have not been observed, although, as men age, the erectile tissue of the penis becomes infiltrated with fibrous tissue. Although the penis can still become erect, the fibrous tissue results in a less stiff erection. The testes decrease slightly in both size and weight, but the extent of decrease is highly variable. Most of the age-related size changes in the testes can be accounted for by loss of cells in the **seminiferous tubules**, the structures responsible for sperm and testosterone production. Age-related anatomical changes in the epididymis, vas deferens (ductus deferens), urethra, and seminal vesicles that lead to changes in sexual function have not been widely observed.

The slight age-related decline in fertility in men most likely reflects changes in the cells of the seminiferous tubules. As a man ages, the sperm-supporting cells, called **Sertoli cells**, are slowly replaced by fibrous tissue, decreasing the total number of sperm per ejaculation. The number of sperm produced per Sertoli cell also appears to decline, although wide variation in this variable has been reported. The number of **Leydig cells**, the cells that produce testosterone, also declines. **Testosterone** is required for the initiation of sperm production. It is also required for the maintenance of healthy gametes. Reduced testosterone production has been suggested as one mechanism for the higher rate of genetic errors observed in sperm from older men.

Although a decline in serum testosterone concentration has been shown to correlate with a decrease in Leydig cell number, some studies suggest that alterations in the hypothalamic-pituitary-testicular axis may also be responsible for the reduced testosterone. Testosterone production in young men follows a **circadian rhythm**, a rhythm based on 24 hours. Production is highest between the hours of 4 and 8 a.m. and lowest around midnight. The circadian rhythm for testosterone corresponds to a similar rhythm for LH, the pituitary hormone primarily responsible for initiating testosterone production. Older men have both lower concentrations of testosterone and a loss of circadian rhythm in its production. Nonetheless, LH levels do not decline in proportion to the decline in testosterone, suggesting that, with aging, the Leydig cells become less responsive to the stimulatory effects of LH. Because a similar response is observed in women during menopause, the decrease in responsiveness of Leydig cells to LH has been named **andropause**.

Old age is not a barrier to sexual activity

The age-related changes in reproductive organs in men and women have only a minor impact on sexual performance. In men, the erectile tissue of the penis accumulates fibrous material that can alter the speed of erection and the stiffness of the tissue. Impotence, the inability to engage in sexual intercourse because the penis is not sufficiently erect, is not a normal age-related condition. It occurs in only about 15–20% of men over the age of 60. Although normal age-related alterations in the testes, seminal vesicles, and prostate do affect fertility and increase the risk of disease, the changes in these organs do not seem to affect sexual performance.

In women, the thickness of the vagina wall decreases with age, resulting in changes that can affect sexual performance and health. The lubricating secretion of the mucosal cells declines, increasing friction during sexual intercourse and the possibility of pain and minor physical trauma. However, recent evidence suggests that women who participate in regular sexual activity show little decline in mucosal cell secretions. Most research shows that age-related decreases in sexual activity in women often reflect psychological problems or the inability to find a suitable partner.

ESSENTIAL CONCEPTS

- Once humans have attained full growth and development (i.e., maturity), changes in body weight reflect changes in the amount of fat stores.

- Energy balance—energy intake minus energy expenditure—can be used to calculate changes in fat storage.

- Energy intake can be determined by calculating the amount of fat, protein, and carbohydrate we eat. Energy expenditure is calculated by measuring respiratory gases; total energy expenditure equals resting energy expenditure plus physical activity plus diet-induced thermogenesis.

- Body weight increases, on average, by approximately 15% between the ages of 20 and 70. Longitudinal studies show that decline in muscle mass during maturity occurs at a rate of about 2% per decade in men and less than 1% per decade in women.

- Near the end of life, body weight declines and this reflects losses in muscle and adipose tissue mass. Decrease in food intake parallels the end-of-life decline in body weight and can lead to a clinical condition known as anorexia of aging.

- Skin wrinkles are caused by three age-related changes: (1) loss in number and function of skin cells; (2) loss of subcutaneous fat; and (3) increase in nonenzymatic cross-links in collagen. The loss in subcutaneous fat in the face, arms, and legs causes irregularities that are covered by a thinner, relatively inelastic epidermis, and thus the creation of wrinkles.

- The majority of skin aging results from overexposure to UV light.

- The age-related decline in hearing, known as presbycusis, results from alterations in the inner ear, but the reasons for these changes are unknown.

- All individuals over the age of 50 have alterations in the optics portion of the eye that affect the ability to focus on close objects, a condition known as presbyopia.

- Presbyopia seems to be primarily caused by two factors: the lens's inability to repair damage and excess nonenzymatic cross-linking.

- Age-related changes in taste buds and olfactory centers are minimal.

- The incidence of atrophic gastritis increases with aging and is the result of infection by *Helicobacter pylori*.

- Age-related changes in the small intestine are minimal and seem not to significantly disrupt the absorption of most nutrients.

- Renal blood flow decreases with age in terms of both total amount and percentage of cardiac output.

- Determining age-related loss in the human immune system is challenging because the acquired immune system develops mainly during the individual's early development period and depends to a great extent on his or her interaction with the environment.

- Age-related functional loss in the innate immune system reflects losses in the phagocytotic function of neutrophils and macrophages and a decrease in cytokine and chemokine release.

- Aging results in significant atrophy of the thymus gland and thus a reduced production of naive T cells.

- The number of peripheral B cells declines with age, reflecting changes in the B-cell progenitor pathway in the bone marrow.

- Women and men experience significant age-related changes in their reproductive processes. Female fertility ends with the cessation of menses and the onset of menopause at about 50–60 year of age. Ovarian production of sex-linked hormones drops dramatically. Male fertility and production of sex hormones decline with age. The genetic quality of sperm also seems to decline with advanced age, resulting in an increased risk for genetic problems in offspring.

DISCUSSION QUESTIONS

Q8.1 Oberta consumes 10,460 kJ of energy per day and has a total TEE of 9623 kJ. Is Oberta in positive or negative energy balance? If Oberta maintains this ratio of energy intake to energy expenditure, how many kilograms of body weight will she gain in 5 years?

Q8.2 What evidence can you provide supporting the thesis that a decrease in energy expenditure rather than an increase in energy intake is the primary cause of overweight and obesity in the older population?

Q8.3 Briefly define atrophic gastritis. Include the primary cause and the treatment for this disorder in the elderly. What effects does atrophic gastritis have on digestion?

Q8.4 Describe how the age-related loss of stereocilia affects hearing. Include in your description how the physical properties of a sound wave are converted into neural signals that the brain uses to interpret sound.

Q8.5 What are advanced glycation end products (AGEs) and how might they cause functional alterations in proteins and other biological structures?

Q8.6 Define the process of accommodation in the mechanism of focusing by the eye. What factors lead to age-related changes in accommodation and what is the visual result of these changes?

Q8.7 Briefly explain how age-related changes in the acquired immune system can result in a less effective response to vaccines.

Q8.8 Briefly describe the factors that lead to an increase in wound healing time in older people.

Q8.9 Define glomerular filtration rate and explain why age-related changes in the kidney's blood flow result in declining GFR.

Q8.10 What are the two age-related changes that trigger the start of menopause?

FURTHER READING

CHANGES IN BODY COMPOSITION AND ENERGY METABOLISM

Byrd-Bredbenner C, Moe G, Beshgetoor D & Berning J (2009) Wardlaw's Perspectives in Nutrition, 8th ed. Boston: McGraw Hill, pp 313–359.

Calle EE, Thun MJ, Petrelli JM et al (1999) Body-mass index and mortality in a prospective cohort of U.S. adults. *N Engl J Med* 341:1097–1105.

Narici MV & Maffulli N (2010) Sarcopenia: characteristics, mechanisms and functional significance. *Br Med Bull* 95:139–159.

Roberts SB & Dallal GE (2005) Energy requirements and aging. *Public Health Nutr* 8:1028–1036.

Robertson RG & Montagnini M (2004) Geriatric failure to thrive. *Am Fam Physician* 70:343–350.

U.S. Department of Health and Human Services (2010) Summary Health Statistics for U.S. Adults: National Health Interview Survey, 2009. Hyattsville, MD: U.S. Department of Health and Human Services. www.cdc.gov/nchs/data/series/sr_10/sr10_249.pdf

CHANGES IN THE SKIN

Buckingham EM & Klingelhutz AJ (2011) The role of telomeres in the ageing of human skin. *Exp Dermatol* 20:297–302.

Prunier C, Masson-Genteuil G, Ugolin N et al. (2012) Aging and photo-aging DNA repair phenotype of skin cells: evidence toward an effect of chronic sun-exposure. *Mutat Res* 736:48–55.

Puizina-Ivic N (2008) Skin aging. *Acta Dermatovenerol Alp Panonica Adriat* 17:47–54.

CHANGES IN THE SENSES: HEARING, VISION, TASTE, AND SMELL

Croft MA, Glasser A & Kaufman PL (2001) Accommodation and presbyopia. *Int Ophthalmol Clin* 41:33–46.

Eye Disease Prevalence Research Group (2004) Causes and prevalence of visual impairment among adults in the United States. *Arch Ophthalmol* 122:477–485.

Huang Q & Tang J (2010) Age-related hearing loss or presbycusis. *Eur Arch Otorhinolaryngol* 267:1179–1191.

Murphy CC (2008) The chemical senses and nutrition in older adults. *J Nutr Elderly* 27:247–264.

Van Eyken E, Van Camp G & Van Laer L (2007) The complexity of age-related hearing impairment: contributing environmental and genetic factors. *Audiol Neurootol* 12:345–358.

Widmaier EP, Raff H & Strang KT (2004) Human Physiology: Mechanisms of Body Functions, 9th ed. Boston: McGraw Hill, pp 231–241.

CHANGES IN THE DIGESTIVE SYSTEM

Byrd-Bredbenner C, Moe G, Beshgetoor D & Berning J (2009) Wardlaw's Perspectives in Nutrition, 8th ed. Boston: McGraw Hill, pp 116–151.

Drozdowski L & Thomson AB (2006) Aging and the intestine. *World J Gastroenterol* 12:7578–7584.

Holt PR (2007) Intestinal malabsorption in the elderly. *Dig Dis* 25:144–150.

Kandulski A, Selgrad M & Malfertheiner P (2008) *Helicobacter pylori* infection: a clinical overview. *Dig Liver Dis* 40:619–626.

CHANGES IN THE URINARY SYSTEM

Esposito C & Dal Canton A (2010) Functional changes in the aging kidney. *J Nephrol* 23 suppl 15:S41–45.

Martin JE & Sheaff MT (2007) Renal ageing. *J Pathol* 211:198–205.

Pannarale G, Carbone R, Del Mastro G et al (2010) The aging kidney: structural changes. *J Nephrol* 23 suppl 15:S37–40.

Widmaier EP, Raff H & Strang KT (2004) Human Physiology: Mechanisms of Body Functions, 9th ed. Boston: McGraw Hill, pp 513–562.

CHANGES IN THE IMMUNE SYSTEM

Gruver AL, Hudson LL & Sempowski GD (2007) Immunosenescence of ageing. *J Pathol* 211:144–156.

Plowden J, Renshaw-Hoelscher M, Engleman C et al (2004) Innate immunity in aging: impact on macrophage function. *Aging Cell* 3:161–167.

Widmaier EP, Raff H & Strang KT (2004) Human Physiology: Mechanisms of Body Functions, 9th ed. Boston: McGraw Hill, pp 695–738.

CHANGES IN THE REPRODUCTIVE SYSTEM

Kuhnert B & Nieschlag E (2004) Reproductive functions of the ageing male. *Hum Reprod Update* 10:327–339.

Widmaier EP, Raff H & Strang KT (2004) Human Physiology: Mechanisms of Body Functions, 9th ed. Boston: McGraw Hill, pp 643–694.

AGE-RELATED DISEASE IN HUMANS

9

"THERE IS NO CURE FOR THE COMMON BIRTHDAY."

-JOHN GLENN, ASTRONAUT AND FORMER
U.S. SENATOR (1921–)

In Chapter 1, we discussed in some detail the distinctions between aging and disease. Recall that disease reflects a process that disrupts the physical and biochemical functions of the cell, whereas biological aging occurs within the bounds of normal cellular function. Defining the difference between aging and disease serves to clarify terms in discussing the biology of aging and helps establish the boundaries for experimental research. However, the strict lines drawn in this textbook between normal age-related functional loss and definable disease are not as clear-cut in the real world of aging humans.

The incidence of most diseases increases with age; this fact cannot be disputed. The real question with regard to aging and disease is, Why does normal age-related functional loss become disease? During the development and maturity phases of the life span, most physiological systems operate at a level where the difference between normal and abnormal function is easily detected. With aging, the difference between normal and abnormal becomes less well defined. For example, as you will learn in this chapter, all individuals over the age of 70 years have a certain amount of functional loss in the cardiovascular system. For most individuals, this decline does not result in significant alterations in overall physiological function or daily living. But for some, the decline in cardiovascular function will progress to a lethal disease called congestive heart failure.

In this chapter, we focus on clearly defined diseases that are limited, for the most part, to the post-reproductive period. To this end, we discuss five of the most common diseases of the elderly: Alzheimer's disease, Parkinson's disease, cardiovascular disease, type 2 diabetes, and osteoporosis. For each of these diseases, you'll learn how normal age-related changes in a particular organ or physiological system progress to an overt disease state. Cancer is not considered here, as this disease affects all age groups.

For some organ systems, such as the liver and respiratory tract, age-related disease occurs more as a consequence of environmental insult

IN THIS CHAPTER . . .

THE NERVOUS SYSTEM AND NEURAL SIGNALS

AGE-RELATED DISEASES OF THE HUMAN BRAIN: ALZHEIMER'S DISEASE AND PARKINSON'S DISEASE

THE CARDIOVASCULAR SYSTEM

AGE-RELATED DISEASES OF THE CARDIOVASCULAR SYSTEM: CARDIOVASCULAR DISEASE

THE ENDOCRINE SYSTEM AND GLUCOSE REGULATION

AGE-RELATED DISEASE OF THE ENDOCRINE SYSTEM: TYPE 2 DIABETES MELLITUS

THE SKELETAL SYSTEM AND BONE CALCIUM METABOLISM

AGE-RELATED DISEASES OF BONE: OSTEOPOROSIS

or lifestyle choices than as a result of aging per se. For example, smoking is the number one cause of age-related disease of the lung.

THE NERVOUS SYSTEM AND NEURAL SIGNALS

The nervous system is the most complex system in the human body. Every movement, every breath, every heartbeat, every sensory perception begins in the nervous system. Our nervous system has three overlapping functions: sensory input, integration, and motor output. These functions are performed by the brain, the spinal cord, the sensory receptors that receive stimuli, the effector cells that carry out the body's responses to stimuli, and a network of nerves that carry information to various parts of the body. The brain and spinal cord make up the **central nervous system (CNS)**, and the nerves make up the **peripheral nervous system (PNS) (Figure 9.1)**. All nerves communicate their messages between the CNS and the rest of the body through a combination of electrical and chemical signals, generated by the exchange of ions across the nerve cell (plasma) membrane. In this section, we

Figure 9.1 Gross anatomy of the human nervous system. The central nervous system (CNS) consists of the brain and the spinal cord. Nerves originating from the CNS make up the peripheral nervous system (PNS).

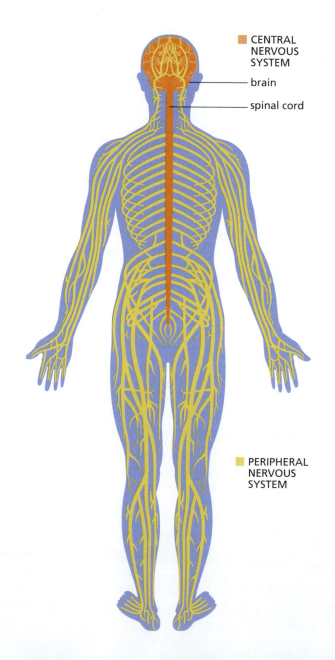

CENTRAL NERVOUS SYSTEM

brain

spinal cord

PERIPHERAL NERVOUS SYSTEM

present an overview of the human nervous system and examine the nature of neural signals in the normally functioning system.

The nervous system is composed of neurons and supporting cells

Neurons, or nerve cells, are the cells that conduct messages along the communication pathways of the nervous system. There are three main types of neurons: **sensory neurons**, which convey sensory input about the external and internal environments to the CNS; **motor neurons**, which convey motor output from the CNS to **effector cells** (muscle or gland cells); and **interneurons**, which integrate sensory input and motor output. Although neurons vary somewhat in structure depending on their function, most consist of three general components: a cell body, dendrites, and an axon **(Figure 9.2)**.

The **cell body** contains the nucleus and a variety of other cellular organelles, which carry out the normal functions of the cell. The **dendrites** are branch-like structures that extend from the cell body, increasing the surface area of the neuron. The dendrites receive signals from other neurons and sensory receptors and convey them to the rest of the neuron. The **axons** are long tubular extensions that carry signals from the cell body to their tips, the axon terminals. The axon terminals connect with the dendrites of other neurons. Many neurons have a single axon, which varies in length. For example, the axons that control fine movements in the fingers originate in cell bodies within the spinal cord and extend all the way to the hand, whereas other axons may be less than 1 mm in length. In addition, the axon may be branched, with each branch giving rise to specialized endings called **synaptic terminals**. The site of contact between a synaptic terminal and another cell is called a **synapse**. As you will see, the synapse is the location where one neuron communicates with another.

Figure 9.2 Anatomy of a neuron.

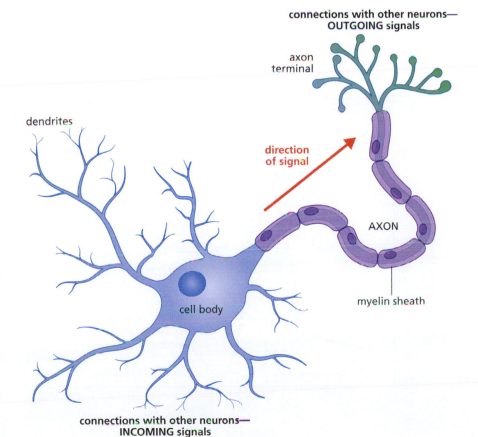

connections with other neurons—
OUTGOING signals

axon
terminal

**direction
of signal**

AXON

myelin sheath

dendrites

cell body

connections with other neurons—
INCOMING signals

Axons are often surrounded by an insulating layer called the **myelin sheath**. In the PNS, the myelin sheath is formed from supporting cells called **Schwann cells**. In the CNS, the myelin sheath is produced by supporting cells called **oligodendrocytes**.

Membrane potentials establish the conditions for neural signal transmission

All nerves generate impulses, or electrical signals, that transmit information to different parts of the body. These impulses depend on the flow of ions across the plasma membrane of neurons. All living cells have a difference in electrical charge across their plasma membranes. The **membrane potential** is the difference in voltage, or electrical potential, between the interior and exterior of the cell. The membrane potential exists because of differences in ion concentrations in the intracellular and extracellular fluids. The extracellular fluid contains a greater concentration of sodium (Na^+) and chloride (Cl^-) ions, while the intracellular fluid (cytosol) contains a greater concentration of potassium (K^+) ions.

Although all cells have a membrane potential, only neurons and muscle cells are excitable, capable of changing the electrical state of their membranes. The size of the membrane potential determines a membrane's readiness or ability to propagate an electrical signal—the greater the membrane's potential, the greater the probability that it can undergo an electrical event. For a neuron in its resting (unexcited) state, the **resting membrane potential** is typically −70 millivolts (mV). By convention, the voltage outside the cell is zero, so a negative membrane potential indicates that there are more negative charges on the inside of the cell than on the outside **(Figure 9.3)**. A change in potential that causes the membrane to become less polarized (less negative on the inside) compared with the resting membrane potential is called **depolarization**. A change that causes an increase in the potential is called **hyperpolarization**.

Recall from Chapters 4 and 8 that the plasma membrane is a phospholipid bilayer. Because ions are electrically charged, they cannot directly diffuse across the lipid of the membrane. The plasma membranes of

Figure 9.3 Membrane potentials and threshold potentials. The resting membrane potential of the neuron is −70 mV, making it highly polarized and ready for changes in electrical activity. When a triggering event occurs, the membrane potential begins to change slowly in the positive direction (less negative inside the cell). At approximately −55 mV, the threshold potential is reached and Na⁺ channels open, causing a rapid depolarization. At +30 mV, depolarization stops and begins to reverse (repolarization), returning the neuron to its resting membrane potential.

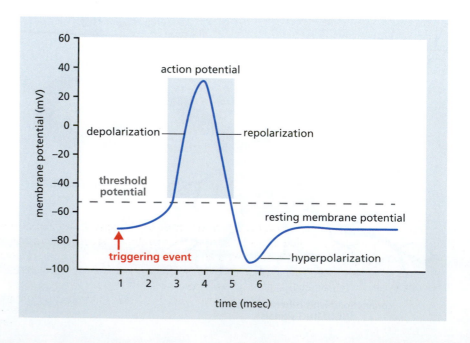

neurons contain **voltage-gated ion channels**, which allow the movement of Na^+ and K^+ ions into and out of the cell. These channels have gates that open and close in response to electrical signals. This allows the cell to change its membrane potential in response to stimuli, or triggers, that the cell receives **(Figure 9.4)**. When a small section of the cell membrane receives a triggering event, such as a thought by your brain to move your index finger, the event initiates an **action potential**. The triggering event causes the membrane to become more permeable to Na^+, which begins to pass through the membrane into the intracellular compartment, starting depolarization. The reversal of charge occurs slowly until it reaches a threshold potential of about −55 mV. The threshold potential causes gated Na^+ channels to open, resulting in an explosive depolarization and the creation of an action potential. When the action potential reaches about +30 mV, K^+ channels open,

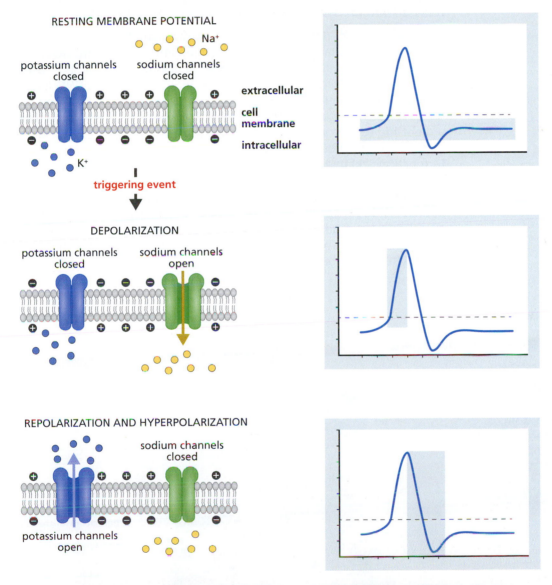

Figure 9.4 The role of voltage-gated ion channels in changes in membrane potential.
A triggering event causes a slow depolarization as the membrane becomes more permeable to Na^+. Sodium channels open when the membrane potential reaches −55 mV. The depolarization continues until the action potential reaches about +30 mV. Then the Na^+ channels close and the K^+ channels open. Potassium exits the cell, causing reversal of the depolarization process (repolarization). When the membrane potential hits −80 mV, the K^+ channels close and the membrane returns to its resting potential.

and the flow of K$^+$ out of the cell begins the repolarization process. The ending of the action potential in one segment of the membrane is the triggering event for initiation of the next action potential.

Neurotransmitters chemically link neurons together at the synapse

Axon terminals do not come into physical contact with dendrites. Rather, the axon terminal of one neuron forms a chemical junction with the dendrites of another neuron **(Figure 9.5)**. The transmitting cell is called the **presynaptic neuron**, and the receiving cell is called the **postsynaptic neuron**. The space between the two neurons is called the **synaptic cleft**. Together, the presynaptic neuron, postsynaptic neuron, and synaptic cleft make up the synapse. The ends of the presynaptic cell, called synaptic knobs, contain synaptic vesicles. Each vesicle contains thousands of neurotransmitter molecules, chemicals that are released into the synaptic cleft in response to an action potential.

There are 50–100 different types of neurotransmitters. Here we discuss briefly two major classes of neurotransmitters that are involved in age-related diseases of the brain: **acetylcholine (Ach)** and the **catecholamines**. Acetylcholine is found throughout the PNS and CNS. It can be inhibitory or excitatory, depending on the type of receptor found on the postsynaptic neurons. One type of acetylcholine receptor, known as a **nicotinic receptor**, propagates a signal by opening Na$^+$ channels on the postsynaptic neuron. Nicotinic receptors in the brain are important in functions related to attention, learning, and memory. Degradation of neurons with nicotinic receptors is involved in the progression of Alzheimer's disease. Another type of acetylcholine receptor, known as a **muscarinic receptor**, uses a G-coupled mechanism to propagate the signal and is found primarily in the PNS. A decrease in the number of muscarinic receptors in the heart may be one reason for the age-related functional decline in this organ.

Figure 9.5 General mechanism of signal propagation between two neurons. (1) Generation of an action potential in the presynaptic cell causes (2) Ca^{2+} channels to open. The influx of Ca^{2+} into the synaptic knob causes (3) synaptic vesicles to move to the membrane and (4) undergo exocytosis, releasing neurotransmitter molecules into the synaptic cleft. (5) Binding of neurotransmitter molecules to receptors on the postsynaptic neuron induces (6) the opening of Na$^+$ channels, which (7) creates the action potential.

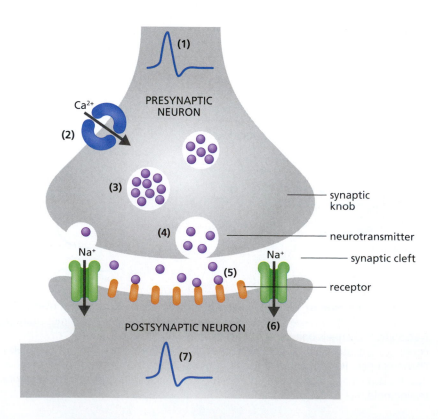

The catecholamines are amines derived from catechol. There are three main types: **dopamine**, **norepinephrine**, and **epinephrine**. Norepinephrine and epinephrine were once thought to be a single neurotransmitter, called adrenaline. Because of this, neurons that release epinephrine or norepinephrine are called **adrenergic neurons** and the receptors that bind these catecholamines are called **adrenergic receptors**. Neurons that release norepinephrine are found primarily in brain centers that control the basic functions of life (see the discussion of the brain below). Epinephrine-releasing neurons are found primarily in the PNS and are responsible for our flight-or-fight response (when you jump when someone scares you, that is epinephrine doing its job). Dopamine-releasing neurons are primarily found in brain centers that coordinate movement. All three catecholamines can bind to two different types of receptors. The alpha-adrenergic receptors propagate the action potential by modulating ion channels. The beta-adrenergic receptors, like the muscarinic receptors, use the G-coupled mode of cellular signaling. As you will see in the next section, loss in function of adrenergic neurons is the primary cause of Parkinson's disease.

The human brain is a collection of separate organs and cell types

The adult human brain weighs approximately 1.3–1.5 kg. While we tend to consider the brain as one organ, it actually consists of a collection of separate neural centers, all having their own specialized function **(Figure 9.6)**. There are hundreds of distinct neural structures and centers. **TABLE 9.1** provides a list of some of the neural structures and their functions that will be discussed later in connection with aging and the development of neuropathology.

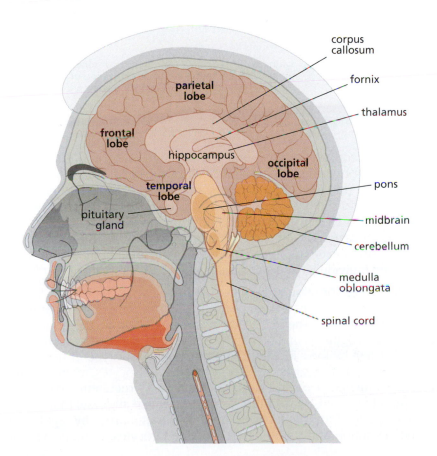

Figure 9.6 Structures of the human brain that are commonly affected by age-related pathology.

TABLE 9.1
STRUCTURES OF THE HUMAN BRAIN WITH POSSIBLE CONNECTIONS TO AGE-RELATED DISEASE

Structure	Function
Cortex (cerebrum)	
Frontal lobe	Reasoning, planning, parts of speech, movement, emotions, problem solving
Temporal lobe	Perception and recognition of auditory stimuli, memory, speech
Parietal lobe	Movement, orientation, recognition, perception of stimuli
Occipital lobe	Visual processing
Limbic system	
Hypothalamus	Temperature regulation, food intake, water regulation, circadian rhythms
Thalamus	Processing center for sensory input and motor activity; distribution of signals to other brain centers
Amygdala	Emotions, fear, memory
Hippocampus	Memory, learning
Entorhinal cortex	Interface between cortex and hippocampus, memory
Fornix	Interface between hypothalamus and memory centers
Basal ganglia/brain stem	
Midbrain	Motor response to sight, eye movement, pupil dilation, hearing
Pons	Sleep and arousal, relay station for various autonomic functions, relay station between cortex and cerebellum; contains the locus coeruleus, a center that produces norepinephrine and regulates stress and panic
Medulla oblongata	Autonomic functions, relay between brain and spinal cord
Substantia nigra	Interface between cortex and brain stem controlling voluntary movement; mood, addiction (possibly)
Other structures	
Cerebellum	Latin for "little brain"; coordination of sensory perception and motor control, fine-tuning of motor movements in response to sensory feedback

Brain tissue contains two general cell types, neurons and **glial cells**, or **glia**. Brain neurons are formed to perform a specific task at a specific location, and once formed, they cannot be replaced. That is, brain neurons are terminally differentiated. The principle of terminal differentiation ensures that the neuron performs the task it was originally designed to carry out and will not be replaced by a different type of neuron. Although the number of neurons in the adult human brain varies among individuals, most estimates place the number at 1×10^{12}.

The glia (Greek for "glue"), the other neurological cell type in the brain, provide support and maintenance for the neurons and outnumber neurons by about 10:1. The glial cells occur in two forms, the **neuroglia** and the **microglia**. Neuroglia can be subdivided into various cell types and classified by their function. The most abundant type of neuroglia is the astrocytes, also known as astroglia. Astrocytes maintain the proper extracellular environment for neurons by regulating ion levels and recycling neurotransmitters. Recent evidence suggests that astrocytes may also form the basis of the blood-brain barrier by regulating vasoconstriction and vasodilation. Oligodendrocytes, another type of neuroglia, produce myelin, the protein that coats axons and improves electrical conduction by providing insulation.

Microglia are mobile neuromacrophages capable of phagocytosis and initiating the inflammatory response. Although microglia make up less than 15% of all cells in the brain, they perform a critical function that may be important to aging—they respond to damage by removing injured or nonfunctional neurons. The failure to remove damaged brain tissue may establish conditions leading to various age-related neuropathologies.

In addition to glia, the brain is protected from insult by the blood-brain barrier, a physiological mechanism that places a barrier between circulating blood and brain tissue. This barrier alters the permeability of brain capillaries so that large bloodborne molecules or particles, such as bacteria, do not pass into the brain tissue (infections of the brain are rare). The altered permeability of brain capillaries arises from structures called tight junctions. Tight junctions increase the points of cohesion between the endothelial cells of the brain capillaries, reducing the space for diffusion through the capillary wall. Recent evidence suggests that alterations to the blood-brain barrier may contribute to the progression of Alzheimer's disease.

AGE-RELATED DISEASES OF THE HUMAN BRAIN: ALZHEIMER'S DISEASE AND PARKINSON'S DISEASE

Compared with what scientists know about other human organs, they have only a rudimentary understanding of how the human brain performs its functions. We know even less about how aging affects brain function. Nonetheless, we do know that the healthy human brain seems to retain significant function and to show little structural change throughout most of the adult life span. In cases where changes have been reported, there is significant variation among individuals with respect to the amount of age-related dysfunction. Moreover, the anatomical location in the brain where age-related functional loss occurs also differs significantly among individuals. Thus, generalities about age-related functional loss in the human brain are difficult to make. You will see, however, that age-related alterations in the human brain, although minor, have the potential to progress to devastating neuropathology.

In this section, we focus exclusively on the two most common age-related diseases that affect the CNS: Alzheimer's disease and Parkinson's disease. (Age-related diseases of the PNS, the nerves that carry information to and from the CNS, have not been well studied and are not covered here.)

Changes in structure and neurotransmission seem to be minor in the aging brain

Structural changes in the brain at advanced ages are consistently found to be minor, to vary among individuals, and to not affect the same anatomical location in all people. Moreover, the ability of the brain to develop new neural connections (that is, its **neuroplasticity**) remains remarkably robust in the aging human brain. In fact, studies using magnetic resonance imaging (MRI) find only small changes in brain size with aging. The lack of brain atrophy reflects two general observations: (1) neural cell number remains fairly constant throughout the adult life span, and (2) glial cell formation slightly increases (in a process called gliosis) with age. When neural cell loss has been reported, it appears to be limited to specific brain centers such as the hippocampus, locus coeruleus, and cerebellum. The importance of age-related neural cell loss to brain function remains obscure, because slight neural cell loss does not correlate well with decreased function.

There also seem to be only minor age-related changes in the structures involved in neurotransmission, including synaptic density, synaptic size, and volume of the synaptic cleft. Although determining these values for large populations of aged individuals is technically difficult, and reports of a post-reproductive decline in synaptic quality have generally been limited to brains derived from autopsy, synaptic quality appears to be highly maintained, except in the hippocampus, where synaptic size and volume decline. However, functional loss due to changes in synaptic quality have not been demonstrated.

Current evidence also suggests that concentrations of neurotransmitters, such as acetylcholine, norepinephrine, epinephrine, and dopamine, do not decrease significantly with age. However, measurements can be done only indirectly in large populations by measuring neurotransmitter concentrations in blood and urine. New techniques using labeled probes in combination with imaging technology are being introduced, but the population of aged individuals in which these techniques have been applied remains too small for definitive conclusions.

Amyloid plaques and neurofibrillary tangles accumulate in the aged brain

The aging human brain appears to have a diminished capacity to adequately rid itself of damaged proteins, and this results in an accumulation of potentially toxic compounds. Two of these neurotoxic compounds, **amyloid plaques** and **neurofibrillary tangles**, have been identified as precursors to neurological diseases such as Alzheimer's disease and Parkinson's disease **(Figure 9.7)**. However, accumulation of these compounds in the healthy aging brain does not seem to have a significant effect on functional capacity in the majority of

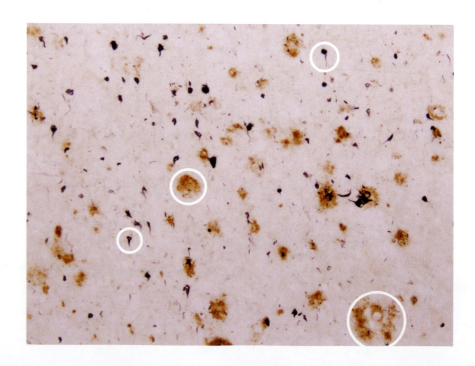

Figure 9.7 Amyloid plaques (*large circles*) and neurofibrillary tangles (*small circles*) in the brain tissues of an individual who had Alzheimer's disease. (From E.G. McGeer and P.L. McGeer, *Mol. Interv.* 1:22–29, 2001. With permission from American Society for Pharmacology and Experimental Therapeutics.)

post-reproductive individuals. Only about 10–20% of aged individuals will experience a transition from normal aging to definable neurological disease as a result of the accumulation of amyloid plaques and neurofibrillary tangles.

More than 150 years ago, the father of modern pathology, Rudolf Virchow, described a "waxy substance" in the brain of older individuals that had the appearance of a starchlike compound. He named this substance amyloid. (The structure of the amyloid protein is not related to the carbohydrate family, but the name remains.) For the next 130 years, detailed descriptions of amyloid plaques were published. Then, in 1984, the amino acid sequence of the protein forming the amyloid plaques, **Aβ protein**, was elucidated. (The β designation refers to the protein's secondary structure, a **β-sheet**, a rigid structure formed by hydrogen bonding, as shown in **Figure 9.8**.) The amyloid fibrils (plaque) are made up of several Aβ proteins wrapped around each other to create a highly insoluble molecule. The formation of amyloid plaques from Aβ protein in brain tissue is a normal part of human aging.

Analysis of the amino acid sequence of the Aβ protein led to the discovery of the gene that encodes the protein. The gene was found to be part of a larger gene, localized to Chromosome 21, that encodes the **amyloid precursor protein (APP)**. APP has a large hydrophilic extracellular domain, a transmembrane domain consisting of 23 amino acids, and a small intracellular domain **(Figure 9.9)**. The expression of APP occurs in both neural and nonneural tissue. Although the function of APP has yet to be fully elucidated, physiological studies suggest that APP supports dendrite outgrowth, synaptogenesis, and inhibition of platelet activation, and it may function as a copper transport protein.

Synthesis of APP in neural tissue takes place in the endoplasmic reticulum, with post-translational modification occurring during transit to its position in the cell membrane. The processing of APP within the membrane can follow one of two proteolytic pathways (Figure 9.9). The non-amyloidogenic, or α, pathway produces the protein, p3 (3 kD), whose function and metabolism remain largely unresolved. The amyloidogenic, or β, pathway produces the Aβ protein. Enzymes known as secretases determine which proteolytic pathway a membrane-bound APP will follow. In nonneural tissue, α-secretase predominates,

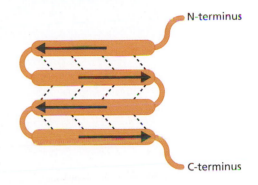

Figure 9.8 β-sheet secondary structure of a protein. *Arrows* indicate the direction (N-terminus to C-terminus) of the amino acid sequence. *Dotted lines* indicate the hydrogen bonds that hold the sheets together. Notice the antiparallel pattern of the hydrogen-bonded sequences. This configuration gives the β-sheet structure its rigidity.

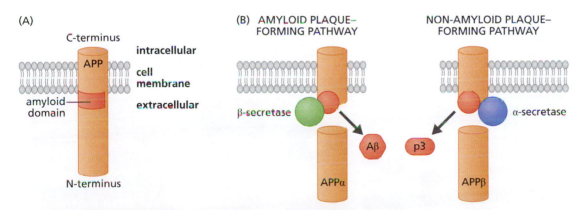

Figure 9.9 The amyloid precursor protein (APP) and formation of the Aβ protein. (A) APP contains an amyloid domain (40–42 amino acid residues) that is mainly located in the extracellular space, with just a small hydrophobic region within the membrane. (B) β-secretase, an enzyme found primarily in neural tissue, cleaves APP just below the amyloid domain. This leaves a small intramembrane protein with a conformation favoring formation of the Aβ protein. Cleavage of APP by α-secretase, the enzyme found primarily in nonneural tissue, favors the formation of the nontoxic p3 protein.

resulting in the production of the nontoxic p3 protein. In contrast, in neural tissue, β-secretase (also known as β-site APP cleavage enzyme 1, or BACE1) predominates, resulting in formation of the potentially neurotoxic Aβ protein.

Production of Aβ in neural tissue leads to the formation of amyloid plaque, a highly insoluble structure with no known catabolic pathway **(Figure 9.10)**. The pathway from Aβ to amyloid plaque has not been described in detail, although a general theory has begun to emerge. It appears that Aβ polymerizes through sequential formation of dimers, tetramers, and long oligomers. The oligomers aggregate into a stacked conformation, known as a β-cross formation, forming a fibril. Several fibrils then aggregate to form the plaque. Enzymes catalyzing Aβ polymerization have not been identified, suggesting that aggregation is nonenzymatic and concentration dependent. Aβ amyloid plaques are found primarily in the extracellular space, as would be predicted from the APP-processing mechanism. However, recent studies have found low concentrations of intracellular Aβ amyloid plaques in the Golgi complex and endoplasmic reticulum.

The limited understanding of the mechanism underlying amyloid plaque formation reflects the unique ability of Aβ to form aggregates. The formation of protein aggregates in biological systems is the exception rather than the rule and most likely reflects an abnormally folded precursor protein—in the current scenario, the Aβ protein. The cell invests considerable energy into ensuring that the folding process occurs without error and independent of other proteins, a critical control mechanism needed to maintain the structure-function relationship of protein activity. Several intracellular proteins, many of which are part of the chaperone family, support the protein-folding process by inhibiting aggregate formation or by molecularly marking for degradation those proteins that have formed an aggregate. Such chaperone proteins do not seem to exist for the Aβ protein, resulting in an aggregate—that is, the amyloid plaque—that resists proteolysis.

Neurofibrillary tangles are insoluble twisted fibers found inside brain cells. The **microtubules** of the **cytoskeleton** perform a vital function by directing the movement and determining the final position of organelles and/or proteins, from the cell body to the axon. Under normal conditions, binding of the **tau protein** to the microtubule provides stability to this cellular structure **(Figure 9.11)**. The protein's function

Figure 9.10 Amyloid plaque formation. *Left to right:* At high concentrations, the Aβ$_{42}$ protein cleaved from the APP complex (see Figure 9.9) forms aggregates by sequential polymerization. The polymerization process leads to the formation of a fibril in the β-cross configuration—a configuration that resists proteolysis. This induces further aggregation into the stacked configuration of the amyloid plaque.

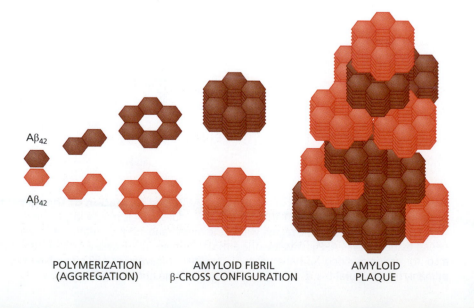

Aβ$_{42}$

Aβ$_{42}$

POLYMERIZATION
(AGGREGATION)

AMYLOID FIBRIL
β-CROSS CONFIGURATION

AMYLOID
PLAQUE

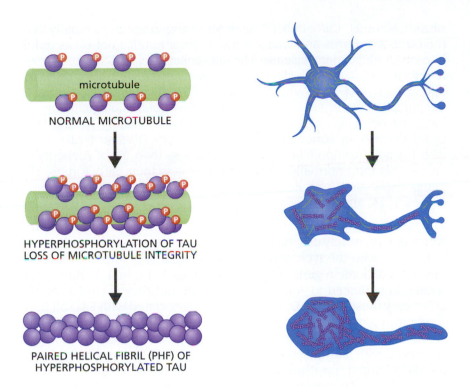

NORMAL MICROTUBULE

HYPERPHOSPHORYLATION OF TAU
LOSS OF MICROTUBULE INTEGRITY

PAIRED HELICAL FIBRIL (PHF) OF
HYPERPHOSPHORYLATED TAU

Figure 9.11 Formation of paired helical fibrils that result in neural fibrillary tangles and neural degeneration. The microtubules of the cytoskeleton maintain their integrity by means of microtubule-associated proteins (MAPs), such as the tau protein shown here. Tau proteins bind to the microtubule surface and perform their function through phosphorylation and dephosphorylation by kinases and phosphatases. If the tau proteins become hyperphosphorylated, the microtubules lose their integrity, leading to neural degeneration. Hyperphosphorylated tau proteins aggregate into insoluble structures called paired helical fibrils (PHFs), which are the primary component of neurofibrillary tangles.

as a support protein depends on its degree of phosphorylation, a process regulated by various kinases and phosphatases. Normal tau has 2–3 moles of phosphate per mole of protein. However, histological studies have shown that brain tissue taken from neurologically healthy aged individuals contains low concentrations of tau protein having phosphate binding at four to five times the normal level. The hyperphosphorylated tau proteins weaken the integrity of the microtubule, increase the risk of neurodegeneration, and lead to the formation of neurofibrillary tangles.

Microscopic observation shows that hyperphosphorylated tau protein aggregates to form **paired helical fibrils (PHFs)**, which aggregate further, leading to a breakdown in microtubule structure and disruption of intracellular transport. In turn, organelles important in maintaining the integrity of the neuron begin to break down, causing cell degeneration and dysfunction. Ultimately, PHFs replace the microtubule network, resulting in complete neural cell degeneration.

Paired helical fibrils are highly insoluble and cannot be degraded by the microglia, and thus they remain permanently affixed to the brain's extracellular matrix. Controversy exists about the effect of tangles on brain function. One theory suggests that tangles are inert and simply represent the end process of age-related neural degradation. Others suggest that tangles have a toxic effect on healthy nerve cells, which, in turn, induces further formation of PHFs.

Alzheimer's disease is an age-related, nonreversible brain disorder

Alzheimer's disease, named after Alois Alzheimer (1864–1915), is a type of age-related dementia that causes problems with memory, thinking, and behavior. There are three primary types of Alzheimer's: early-onset, late-onset, and familial. Early-onset is a rare form of Alzheimer's in which people are diagnosed with the disease before age 65. Most cases occur in 40- to 50-year-old patients with Down's syndrome, an observation implicating the involvement of genes on

Chromosome 21. Early-onset Alzheimer's progresses more rapidly than the other two forms and leads to brain abnormalities not observed in late-onset Alzheimer's disease. The late-onset disease, the most prevalent form (80–90% of all Alzheimer's cases), normally begins after the age of 65 **(TABLE 9.2)**. Whether late-onset Alzheimer's disease has a genetic component remains unknown. Most epidemiological surveys have not been able to show a significant family history for the late-onset disease. In contrast, familial Alzheimer's disease (FAD) has a direct genetic component; it accounts for less than 5% of Alzheimer's cases. This form normally affects individuals in their early forties. The genes associated with FAD appear to be located on three chromosomes: 1, 14, and 21.

The principal features of all three types of Alzheimer's disease are the same and are widely described, but the cause appears to be multifactorial, varies with the type of disease, and remains unknown. As noted above, mutations in genes on Chromosomes 1, 14, and 21 have been shown to be linked to Alzheimer's disease, but they seem to directly affect only those with FAD. The mutations associated with FAD all have a role in the processing of APP and may lead to the accumulation of amyloid plaques. Neither genetics nor family history has been strongly correlated with late-onset Alzheimer's. Nonetheless, a **polymorphism**, an allele variant, on Chromosome 19 has been identified as increasing the risk of developing Alzheimer's. Inflammation, oxidative stress, and disruption in the production of neurotransmitters have also been suggested as possible causes of Alzheimer's disease. The evidence supporting a direct link between these potential causes and Alzheimer's is not yet strong enough to warrant inclusion here.

Alzheimer's disease begins in the entorhinal cortex and progresses into the cortex

All types of Alzheimer's disease show a similar progression of symptoms and brain pathology **(Figure 9.12)**. The early or pre-clinical pathology begins in the entorhinal cortex, a small structure at the base of the hippocampus that is responsible for relaying information between the cortex and hippocampus. Some estimates suggest that pre-clinical pathology can begin 10–15 years before symptoms appear. The mild loss of function in the hippocampus during the pre-clinical stage leads to some memory loss, a symptom that may be noticed only in retrospect, after diagnosis. As the disease progresses into the mild or clinical stage, cortical shrinkage occurs in the frontal, temporal, and occipital lobes. The symptoms that begin during this stage reflect effects

TABLE 9.2

ESTIMATES OF NUMBER OF INDIVIDUALS WITH LATE-ONSET ALZHEIMER'S DISEASE, UNITED STATES, 2002–2005

Age	Total	Men	Women
71–79	332,000 (2.32)	148,000 (2.30)	184,000 (2.33)
80–89	1,493,000 (18.1)	409,000 (12.33)	1,084,000 (21.34)
≥90	558,000 (29.7)	190,000 (33.89)	368,000 (28.15)
Total (≥71)	2,383,000 (9.7)	747,000 (7.05)	1,636,000 (11.48)

From B.L. Plassman, K.M. Langa, G.G. Fisher et al., *Neuroepidemiology* 29:125–132, 2007.

Note: Values in parentheses are percentages of the total population having Alzheimer's disease. Average age at diagnosis, 75 years; average length of life after diagnosis, 10 years; estimated yearly cost for care of all Alzheimer's patients, $500 million to $1 billion; estimated yearly cost for an individual with Alzheimer's, $18,000–36,000; predicted number of individuals with Alzheimer's in 2040, 6–7 million.

PRE-CLINICAL TO MILD   MILD TO MODERATE   SEVERE

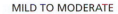

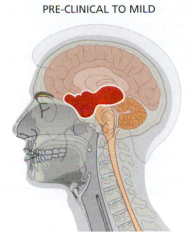
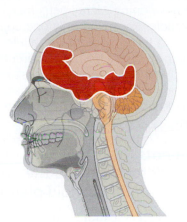
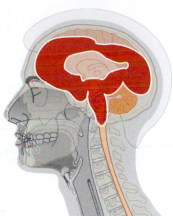

in areas of the cortex (see Table 9.1). That is, loss of neurons in the cortical areas results in declining language skills (temporal lobe), mild loss in reasoning (frontal lobe), and hallucinations (occipital lobe). The patient also experiences changes in circadian rhythm that may lead to day-night reversal, such as sleeping during the day and being awake at night. The day-night reversal disappears in many Alzheimer's patients as the disease progresses to more advanced stages. The patient's sense of personality and relationship to the world also begins to deteriorate during the mild stage. Amyloid plaques and neurofibrillary tangles spread into the brain stem, affecting autonomic function. In particular, the locus coeruleus, located in the pons, becomes affected, causing a decline in the ability to regulate stress and panic.

In the final stage of Alzheimer's disease, also referred to as the demented stage, the hippocampus is half the size of that found in a normal brain, resulting in complete loss of memory. Severe neurodegeneration in the cortex leads to an inability to speak or understand any form of communication. The patient loses all sense of self. The cerebellum atrophies, and the patient can no longer control his or her movements, resulting in confinement to bed. In the 2–3 months prior to death, the pathology spreads into the medulla oblongata, affecting basic autonomic functions such as bowel and bladder control, heart rate, swallowing, and respiration. The patient's inability to breathe properly and expel material from the lungs often leads to pneumonia and death.

Figure 9.12 Progression of Alzheimer's disease from pre-clinical to severe. Alzheimer's begins subtly, and without symptoms, in the entorhinal cortex, which lies just below the hippocampus. With the transition from mild to moderate Alzheimer's disease, the hippocampus shrinks considerably and neurofibrillary tangles and amyloid plaques begin to appear in the cortex. Areas in the upper brain stem are also affected during the moderate stage. In the final stage of Alzheimer's, large areas of the cortex are involved. The disease spreads into the autonomic centers of the medulla oblongata, causing difficulties in respiration. Death due to pneumonia or heart failure is common.

The ε4 allele of the apolipoprotein E gene is a risk factor for late-onset Alzheimer's disease

Although the cause of Alzheimer's disease remains unknown, a genetic variant of the lipid-binding protein **apolipoprotein E (ApoE)** has been found to be more frequent in patients with late-onset Alzheimer's. The *APOE* gene resides on Chromosome 19 and has three common alleles, ε2 (protective against Alzheimer's), ε3 (the most frequent in Caucasians), and ε4 (a risk factor for Alzheimer's). Synthesis of ApoE occurs in the liver, and its function has been primarily associated with the transport of cholesterol in the blood (the importance of this to cardiovascular health is discussed later in the chapter). In the brain, ApoE is synthesized by astrocytes and microglia, and here it has a major role as an extracellular lipid carrier and in the maintenance of the microvasculature.

Approximately 40–60% of individuals who have at least one copy of the ε4 allele develop late-onset Alzheimer's disease. That is, having the ε4 allele does not *cause* Alzheimer's disease, it only increases the risk. The risk of developing Alzheimer's increases with the number of copies of the ε4 allele. Receiving only one copy increases the risk by 30% compared with those without this allele variant. A person who receives a copy of the ε4 allele from both parents has a 50–60% chance of developing the disease. These risk factors are striking when compared with the 9% lifetime risk for Alzheimer's disease in people without the ε4 allele. The biochemical mechanism underlying ApoE's impact on Alzheimer's remains, by and large, a mystery.

Treatments for Alzheimer's disease target neurotransmission and the prevention and degradation of amyloid plaques

In general, medications currently in use for treating patients with Alzheimer's focus on helping to delay symptoms or prevent them from becoming worse for a limited period of time. Current medications do not cure or prevent the progression of the disease. The medications currently prescribed for mild to moderate Alzheimer's are known as cholinesterase inhibitors and most likely work by inhibiting the breakdown of acetylcholine, the neurotransmitter important for memory and cognition. The three most widely prescribed cholinesterase inhibitors are galantamine (Razadyne), rivastigmine (Exelon), and donepezil (Aricept®)). Because Alzheimer's is characterized by a progressive loss in the brain's ability to synthesize acetylcholine, cholinesterase inhibitors are effective for just a limited time. The medication used for the treatment of moderate to severe Alzheimer's is memantine (Namenda®). The precise reason that memantine helps Alzheimer's patients think more clearly remains obscure, but the chemical is known to prevent accumulation of glutamate in the synaptic cleft. Glutamate, a neurotransmitter normally found in areas of the brain associated with cognition, can be toxic to neurons in abnormally high amounts.

Ongoing research into new treatments and therapies for Alzheimer's disease has focused primarily on preventing the formation of protein aggregates or degrading them once they have formed. One hypothesis posits that plaque formation may be associated with altered immune function. This hypothesis was tested by injecting antibodies to Aβ protein into transgenic mice that overexpress Aβ protein and accumulate amyloid plaques. Remarkably, the plaque concentration decreased significantly in mice receiving the antibodies compared with those receiving a placebo injection. Clinical trials in humans using a similar approach are currently underway.

Some of the most recent research has linked two important questions concerning the development of Alzheimer's disease: (1) Why do all aging brains show some accumulation of amyloid plaques, but only a small percentage experience the widespread accumulation associated with Alzheimer's? (2) What is the mechanism underlying the increased risk of Alzheimer's associated with the *APOE* ε4 (*APOE4*) gene variant? The answers to these questions may lie in an abnormal disruption in the blood-brain barrier. Recall that the blood-brain barrier protects the brain by preventing the entry of large molecules. We also know that the most common *APOE* variant, *APOE2*, expresses a protein that supports vessel integrity, whereas the protein expressed by the *APOE4* variant seems to injure capillary cells. Using this knowledge, researchers found that transgenic mice overexpressing *APOE4* showed increased damage to cells of the tight junctions that form the blood-brain barrier when compared with *APOE2* transgenic mice. The resulting damage would cause holes in the tight junctions and allow large molecules in the blood, such as the Aβ protein, to breach the blood–brain barrier. If correct, these findings would help to solve the mystery of why excess amyloid plaque accumulation occurs in Alzheimer's disease but not in normal aging. Most important, this research found that the damage caused by the protein product of *APOE4* responded to pharmacological treatment in the transgenic mice, suggesting that similar treatments could be developed for humans.

Parkinson's disease is associated with loss of dopaminergic neurons

Parkinson's disease is an age-related, motor system disorder and usually affects people over the age of 50. The primary symptoms of Parkinson's—tremor or trembling of the hands, arms, legs, jaw, and face; **bradykinesia**, or slowness of movement; muscular rigidity; and postural instability—reflect the loss of dopamine-producing neurons in the substantia nigra region of the basal ganglia **(Figure 9.13)**. Patients often experience weakness and tremor on one side of the body only. Moreover, some individuals notice a "feeling of tremor" within the deep portions of large muscle groups. Tremor becomes greater during times of strong emotions, such as sexual arousal or anxiety, but subsides after the emotion has passed and returns to its normal level. An inability to create facial expressions that match the person's emotional state (smile, frown, and so on.) begins to appear during the early stages of Parkinson's disease, as does a slight change in the voice. Both effects are caused by rigidity and slowing of muscles in the head and neck. A slight stooped posture occurring during the later stage of early Parkinson's results in a feeling of being out of balance and creates difficulty in walking.

As Parkinson's disease progresses to the moderate stage, the motor function impairments become more severe. The muscle rigidity and cramps that accompany loss of motor function can cause considerable

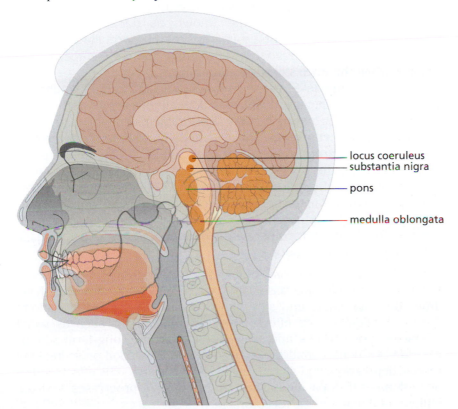

locus coeruleus
substantia nigra

pons

medulla oblongata

Figure 9.13 Location of the substantia nigra and locus coeruleus. Note the close proximity of the locus coeruleus to the substantia nigra. Declining synthesis of the neurotransmitter dopamine in the substantia nigra results in the early symptoms of Parkinson's disease. Loss of norepinephrine-producing neurons in the locus coeruleus occurs in Alzheimer's disease. Many Alzheimer's patients develop Parkinson's disease, and many Parkinson's patients develop Alzheimer's disease.

pain. While many of the effects caused by the declining ability of neurons in the substantia nigra to synthesize dopamine can be significantly reduced through drug therapy, the autonomic centers in the brain stem that synthesize acetylcholine and/or norepinephrine begin to show dysfunction. Parkinson's patients often suffer from constipation, overproduction of saliva (drooling), lack of bladder control, and an inability to regulate body temperature. Finally, some individuals might suffer from what clinicians call day-night reversal. That is, the patient is awake during the night and sleeps during the day.

Although large-scale epidemiological investigations of Parkinson's disease have not been completed, estimates from several small population studies indicate that the **prevalence** of Parkinson's worldwide increases after age 50. Indeed, age remains the only widely accepted risk factor for Parkinson's. Occurrence of the disease in people under 50 years of age reflects a rare genetic/family form of Parkinson's that is caused by mutations in genes found primarily on Chromosomes 1, 4, and 6. Controversy exists as to the variation of disease prevalence with respect to ethnicity/race and gender. For example, African Americans seem to have the highest rate of Parkinson's disease in the United States. However, the prevalence of Parkinson's in Nigeria is among the lowest in the world, suggesting that factors other than race account for the high rate of Parkinson's in African Americans. No clear difference has been shown in the prevalence of Parkinson's in men and women of similar age.

Increasing the brain's concentration of dopamine is the primary objective in treatment of Parkinson's disease

The cause of Parkinson's disease is loss of the neurotransmitter dopamine, and there are no treatments for reversing or curing this loss. Increasing the level of dopamine in the brain has proved effective in reducing the symptoms of Parkinson's, and there are two general approaches to this treatment. During the early stages of the disease, many physicians prescribe drugs known as dopamine agonists, which stimulate the receptors in nerves that are normally stimulated by dopamine. These drugs work only during the early stages of the disease, when there are still sufficient numbers of dopaminergic receptors/nerves to stimulate. As the disease progresses and dopaminergic receptors decline, the effectiveness of dopamine agonists declines. At this point, the focus of drug therapy shifts to increasing the concentration of dopamine in the brain. However, dopamine cannot pass through the blood-brain barrier. A precursor to dopamine, called **levodopa** or L-dopa, can cross the blood-brain barrier and is converted to dopamine in the brain, through a reaction catalyzed by DOPA decarboxylase. This reaction also occurs in the peripheral nervous system, however, resulting in unwanted side effects such as nausea and vomiting. In most cases, L-dopa is given together with a peripheral DOPA decarboxylase inhibitor known as **carbidopa**.

While the L-dopa–carbidopa combination is extremely effective at reducing the symptoms associated with Parkinson's disease, these drugs do have limits and side effects that are progressive. Among the most frequent side effects are jerky movements such as facial grimacing, postural tics, and exaggerated chewing. Long-term administration of L-dopa–carbidopa often results in low blood pressure, skin rashes, depression, and alterations in sleep patterns. The effectiveness of L-dopa–carbidopa diminishes as Parkinson's progresses and the number of dopaminergic receptors/nerves decreases.

Lewy bodies are the pathological hallmark of Parkinson's disease

The cause of Parkinson's disease—the reason for the loss of dopaminergic neurons—remains unknown, and early diagnosis is difficult, due in large part to the slow progress of the disease and clinical signs that can be confused with other neurological disorders. We have discussed how the accumulation of damaged or misfolded proteins in brain tissue is often associated with aging and age-related disease. Such proteins are also found in association with Parkinson's disease. These proteins, known as **Lewy bodies**, are highly insoluble aggregates of fibrous proteins that appear in the cytoplasm of neurons. Although Lewy bodies can be found in aged humans without Parkinson's, their accumulation in the substantia nigra and locus coeruleus is a pathological hallmark of the disease. The major components of Lewy bodies are two proteins that normally participate in the maintenance of protein structure, **ubiquitin** and **α-synuclein**. Ubiquitin, a small (76 amino acid) heat shock protein, attaches to misfolded or damaged proteins to mark them for degradation. The precise function of α-synuclein remains unclear, but most researchers agree that this protein plays a vital role in the maintenance and regulation of dopamine vesicles at the synaptic terminals. Its β-sheet conformation involving specific amino acid residues also promotes the formation of β-sheet aggregates, if the protein does not undergo degradation. These aggregates form the primary component of Lewy bodies. As in amyloid plaque accumulation in Alzheimer's disease, the accumulation of Lewy bodies in the brain in Parkinson's does not cause the disease but seems to result from it.

Several genes are associated with early-onset Parkinson's disease

Early-onset Parkinson's disease, occurring before the age of 50, accounts for less than 1% of all Parkinson's cases and may be directly linked to family history and mutations in several genes. Mutations linked to early-onset Parkinson's have been discovered at several loci, but genes associated with the proteins that constitute the majority of Lewy bodies, ubiquitin (the *Parkin* gene, on Chromosome 6) and α-synuclein (a gene on Chromosome 4), have drawn the most attention. The protein product of the *Parkin* gene functions as a component of ubiquitin ligase, the enzyme that adds ubiquitin to other proteins. A mutation in the *Parkin* gene disrupts the normal pathway that marks proteins for degradation, thereby allowing damaged or misfolded proteins to accumulate and protein aggregates to form. Interestingly, the primary protein aggregates found in Parkinson's disease, the Lewy bodies, are not found in the neurons of patients with the *Parkin* gene mutation. Rather, current research suggests that the mutation in the *Parkin* gene causes mitochondrial dysfunction and overproduction of reactive oxygen species, leading to cell death.

Regardless of the type of mutation, in the *Parkin* gene or in the α-synuclein gene, only about 50% of cases of early-onset Parkinson's disease can be linked to these mutations. Moreover, people with the late-onset disease, the type accounting for 99% of all cases of Parkinson's, rarely have this mutation.

Several factors may predispose individuals to Parkinson's disease

Parkinson's disease, as we've seen, does not have a clear and unequivocal cause. It is a multifactorial disorder involving both genetic and environmental factors. Mutations of genes leading to early-onset Parkinson's strongly suggest a genetic component. The first clue that the environment may contribute to the disease came when drug users, in an attempt to synthesize the morphine-like drug meperidine, instead produced a neurotoxin, methyl-phenyl-tetrahydropyridine (MPTP). MPTP selectively kills dopaminergic neurons in the substantia nigra and other parts of the brain stem. Thus, when these drug users took what they believed to be meperidine, they developed an irreversible Parkinson's-like condition. Subsequent studies investigating the connection between MPTP and Parkinson's suggested that this compound inhibits a biochemical pathway in the mitochondria that protects neurons from oxidative damage. While the association between oxidative damage and Parkinson's remains only correlative, exposure to other agents that increase oxidative stress, such as Paraquat or high levels of iron and manganese, has been shown to increase the risk of developing Parkinson's.

THE CARDIOVASCULAR SYSTEM

In this section, we consider a few basic physiological properties that govern blood flow and the subsequent oxygenation of tissue in the normally functioning cardiovascular system. We are taking this approach because impedance to blood flow (and thus oxygenation) is the major factor underlying age-related and/or disease-related decline in many organs and tissues.

The cardiovascular system is a closed system of fluid transport

The **cardiovascular system**, like all closed systems of fluid transport, consists of a central pump (the heart) and conduits (arteries and veins) that carry the fluid to target structures (cells) and back to the pump. The heart consists of two separate pumps: the right side delivers oxygen-poor blood to the pulmonary system (lungs), and the left side supplies oxygen-rich blood to the body **(Figure 9.14)**. Each side of the heart has two chambers, the **atrium** and the **ventricle**. The atrium acts as a primer or regulator of blood volume for the ventricle. The right atrium receives oxygen-poor blood from the **superior vena cava**, and the left atrium receives oxygen-rich blood from the **pulmonary vein**. The atria are separated from the larger and more powerful ventricles by one-way valves (the **tricuspid valve** on the right side and the **mitral valve** on the left side). The ventricles eject blood from the heart through the **aortic valve** on the left and the **pulmonary valve** on the right.

The circulatory system, like the heart, consists of two separate but connected systems, the **pulmonary circulation** and the **systemic circulation (Figure 9.15)**. Vessels leaving the heart are called **arteries** and vessels entering the heart are called **veins**. The circulatory system is organized so that, in the direction of blood flow, arteries become progressively smaller and veins become progressively bigger. This arrangement has evolved so that the exchange of gases between the red blood cells and tissue cells occurs efficiently. The lumen of the capillaries (the smallest vessels), where gas exchange occurs through the semi-permeable walls, has a diameter equal to that of one red blood cell, resulting in a very low ratio of dead space to surface area.

The architecture of arteries is significantly more complex than that of veins, because arteries help maintain blood pressure and flow to the tissue **(Figure 9.16)**. Arteries have a significant amount of circular

Figure 9.14 Anatomy of the human heart.

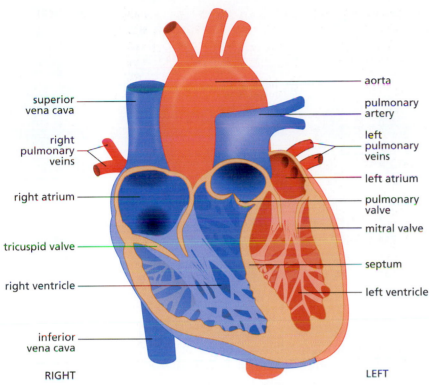

superior vena cava

aorta

pulmonary artery

right pulmonary veins

left pulmonary veins

left atrium

right atrium

pulmonary valve

tricuspid valve

mitral valve

septum

right ventricle

left ventricle

inferior vena cava

RIGHT

LEFT

smooth muscle that, when it contracts, causes a narrowing of the lumen. The smooth muscle responds to neural signals from the brain when systemic blood pressure needs to be increased or blood flow to the extremities needs to be shut down (such as during cold exposure; see Chapter 8). Veins do not use pressure to move the blood back to the heart. Rather, they maintain the flow by one-way valves. When the heart beats and ejects blood, the pressure in the arteries, called **systolic pressure**, causes the valves to open for blood flow. When the heart is at rest, the drop in pressure in the arteries, called **diastolic pressure**, causes the valves to close so that the blood in the veins does not flow backward. Skeletal muscle contractions also help veins "pump" blood from areas below the heart. Abnormal age-related alterations in the smooth muscle of the arteries and in the valves of the veins in the extremities result in an age-related disease known as congestive heart failure.

The heart and arteries are excitable tissues

Both the heart and the arteries use the contraction of muscle tissue to generate pressure and flow. Muscle tissue, like nerve tissue, is an excitable tissue that generates action potentials to propagate the contraction from muscle cell to muscle cell. The heart muscle (myocardium) must follow a precise pattern of contraction to operate at peak performance. This pattern begins when the **sinus (sinoatrial) node**, a small group of specialized cells at the top of the right atrium, generates an action potential **(Figure 9.17)**. The action potential spreads into the adjacent heart muscle cells (myocytes) and causes simultaneous contraction of the right and left atria. The **atrioventricular (AV) node** picks up the electrical discharge and transmits the signal through the AV bundle and into the AV-bundle branches. Spreading the action potential via the AV node rather than by direct contact between muscle fibers of the atria and ventricles causes a delay of about one-sixth of a second. This delay

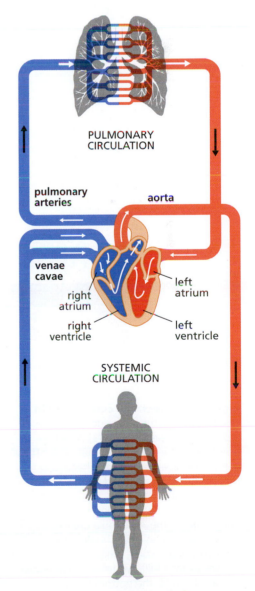

PULMONARY CIRCULATION

pulmonary arteries

aorta

venae cavae

right atrium

left atrium

right ventricle

left ventricle

SYSTEMIC CIRCULATION

Figure 9.15 The systemic and pulmonary circulations. Oxygenated blood is represented by *red*; deoxygenated blood, by *blue*.

Figure 9.16 Anatomy of arteries and veins. The arteries contain significantly more elastic and connective tissue than the veins. This type of tissue, along with the considerable amount of smooth muscle, has two functions: (1) it allows the arteries to participate with the heart in maintaining blood pressure and flow; and (2) it maintains the structure of the vessels. The connective tissue and smooth muscle of the veins maintain the structure of these vessels.

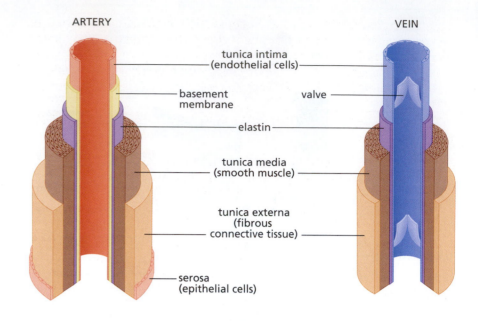

ARTERY

VEIN

- tunica intima (endothelial cells)
- basement membrane
- valve
- elastin
- tunica media (smooth muscle)
- tunica externa (fibrous connective tissue)
- serosa (epithelial cells)

Figure 9.17 Electrical conduction in the heart. Electrical conduction follows a pattern that begins with (1) an action potential in the sinus (sinoatrial) node. The action potential spreads simultaneously throughout both atria and into (2) the atrioventricular (AV) node. Propagation of the action potential through the AV node and (3) down the AV bundle branches delays contraction of the ventricles by one-sixth of a second, long enough for the ventricles to fill. As the action potential spreads from (4) the bottom of the ventricles to (5) the top, the high speed of transmission causes all fibers to contract at virtually the same moment.

sinus node

atrioventricular node

AV bundle

right AV-bundle branch

left AV-bundle branch

allows the ventricles to fill with blood prior to contraction. Control of cardiac tissue contraction—the heartbeat—occurs primarily through nerve fibers from the brain stem but can be modulated by hormones such as catecholamines.

Whereas cardiac muscle contraction is controlled by neural stimulation, contraction of the smooth muscle of the arteries is controlled primarily by hormones and events occurring locally in the arteries **(TABLE 9.3)**. Contraction of the smooth muscle of an artery, called **vasoconstriction**, results in a decrease in lumen diameter. Relaxation of the smooth muscle, called **vasodilation**, causes an increase in diameter. We can use the example of an increase in metabolic activity to demonstrate how vasoconstriction and vasodilation work together to supply the necessary blood to an organ. At rest, the arteries of a skeletal muscle are partially vasoconstricted—that is, they have a relatively low need for blood flow. When exercise begins and the muscle begins to contract, catecholamines are released and bind to alpha-adrenergic receptors, causing vasodilation of the arteries (increasing blood flow). This neurohormonal response lasts only for the first few seconds to minutes of the exercise. Vasodilation and increased blood flow during sustained

TABLE 9.3

SOME VASODILATORS AND VASOCONSTRICTORS THAT AFFECT ARTERIAL BLOOD FLOW

Agent	Type	Effect
Sympathetic nerves	Neural	Vasoconstrictor
Nitric oxide	Neural	Vasodilator
Epinephrine	Hormonal	Vasodilator
Angiotensin II	Hormonal	Vasoconstrictor
Norepinephrine	Neural	Vasoconstrictor
Decreased oxygen level	Local	Vasodilator
Decreased blood pressure	Local	Vasoconstrictor
Carbon dioxide	Local	Vasodilator
Increased K+ level in blood	Local	Vasodilator

exercise are maintained by local events such as the decreased oxygen content across the arterial-venous bed and the increased blood levels of metabolites of cellular respiration. As exercise comes to a halt, vaso-constricting agents, such as the drop in blood pressure and the binding of catecholamines to beta-adrenergic receptors, reduce blood flow.

The heart controls blood flow and pressure by adjusting cardiac output

The amount of oxygen-rich blood that a heart pumps into the circula-tion is called **cardiac output** and can be expressed as the product of the **stroke volume**, the amount of blood ejected during contraction (**systole**), and the **heart rate**, the number of beats per minute (Equation 9.1). Cardiac output corresponds to the sum of all tissue needs. That is, if the tissues need 100 units of blood, then the heart will release 100 units of blood. Tissue needs are determined by the amount of blood returned to the heart from the veins (**venous return**).

Cardiac output = stroke volume × heart rate   (9.1)

Where

stroke volume = amount of blood (ml) ejected from the left ventricle during one contraction of the heart (one heartbeat)

heart rate = number of beats per minute

As we begin to move our muscles, such as when we start to climb stairs, neurohormonal factors increase heart rate to supply the body with the extra oxygen needed for muscular work. The increase in car-diac output also causes greater venous return, with more blood filling the ventricles. Stretching of the muscle around the ventricles lengthens the myocytes, a property known as **compliance**, and causes a greater force of contraction, or greater **contractility**—that is, more blood can be ejected per beat. Contractility matches compliance over a wide range of **end diastolic volumes**, the amount of blood in the ventricle before it contracts and a measure of venous return. That is, the amount of blood ejected from the heart matches venous return. However, the myocyte has an optimal length, and any further stretching of the ventricle will decrease its contractility and, in turn, the stroke volume. The length-tension relationship of compliance and contractility in cardiac muscle, known as the Frank-Starling mechanism, does not operate beyond the optimal myocyte length under normal physiological conditions **(Figure 9.18)**. Later in the chapter, we will see how congestive heart failure can be explained, in part, by the stretching of myocytes beyond their optimal length.

Principles of fluid dynamics govern overall blood flow

As we have seen, both the heart and the vascular system have mecha-nisms that control the volume of blood flow: cardiac output for the heart; vasodilation and vasoconstriction for the arteries. These tissue- and organ-specific mechanisms for blood flow regulation work together to adjust overall blood flow, through some basic principles of fluid dynam-ics. The most important is the relationship among flow, pressure, and resistance. Blood flow is directly proportional to the pressure difference across the system and is inversely proportional to resistance (Equations 9.2, 9.3, and 9.4). For example, using Equation 9.3, we can see that an increase in either resistance (*R*) or flow (*F*) will increase blood pressure (Δ*P*). Pressure can be returned to its basal level by reducing cardiac out-put or decreasing vasoconstriction (or increasing vasodilation).

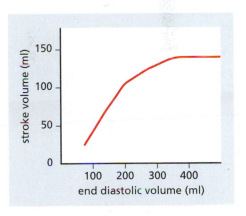

Figure 9.18 Graphical representation of the Frank-Starling mechanism. When venous return increases, so does the end diastolic volume of the ventricles. An increase in end diastolic volume stretches the ventricles, causing the myocytes to lengthen and generate a greater force (increased contractility). In turn, greater contractility generates the necessary force to eject a larger volume of blood. However, the length-tension relationship has its limits; eventually, contractility will begin to fail as myocytes stretch beyond the point at which they can generate the appropriate force of contraction.

TABLE 9.4
MEAN PRESSURE IN SELECTED VESSELS

Artery or vein	Mean pressure (mm Hg)
Aorta	100
Large arteries	100
Small arteries	80
Arterioles	60
Capillaries	20
Venules	10
Small veins	0
Venae cavae	0

$$F = \Delta P/R \qquad (9.2)$$

$$\Delta P = F \times R \qquad (9.3)$$

$$R = \Delta P/F \qquad (9.4)$$

Where

F = blood flow

ΔP = difference in pressure between the two ends of a vessel

R = resistance

Recall that the arterial vessels get progressively smaller until they reach the capillaries, which are just one red blood cell in diameter. As the vessel gets smaller, pressure decreases **(TABLE 9.4)** to prevent damage to the vessel. That is, while the large arteries can withstand the 100 mm Hg of pressure that occurs close to the heart, small vessels, especially the arterioles (the smallest arteries) and capillaries, cannot. So how can the cardiovascular system adhere to the laws of fluid dynamics and, at the same time, decrease pressure as the vessel diameter becomes smaller?

The answer lies in the vessel's ability to increase its lumen diameter by stretching the wall and taking advantage of another basic physical law of fluid dynamics: the fourth-power law. The fourth-power law states that blood flow in a vessel increases or decreases in proportion to the fourth power of the diameter. This means that small changes in the diameter of a vessel result in large changes in flow. As an example, let's assume that cardiac output has increased, resulting in greater flow to the periphery. The artery wall expands in order to accommodate the increased volume of blood. As the diameter of the lumen increases, resistance decreases, allowing greater flow and the maintenance of blood pressure in the artery. The relevance of the artery's ability to increase or decrease its lumen size will become more apparent when we discuss atherosclerosis.

AGE-RELATED DISEASES OF THE CARDIOVASCULAR SYSTEM: CARDIOVASCULAR DISEASE

Like other physiological systems discussed in this chapter and in Chapter 8, the cardiovascular system undergoes slight to moderate functional loss with age. In most cases, the functional loss does not impede daily activities, but it can reduce a person's ability to respond to stress or overload. As we discussed earlier in the chapter for neural pathologies, an important question to ask in relation to aging and the cardiovascular system is not so much what changes occur but why these changes progress to disease in some individuals but not in others. And, as for the neural pathologies, biogerontologists have yet to answer that question.

In this section, we explore three common age-related diseases of the cardiovascular system: coronary artery disease **(atherosclerosis)**, stroke **(cerebrovascular incident)**, and high blood pressure **(hypertension)**. We begin with a very brief discussion of environmental factors that affect the cardiovascular system.

Environmental factors influence age-related decline in the cardiovascular system

Defining nondisease, age-related alterations in the cardiovascular system has proved difficult. This problem arises because environmental factors such as diet, exercise, and smoking have significant effects on the rate of aging of the cardiovascular system. For example, the elasticity and compliance of the aorta and other large arteries decrease with age, often resulting in an increase in systolic blood pressure and hypertension. However, several studies have shown that physically active, aged individuals have significantly lower systolic blood pressure than age-matched sedentary people. Smoking exacerbates the loss of distensibility of the aorta and increases systolic blood pressure to a greater degree than is observed in nonsmoking, sedentary older individuals.

The physiological response of the heart to an increased demand for systemic flow does change with age and might not be influenced by environmental factors. Recall that the heart regulates its output based on the demands of the tissues for oxygen. During the initial demand for greater blood flow, such as during exercise, cardiac output increases as a result of an increased heart rate, while stroke volume remains the same. The increased heart rate is caused by stimulation of the sinus node by neurohormonal factors, which induces an increased rate of action potential generation. With age, the number of adrenoreceptors (receptors that bind epinephrine and norepinephrine) in the sinus node seems to decrease. This results in two age-related alterations in heart function. First, the increase in heart rate during the initial stage of increased metabolic demand occurs more slowly as we grow older. Thus, it takes longer for older individuals to adjust to increased metabolic demand. Second, the decrease in adrenoreceptors with aging results in a decrease in the maximal heart rate that can be achieved. This means that total cardiac output during high levels of exercise declines—that is, the maximum amount of work performed during exercise declines with age. The heart can compensate for the age-related decline in maximal heart rate by increasing stroke volume, although, because of the Frank-Starling mechanism, the absolute quantity of stroke volume has its limits. This compensation does not, however, offset the loss in total cardiac output caused by the decline in maximal heart rate.

Age-related changes in arteries and veins are also influenced by environmental factors. Recall the discussion on age-related changes in skin in Chapter 8. You learned that many of these changes result from increased cross-linking of collagen fibers. Cross-linking decreases the elasticity of connective tissue, resulting in more rigid structures. Now, look at Figure 9.16 and notice that arteries and veins have a substantial amount of connective tissue, containing mostly collagen. Moreover, the internal and external elastic membranes are composed of elastin, which can also be heavily cross-linked, with a similar decrease in distensibility. Since arteries help to regulate blood flow and pressure in direct proportion to their ability to expand and contract, a decrease in distensibility will affect blood flow and pressure. While a slight loss in vessels' ability to expand and contract is widely accepted as a normal age-related event, a lack of physical activity has been shown to exacerbate the loss. Smoking and poor diet can also have a significant effect on the functioning of blood vessels.

Arterial plaques can lead to atherosclerosis and ischemic events

Most individuals over the age of 60 have some buildup of **arterial plaques**, fatty deposits inside arterial walls. As autopsies have shown, the formation of plaques begins as fatty streaks in the major arteries of the heart as early as six months of age. For the majority of individuals, arterial plaques do not become large enough to disrupt blood flow. However, for reasons still to be determined, arterial plaques can

HEALTHY ARTERY

BUILDUP BEGINS

PLAQUE FORMS

PLAQUE CREATES BLOCKAGE;
blood clot forms

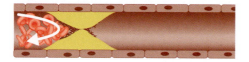

Figure 9.19 Stages of atherosclerotic blockage. The formation of plaques begins as fatty streaks in the major arteries. As plaques form, they cause an occlusion at the primary site of the lesion. To cause an ischemic event, the plaque must cause at least an 85% occlusion. Occlusion of a vessel that results in an ischemic incident usually results from rupture of the plaque and release of a thrombus.

progress to atherosclerosis and disrupt blood flow. If the atherosclerosis is severe enough, blood flow can be blocked, resulting in decreased oxygen delivery to tissues, a condition known as **ischemia (Figure 9.19)**. This can lead to the death of the tissue.

An atherosclerotic lesion typically forms just downstream of an arterial branch point and can block an artery by two means: (1) a thrombus at the primary site (also known as the focal point) of the lesion, or (2) an **embolus**, a piece of the thrombus that has broken loose, blocking a narrower part of the artery downstream from the original lesion **(Figure 9.20)**. Not all atherosclerotic lesions result in ischemia, however. For example, autopsies have shown that some lesions rupture and release an embolus too small to induce an ischemic event. The plaque is repaired and may rupture again, releasing a larger embolus. In some cases, the plaque causes the affected artery to "grow" a new vessel around the occluded area, resulting in a clinically silent lesion.

The formation of an atherosclerotic plaque large enough to cause an ischemic event takes 40–50 years in humans. The fatty streaks in arterial walls are composed primarily of cholesterol in the form of oxidized **low-density lipoproteins (LDLs)**. The invasion of LDLs into the tunica intima layer of the vessel damages the endothelial cells, and this initiates an inflammatory response. The mixture of immune cells, mostly T cells and macrophages, and cholesterol deposits forms new structures known as foam cells. The foam cells weaken the arterial wall, allowing smooth muscle cells to invade the fatty plaque and form a fibrous cap over the foam cells (something like a scab forming over a wound on your skin). As the plaque grows during a 20- to 30-year period, it calcifies and forms its own matrix, consisting primarily of collagen. That is, the atherosclerotic plaque becomes highly thrombotic. The reason for rupture of a fibrous cap and release of the embolus remains largely unknown, but the rupture mostly involves a weakening of the area due to repeated exposure to macrophages, as well as cell death of the smooth muscle.

Ischemic heart disease occurs when a large artery of the heart, one of four coronary arteries **(Figure 9.21)**, becomes blocked by an

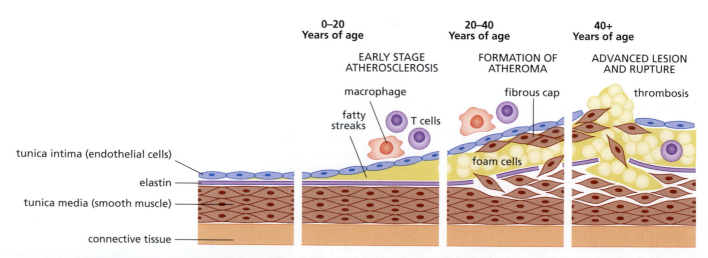

Figure 9.20 Pathogenesis of atherosclerosis over the life span. In the early stages of pathogenesis, observed in individuals as young as six months of age, the tunica intima shows an accumulation of fatty substances, appearing as fatty streaks, under the endothelial cells. The immune system recognizes the wall thickening as an antigen and directs T cells and macrophages to the site. With attack by the immune cells, the elastic lamellae between the tunica intima and media rupture, and smooth muscle cells invade the media to create a fibrous cap over the highly thrombotic foam cell component. The swelling of the artery wall is called an atheroma.

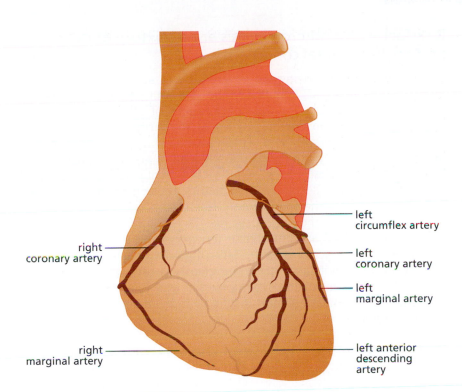

left circumflex artery

right coronary artery

left coronary artery

left marginal artery

right marginal artery

left anterior descending artery

Figure 9.21 The coronary arteries.
Note that the coronary arteries descend directly from the aorta and thus carry highly oxygenated blood. A blockage in one of these arteries can lead to a myocardial infarction (death of tissue due to lack of oxygen), leading to tissue death and decreased cardiac output.

atherosclerotic lesion, and the blood flow to heart tissue decreases. If the blockage reduces blood flow by more than 85% of normal, the lack of oxygen causes cardiac tissue death; one would experience this as a heart attack, or **myocardial infarction**. In turn, the normal electrical conductivity of the heart becomes disrupted and the heart fails to contract, causing a significant drop in cardiac output and damage to other areas of the body. If too much cardiac tissue is affected by the blockage in blood flow, the beating rhythm of the heart is disrupted, leading to **fibrillation**, the uncontrolled and dis-synchronous beating of the heart. Atrial or ventricular fibrillation causes a complete loss of cardiac output and death.

The mechanism leading to a stroke, or cerebrovascular incident, is similar to that causing ischemic heart disease, except that the ischemic event occurs in the brain. In most cases, the blockage resulting in an infarction is caused by an embolus that has broken off from an atherosclerotic lesion in the carotid arteries. However, some strokes can arise from the rupture of an internal brain artery and subsequent hemorrhage. The type of loss in systemic function resulting from a stroke depends on the area of the brain affected. Strokes can affect any part of the brain, but they are typically described in terms of injury to the right or left hemisphere. Right-hemisphere stroke affects motor function on the left side of the body, often causing some paralysis to the arm and leg; it also causes difficulty with judging distance and other spatial relationships, as well as impairment in the ability to understand how parts are connected to the whole. As one would expect, left-hemisphere stroke causes loss of motor control on the right side of the body. Right-hemisphere stroke survivors also have a wide range of speech and language problems and sometimes develop slow and cautious behavior. Strokes can also occur in the cerebellum and brain stem. Since the cerebellum coordinates movement, strokes in this area affect balance, coordination, and reflexes. The brain stem controls most of our autonomic functions, including breathing, blood pressure, and heart rate, and strokes in the brain stem are often fatal.

Risk factors for atherosclerosis are a mixture of genetic and environmental conditions

The mechanism underlying the progression of arterial plaques to the disease atherosclerosis remains to be determined. Four factors are considered major risks for the development of atherosclerosis: age, smoking, high serum cholesterol levels (**hyperlipidemia**), and high blood pressure (hypertension). These four are known as primary risk factors, because they can independently lead to atherosclerosis. Other factors such as obesity, lack of physical activity, and psychological stress are known as secondary risk factors; these must occur in association with one of the four primary risk factors in order to promote the development of atherosclerosis. Risk factors for atherosclerosis are also often classified as controllable (environmentally based) or non-controllable (genetically based). Smoking, obesity, lack of physical activity, and psychological stress are clearly controllable risk factors. Age cannot be controlled. Hyperlipidemia and hypertension are the consequence of environmental insults to individuals who are genetically susceptible.

Statins reduce the synthesis of cholesterol in the liver and lower serum cholesterol

Recall that atherosclerotic plaques begin as fatty streaks in the arteries, containing primarily cholesterol. Several investigations have shown that the higher the concentration of serum cholesterol, the greater the risk of developing atherosclerosis and dying from the disease **(Figure 9.22)**. Since food contains cholesterol, mostly in meats, reducing dietary cholesterol or total amount of dietary fat has been seen as an effective way of reducing serum cholesterol. (Remember, it is serum cholesterol, not dietary cholesterol, that is the risk factor.) Several studies during the 1970s and 1980s supported this view. Subsequent

Figure 9.22 Serum cholesterol and the risk of death from coronary heart disease (CHD). (Data from J.I. Cleeman. and C. Lenfant, *JAMA* 280:2099–2104, 1998. With permission from the American Medical Association.)

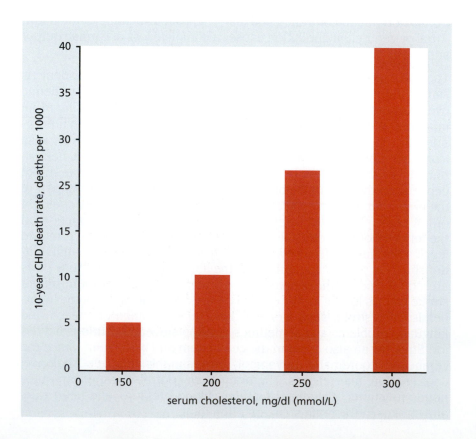

studies have refined this view, however. The results suggest that dietary management of serum cholesterol may be effective only for individuals genetically predisposed to high serum cholesterol—individuals who are known to oversynthesize LDLs, the type of cholesterol most often found in atherosclerotic plaques. The primary role of LDLs is to transport cholesterol in the blood from the liver to body cells.

Cholesterol is a vital component of cellular membranes that provides stability to the phospholipid bilayer. While all cells can synthesize cholesterol, 70–80% of LDL synthesis occurs in the liver. Thus, for genetically predisposed individuals with high serum cholesterol, prevention of the synthesis of liver LDLs by drugs known as **statins** (the most commonly prescribed is Lipitor®) has proved effective in lowering the risk for atherosclerosis. Statins work by inhibiting an enzyme involved in the synthesis of cholesterol. In this synthetic pathway, acetyl-CoA, the end product of fatty acid breakdown (see Figure 4.19), is converted to 3-hydroxy-3-methylglutaryl coenzyme A (HMG-CoA) in a series of reactions **(Figure 9.23)**. The enzyme HMG-CoA reductase catalyzes the rate-limiting step of the pathway, the reduction of HMG-CoA to mevalonate, a precursor of cholesterol. Inhibiting the enzyme HMG-CoA reductase prevents cholesterol production.

Statins have two other effects. First, reduced cholesterol synthesis in the liver up-regulates the synthesis of the liver's LDL receptors. Thus, the uptake of serum LDL by the liver increases and serum levels of LDL fall. Second, statins raise **high-density lipoprotein** (HDL) levels by about 5%, though the reason for this is unclear. Since HDL removes cholesterol from the blood and transports it to the liver for catabolism, the lowering of total serum cholesterol by statins may involve this HDL-related mechanism.

Like all drugs, statins have side effects, but they tend to be more annoying than serious physiological problems and affect only about 5% of users. These side effects include headaches, memory problems, muscle pain or weakness, and weight gain. More serious side effects occur in less than 1% of users and vary from one individual to another.

Hypertension is the most common chronic condition in the aged

More than 50% of individuals 65 years and older in the United States have hypertension, which is defined as a consistent blood pressure of ≥140 mm Hg for systolic or ≥90 mm Hg for diastolic (140/90). The mechanisms underlying the development of hypertension are not known. We can conclude, however, that the increase in pressure reflects a resistance to flow (arterial stiffening), since stroke volume does not increase with age. Although the etiology of hypertension has yet to be described, several physiological and environmental factors are known to be associated with this disease. These include decreased compliance of the arteries (increased stiffness), declining adrenergic function, kidney disease, obesity, lack of exercise, and smoking. While mortality from hypertension remains low (58/100,000 in the United States), hypertension can lead to an increased risk for other forms of heart disease. For example, most studies find that a reduction of 10–15 mm Hg in systolic blood pressure can reduce the risk of an ischemic event by as much as 45%.

Treatments for hypertension include a reduction in salt intake, increased exercise, and drugs that reduce vasoconstriction and blood water volume. The effectiveness of reduced salt intake and increased exercise in reducing hypertension remains somewhat uncertain. While several

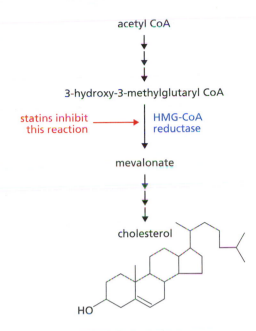

Figure 9.23 Cholesterol synthesis and the action of statins in reducing cholesterol synthesis. The *broken arrows* indicate steps in the pathway that are not shown.

studies have shown that both of these interventions can be effective in some people, there are many people who do not respond. Moreover, the reduction in blood pressure resulting from increased physical activity and/or decreased salt intake is modest, often no more than a 10% reduction in systolic and diastolic pressures. The limited ability of these interventions to significantly reduce blood pressure in the entire population suggests that hypertension has a strong genetic component.

Drug therapies have proved to be the most beneficial way of controlling blood pressure. There are several classes of drugs used to treat hypertension; the three most common are **beta-blockers**, **angiotensin-converting enzyme (ACE) inhibitors**, and **diuretics**. Beta-blockers block beta-adrenergic receptors. ACE inhibitors inhibit the conversion of angiotensin I to angiotensin II. Recall that both norepinephrine and angiotensin II are strong vasoconstrictors (see Table 9.3). Diuretics typically block the formation of antidiuretic hormone (vasopressin), a hormone that prevents passage of water from blood into the urine. Thus, diuretics increase water loss (in the urine) and reduce blood volume. All three types of treatment lower the resistance to blood flow and thus act in accordance with the laws of flow described earlier in the chapter. Recall that a decrease in resistance decreases pressure (see Equation 9.3).

Heart failure results in a decline in cardiac output

The causes of heart failure (also known as **congestive heart failure**) cannot be defined as clearly as the causes of ischemic-related disorders. Heart failure results from a number of conditions that range from infection of the myocardium to abnormal retention of salt and water. Nonetheless, the deterioration in the physical properties of the myocardium that leads to heart failure is the same, regardless of the underlying cause. In all cases, heart failure can be defined as the inability of the myocardium to generate the force of contraction needed to eject the blood volume required to meet the oxygen needs of the body—that is, an abnormal decrease in cardiac output.

Heart failure is classified into two types: difficulty in filling the heart (diastolic dysfunction) and difficulty in ejecting blood from the heart (systolic dysfunction). Diastolic dysfunction occurs primarily when the ventricle wall has become stiff, with a consequent decrease in compliance and end diastolic volume; the contractility of the heart often remains unaffected. That is, although the heart cells are working normally, the heart muscle cannot expand sufficiently to accommodate the volume of venous return. The declining end diastolic volume causes a decrease in the force of contraction and the amount of blood ejected during systole. In turn, peripheral tissues do not receive sufficient blood to adequately oxygenate the cells.

Systolic dysfunction results primarily from damage to heart muscle—that is, myocardial infarction. Systolic dysfunction, like diastolic dysfunction, results in decreased stroke volume, regardless of the end diastolic volume. However, the mechanism underlying the decrease in stroke volume differs from that in diastolic dysfunction. The loss of heart tissue due to an infarction reduces the total number of myocytes that participate in contraction during systole, so the contractility of the heart is reduced; compliance may remain unchanged.

The physiological response to heart failure, diastolic or systolic, is progressive and results, for the most part, in fluid retention, or **edema**. The mechanisms leading to edema caused by heart failure are complex, but the simple explanation is that blood in the veins backs up. The resulting increase in osmotic pressure causes fluid to leave the veins and enter the interstitial spaces. The most serious location for edema is the lungs, in a condition known as pulmonary edema. An increase in fluid in the lungs impairs gas exchange. Individuals with heart failure literally begin to drown in their own body fluids.

Prevalence is a better descriptor of cardiovascular disease than is mortality

In the United States between 1950 and 2007, deaths caused by diseases of the heart in individuals over the age of 65 dropped by 41% for males and 39% for females **(Figure 9.24)**. Nonetheless, ischemic heart disease remains the number one cause of death in older individuals in the United States and other developed countries, and the rate of mortality increases with advancing age. Mortality rate was once considered the defining measure of the age-related occurrence of cardiovascular disease, because medical science had not yet developed procedures to prevent death in patients with ischemic heart disease. Today, however, improved diagnostic techniques significantly reduce deaths due to cardiovascular disease. With the introduction of bypass surgery, angioplasty, and insertion of stents, along with a greater awareness of the impact of diet on disease prevention, cardiovascular disease is now, to a large degree, a chronic rather than a lethal condition. Thus, prevalence rather than mortality may be a better estimate by which to gauge the effect of cardiovascular disease in the aging population **(Figure 9.25)**.

THE ENDOCRINE SYSTEM AND GLUCOSE REGULATION

The **endocrine system** is a system of nonconnected organs and glands, each of which secretes hormones directly into the blood **(Figure 9.26)**. The hormones have specific target cells, and the response of these cells to a particular hormone causes the tissue to perform specific functions **(TABLE 9.5)**. The endocrine system is responsible for regulating many human functions, including metabolism, growth and development, various tissue functions, and maintenance of blood glucose levels. In addition to the traditional endocrine organs and glands illustrated in Figure 9.26, several others with mixed functions are now recognized as part of the endocrine system. These include the liver, heart, small intestine, adipose tissue, and skin.

Figure 9.24 Death rate (per 100,000) for diseases of the heart (ischemic heart disease) in men and women aged 65 and older, United States, 1950–2007. (Data from National Center for Health Statistics, Health, United States, 2010: With Special Feature on Death and Dying, Hyattsville, MD: Centers for Disease Control and Prevention, 2011.)

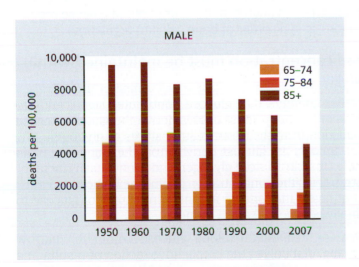

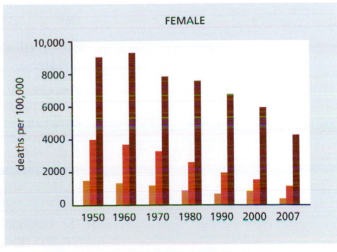

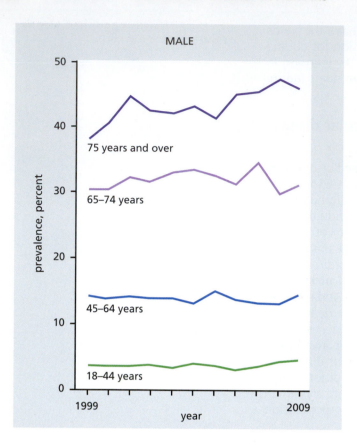

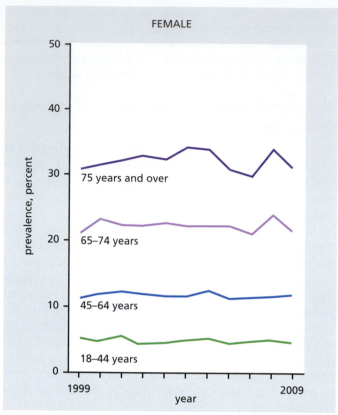

Figure 9.25 Prevalence of heart disease in men and women, United States, 1999–2009. The prevalence of heart disease has remained stable or increased slightly at the same time that death rate was dropping (see Figure 9.24). Together, decreasing death rates with increasing prevalence indicates that medical advances have extended longevity for individuals with this disorder. (Data from National Center for Health Statistics, Health, United States, 2010: With Special Feature on Death and Dying, Hyattsville, MD: Centers for Disease Control and Prevention, 2011.)

Animal studies have shown an age-related decline in function in virtually all endocrine organs and glands. The results from highly controlled experiments using laboratory animals, however, have not generally been corroborated in human trials. The reasons for the different findings in experimental animals and in humans are the same as those we have discussed previously: an inability to control for environmental influences in human studies, inclusion of individuals with disease in the sample population, the small number of studies, the small number of individuals included in each study, intrinsic genetic differences among species, and so on. We therefore limit our discussion here to two endocrine systems—the regulation of blood glucose and of bone calcium—for which there are sufficient data to suggest that age-related changes will progress to disease. In this and several following sections, we discuss the biochemical and physiological mechanisms involved in regulating blood glucose and bone calcium and the two diseases associated with altered regulation: type 2 diabetes (a disorder of blood glucose regulation) and osteoporosis (a disorder of bone calcium regulation).

Blood glucose concentration must be maintained within a narrow range

You saw in earlier chapters that glucose contributes significantly to the energy needs of the body. Most tissues derive the glucose needed for energy production from their intracellular stores of **glycogen**. The brain, however, can only use glucose derived from the blood. The brain does not have the capacity to store glycogen and cannot easily metabolize fat for energy. Thus, the concentration of glucose in the blood must be maintained at levels that ensure a consistent supply of glucose to the brain, normally between 90 and 120 mg/dl.

The mechanism for maintaining blood glucose within the narrow range of 90–120 mg/dl involves the competitive action of **insulin** and

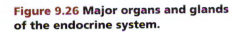

Figure 9.26 Major organs and glands of the endocrine system.

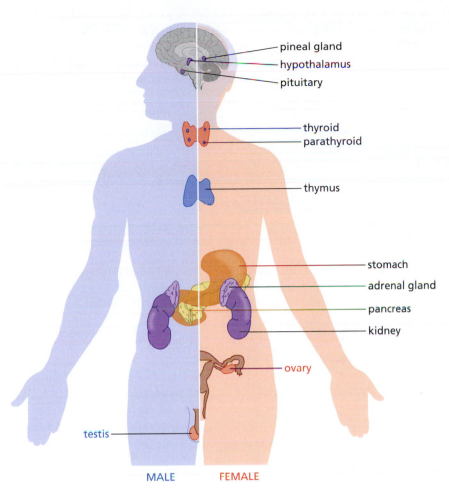

pineal gland
hypothalamus
pituitary
thyroid
parathyroid
thymus
stomach
adrenal gland
pancreas
kidney
ovary
testis

MALE   FEMALE

glucagon, two hormones produced by the pancreas. The endocrine cells of the pancreas are localized in the **islets of Langerhans (Figure 9.27)**. Insulin removes glucose from the blood when the glucose concentration exceeds ~120 mg/dl and stimulates **glycogenesis** (glycogen synthesis) in the liver and muscle. Conversely, glucagon adds glucose to the blood by stimulating the breakdown of glycogen to glucose in the liver when the blood glucose concentration falls below ~90 mg/dl. However, the human body has only a 24-hour supply of glycogen that can be used for energy. Glucose must be supplied by food to replace the glycogen stores broken down and used for energy between meals. **Gluconeogenesis**, the synthesis of glucose from non-carbohydrate carbon sources, does occur in mammals, but only during extended periods of starvation.

Insulin facilitates glucose uptake into liver, muscle, and adipose cells

The cell membranes of liver, muscle, and adipose tissue are, for the most part, impermeable to glucose; the cell membranes of neural tissue and the kidney allow passive diffusion of glucose. Thus, glucose requires carrier proteins for passage into liver, muscle, and adipose tissue cells. A specialized carrier protein for glucose, called glucose transporter 4 (GLUT4), facilitates glucose transport only in the presence of insulin—this ensures that blood glucose remains available for the brain even when blood glucose concentrations are below the level that stimulates insulin secretion (~120 mg/dl). The primary actions of insulin on muscle, liver, and adipose tissue are shown in **TABLE 9.6**.

In the presence of increased blood glucose, such as following a meal, the beta cells of the pancreas secrete insulin **(Figure 9.28)**. Binding of insulin to tyrosine kinase-linked receptors on the surface of cells initiates a series of signal-propagation events, involving phosphorylation, that induce glucose uptake. Binding of insulin to the cell surface side

TABLE 9.5
MAJOR GLANDS AND ORGANS OF THE ENDOCRINE SYSTEM

Endocrine gland or organ	Hormone(s) secreted	Location of target cells	Primary physiological function affected
Pineal gland	Melatonin	Pituitary, reproductive organs, immune system	Regulates body's biological rhythm
Hypothalamus	7 different releasing and inhibitory hormones	Anterior pituitary	Control the release of 7 different anterior pituitary hormones
Posterior pituitary	Vasopressin (antidiuretic hormone, ADH)	Kidney tubules	Increases water reabsorption
	Oxytocin	Arterioles Uterus	Causes vasoconstriction Increases contraction
Anterior pituitary	Thyroid-stimulating hormone (TSH)	Thyroid	Stimulates release of thyroid hormone
	Adrenocorticotropic hormone (ACTH)	Adrenal cortex	Stimulates cortisol secretion
	Growth hormone (GH)	Nearly all tissues	Stimulates growth
	Follicle-stimulating hormone (FSH)	Ovarian follicles and testes	Stimulates follicular growth, estrogen secretion, sperm production
	Luteinizing hormone (LH)	Ovarian follicles and testes	Stimulates ovulation and testosterone secretion
	Prolactin	Mammary glands	Stimulates milk secretion
Thyroid gland	Thyroid hormone	Nearly all tissues	Increases metabolic rate; essential for normal growth
	Calcitonin	Bone	Decreases plasma calcium concentration
Parathyroid gland	Parathyroid hormone (PTH)	Bone, kidneys, intestine	Increases blood calcium concentration
Thymus	Thymopoietin	T cells	Affects T-cell function (role poorly understood)
Stomach	Gastrin	Pancreas, liver, gallbladder	Stimulates secretion of digestive enzymes and bile
Pancreas—islets of Langerhans	Insulin	Nearly all tissues	Increases glucose storage
	Glucagon	Nearly all tissues	Stimulates release of stored glucose into blood
	Somatostatin	Digestive system	Inhibits digestion and absorption of nutrients
Adrenal cortex	Aldosterone	Kidney tubules	Increases sodium reabsorption and potassium secretion
	Cortisol	Nearly all tissues	Converts protein and fat into glucose
	Dehydroepiandrosterone	Bone and sex-linked tissues	Stimulates pubertal growth and sex drive in females
Adrenal medulla	Epinephrine and norepinephrine	Neural receptors throughout body	Affects stress adaptation and blood pressure regulation
Kidneys	Renin	Adrenal gland	Stimulates aldosterone secretion
Ovaries	Estrogen	Female sex-linked tissues	Stimulates follicular development, growth; regulation of secondary sexual characteristics
Testes	Testosterone	Male sex-linked tissues	Stimulates sperm production, growth; regulation of secondary sexual characteristics

(A)

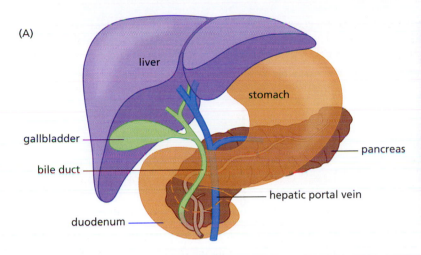

(B)

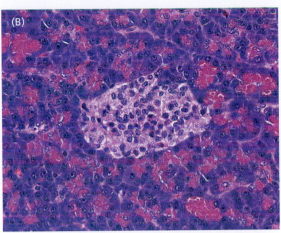

Figure 9.27 The pancreas and islets of Langerhans. (A) Diagram showing the anatomical location of the pancreas. (B) Micrograph of pancreatic tissue. The *blue* structures surrounding the islets of Langerhans (*lighter area*) are pancreatic acini, cells that secrete digestive enzymes into the small intestine. (B, courtesy of vetpathologist/Shutterstock.)

of its transmembrane receptor activates tyrosine kinase. The tyrosine kinase phosphorylates a tyrosine residue on proteins that induce GLUT4 molecules to move from the cytosol to the cell surface. Glucose can then pass through the GLUT4 membrane channels. Binding of insulin to its receptor also activates several other kinases that phosphorylate insulin-related substrates, thus activating several glucose metabolic pathways (**glycolysis** and glycogenesis) as well as the expression of genes needed for glucose-related metabolism.

We generally think of insulin as a hormone that regulates carbohydrate metabolism, but it also influences the metabolism of protein and fat, with the effect of sparing these energy substrates during times of glucose excess. For example, insulin inhibits the action of **hormone-sensitive lipase** in adipose tissue, thus preventing the breakdown of **triglycerides** and release of free fatty acids into the blood. This limits the amount of energy from fat that is available to other tissues and "forces" cells to use glucose for energy. In the liver, excess insulin-stimulated

TABLE 9.6
SUMMARY OF INSULIN'S EFFECTS ON THE METABOLISM OF GLUCOSE, FATS, AND PROTEINS

Nutrient	Tissue	Effect
Glucose	Muscle	Stimulates glucose uptake Increases glycogenesis Inhibits fatty acid oxidation
	Adipose tissue	Promotes conversion of glucose to glycerol used in triglyceride formation; small amounts converted to fatty acids
	Liver	Stimulates glycogenesis
Fatty acids	Adipose tissue	Inhibits fatty acid release by inhibiting hormone-sensitive lipase
	Liver	Stimulates fatty acid synthesis and triglyceride formation after glycogenesis has reached maximum
Protein	Muscle and liver	Stimulates transport of amino acids Stimulates ribosomal activity Inhibits protein catabolism
	Liver	Spares amino acids by inhibiting gluconeogenesis

Figure 9.28 The facilitation of glucose uptake by insulin. (1) Increasing blood glucose concentration stimulates the secretion of insulin by pancreatic beta cells. (2) Insulin binds to a transmembrane tyrosine kinase-linked receptor, which (3) activates tyrosine kinase and phosphorylates protein substrates. (4) Phosphorylation of these substrates induces the movement of glucose transporter 4 (GLUT4) molecules from the cytosol to the membrane surface, allowing the uptake of glucose. (5) Tyrosine kinase also induces the phosphorylation of a number of other insulin-receptor substrates, resulting in activation of glucose metabolic pathways and gene expression.

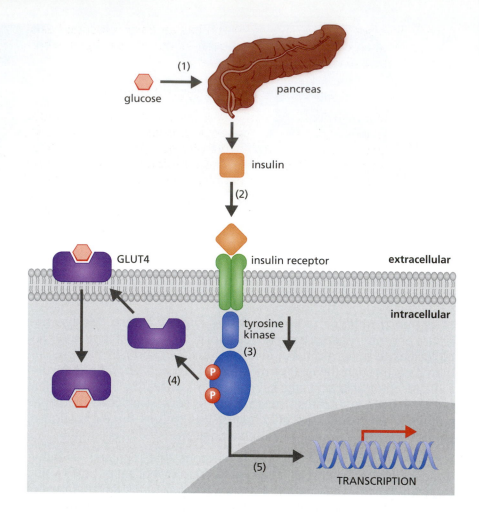

glucose uptake (exceeding the liver's capacity for glycogen storage) results in the accumulation of acetyl-CoA, a precursor of fatty acids and cholesterol. Thus insulin can stimulate fatty acid synthesis in the liver. The newly formed fatty acid is generally packaged as triglycerides, released from the liver into the blood, and transported to adipose tissue for storage.

Insulin can stimulate the synthesis of proteins in various cells by several mechanisms. First, insulin stimulates the uptake of amino acids following a meal, providing cells with the amino acids needed for protein synthesis. Second, insulin can stimulate translation of mRNA by increasing ribosomal activity. Third, insulin increases the rate of transcription of DNA coding for enzymes that stimulate the storage of carbohydrates, fats, and proteins. Finally, insulin inhibits the catabolism of proteins and thus limits the quantity of amino acids that can be used for gluconeogenesis.

AGE-RELATED DISEASE OF THE ENDOCRINE SYSTEM: TYPE 2 DIABETES MELLITUS

Diabetes mellitus is caused by the inability of cells to take up glucose. Alterations in glucose uptake are due either to insufficient insulin secretion by the pancreas **(type 1 diabetes mellitus)** or to cells developing a resistance to the action of insulin **(type 2 diabetes mellitus)**. Type 1 diabetes develops primarily in individuals under the age of 10 and reflects, for the most part, genetic and viral causes. Type 2 diabetes usually develops after the age of 40 and has an environmental basis, although some genetic factors have been identified. We focus exclusively on type 2 diabetes.

Insulin resistance is a precursor to type 2 diabetes

Insulin resistance reflects the inability of insulin to effectively induce glucose uptake by liver, muscle, and adipose tissue when insulin secretion remains normal. The American Diabetes Association defines **insulin resistance** (also known as **glucose intolerance**) as a condition in which an individual has a normal blood glucose concentration at rest but values above 140 mg/dl at the end of a two-hour **oral glucose tolerance test (OGTT) (Figure 9.29)**. The OGTT consists of measuring blood glucose concentration in a fasted state and then administering an oral dose of glucose (75 g) mixed in flavored water. Six to ten additional blood glucose measurements are taken over the next two hours. Insulin-resistant individuals show a slightly steeper rise in the initial blood glucose concentration, to values that are typically greater than those observed in non-insulin-resistant individuals. The drop in blood glucose concentration after one hour is slower in the insulin-resistant than in the non-insulin-resistant individual, and the concentration does not reach the resting level within two hours.

In recent years, a second diagnostic criterion for type 2 diabetes, the level of glycosylated **hemoglobin A1c** (or just A1c), has been added as an indication of plasma glucose concentration over extended periods. Recall our discussion of the nonenzymatic glycosylation of proteins in Chapter 8. Since nonenzymatic glycosylation has no catabolic pathway, once a protein has been glycosylated, it remains that way over its life span. Nondiabetic individuals with an average resting blood glucose level below 120 mg/dl have blood concentrations of glycosylated **hemoglobin** below 6.5% of the total blood hemoglobin concentration. Individuals whose long-term average resting glucose concentration exceeds 120 mg/dl have an A1c level above 6.5%. For people with diabetes, A1c level also provides a convenient and easy biomarker for evaluating the effectiveness of treatment aimed at regulating blood glucose levels.

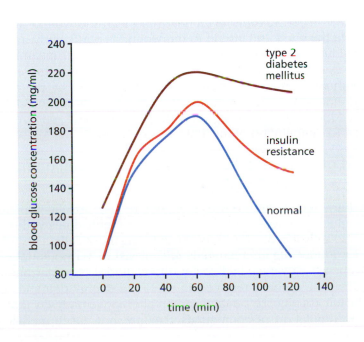

Figure 9.29 Oral glucose tolerance test results for individuals with normal insulin function, insulin resistance, and type 2 diabetes.

The underlying mechanisms of insulin resistance have not been fully described, and most of our current knowledge about the effect of age on impaired glucose tolerance comes from animal studies. In general, the number and affinity of insulin receptors seem to remain unchanged throughout the life span. Only limited information exists on the effect of aging on post-receptor signal transduction. Although the intracellular concentration of GLUT4 does not change, the amount of this transporter found at the membrane during periods of increased blood glucose concentration declines significantly in older animals. Decreased translocation of the GLUT4 protein to the cell membrane surface attenuates the uptake of glucose into the cell and thus results in increased blood glucose concentration. Without therapeutic intervention or alteration in lifestyle, insulin resistance can progress to type 2 diabetes.

Type 2 diabetes impairs microvascular blood flow

The American Diabetes Association defines type 2 diabetes mellitus as a fasting blood glucose concentration of >126 mg/dl. The diagnosis of type 2 diabetes can also be made on the basis of an OGTT as a blood glucose level of >200 mg/dl after two hours. Uncontrolled type 2 diabetes leads to several pathologies, including cardiovascular disease, neuropathy, nephropathy, and retinopathy. The mechanisms underlying development of the pathologies associated with type 2 diabetes have yet to be established, but a reduction in the amount of blood reaching cells is widely accepted as a primary cause. Loss of appropriate blood flow to tissue reflects, in large part, a decrease in the microvasculature's compliance properties that is due to increased glycosylation of connective tissue. Without the ability of blood vessels to contract and expand in response to changes in cardiac output, tissue does not receive the appropriate amount of blood and thus oxygenation of the tissue is impaired. Attenuated oxygenation lowers the rate of ATP production, which, in turn, limits many intracellular reactions, leading to cell dysfunction and/or cell death.

The loss of blood flow and subsequent damage to sensory neurons provides a good example of how type 2 diabetes can result in significant tissue damage. Imagine that an individual with type 2 diabetes and sensory neuron damage steps on a small piece of glass. The sensory neurons fail to signal the brain that a wound has occurred on the bottom of the foot. The individual does not perceive pain and fails to take appropriate action to remove the glass or clean the wound. An effort to remove the glass and attend to the wound is particularly important for individuals with type 2 diabetes, because normal inflammation/immune system mechanisms are diminished due to decreased blood flow. Without appropriate "cleansing" of the wound by the immune system, the individual is at significantly greater risk for infection. An untreated infection could result in widespread necrosis, gangrene, and the need for amputation of the limb.

Altered glucose metabolism may increase cell damage in people with type 2 diabetes

The kidney, retina, and neurons do not require insulin for glucose uptake; glucose moves freely into the cells. In individuals with normal blood glucose concentration, glucose entering these cells is used exclusively for immediate energy needs (glycolysis) or for potential energy (glycogenesis). However, in diabetes, blood glucose concentrations are well above the amount required to meet the energy demands of the cell. The excess glucose can activate an alternative pathway that leads to an accumulation of compounds that are not easily metabolized.

This pathway, the polyol pathway, converts glucose to fructose via the production of sorbitol **(Figure 9.30)**. Sorbitol and fructose have the potential to alter cellular metabolism and increase cellular damage. Given its hydrophilic nature, sorbitol (an alcohol) cannot diffuse through the membrane and it accumulates in the cell. The resulting osmotic stress disrupts normal membrane potential and slows or stops many intracellular reactions. Moreover, the reduction of glucose to sorbitol reduces the concentration of reduced nicotinamide adenine dinucleotide phosphate (NADPH + H$^+$), an electron carrier important in maintaining the intracellular concentration of reduced glutathione (GSH). Reduced glutathione protects cell membranes against the formation of lipid peroxides by scavenging the hydroxyl radical ($^\bullet$OH). Fructose, a sugar normally found at very low concentrations in the cell, forms advanced glycation end products 100 times more effectively than does glucose. Thus, the polyol pathway has the potential to increase the already high rate of protein glycosylation observed in type 2 diabetes.

The increased activity of the polyol pathway has been shown to induce retinopathy in type 2 diabetic dogs, an excellent model for human diabetes. These findings led many clinical research teams to test drugs that inhibit aldose reductase as a means of controlling sorbitol/fructose-induced damage. Unfortunately, investigations in laboratory animals and humans have produced only limited success. That is, aldose reductase inhibitors reduced the activity of the polyol pathway, but retinopathy did not decline. Research is currently underway to evaluate other drugs that inhibit the formation of fructose through the polyol pathway.

Risk factors for diabetes include increasing age, obesity, and genetic background

The cause of type 2 diabetes in the elderly population remains unknown. Several factors have been identified that place elderly individuals at risk for developing type 2 diabetes. These factors include age, obesity, lack of physical activity, and genetics/family history. All appear to play primary roles and often occur together. The percentage of individuals diagnosed with type 2 diabetes becomes greater with age, and in recent decades, new cases have increased significantly in all age groups **(Figure 9.31)**. Although the reason for the rise in new cases of type 2 diabetes remains unknown, the increase in incidence of obesity over the same period must be suspected as contributing significantly to the incidence of this disease **(Figure 9.32)**. Indeed, most experts now believe that obesity accounts for 80% of all new cases of type 2 diabetes **(BOX 9.1)**.

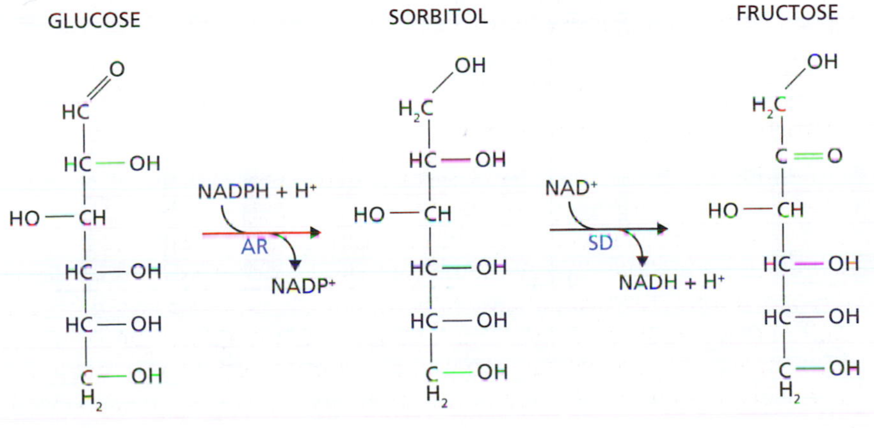

Figure 9.30 The polyol pathway. When the intracellular glucose concentration exceeds the cell's energy requirements, aldose reductase (AR) can catalyze the reduction of glucose to sorbitol, with hydrogen ions donated by reduced NADP (NADPH + H$^+$). Sorbitol dehydrogenase (SD) catalyzes the oxidation of sorbitol to fructose by donating the hydrogen ions to NAD$^+$.

Figure 9.31 Prevalence of type 2 diabetes in the adult population, United States. (Data from National Center for Health Statistics, Health, United States, 2010: With Special Feature on Death and Dying, Hyattsville, MD: Centers for Disease Control and Prevention, 2011.)

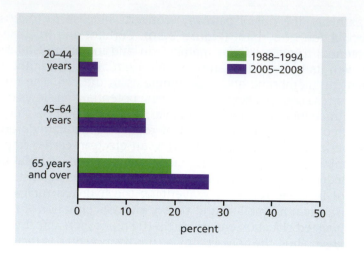

There is also a genetic link. Several studies have shown that where obesity is not involved, the development of type 2 diabetes runs in families, especially in individuals of Asian or African descent. Estimates of the genetic contribution to type 2 diabetes vary, but most experts agree that 10–20% of type 2 diabetes cases have a genetic link. The genetic link might be part of a more general condition known as metabolic syndrome, which increases the risk for several diseases, including coronary heart disease, stroke, and diabetes. Metabolic syndrome is characterized by high levels of blood triglycerides and LDLs, low levels of HDLs, hypertension, and a resting blood glucose level of 120–135 mg/dl. It is unlikely that metabolic syndrome reflects a mutation in a gene or gene group. Rather, like type 2 diabetes, metabolic syndrome most often appears in individuals who have some defect in regulating fat metabolism (obesity) and a sedentary lifestyle.

THE SKELETAL SYSTEM AND BONE CALCIUM METABOLISM

Human bone has three major functions: (1) it provides strength and structure to the human form; (2) it is the site of blood cell production; and (3) it stores calcium and is involved in the regulation of serum

Figure 9.32 Prevalence of obesity in men and women, United States. Obesity is defined as BMI > 30. Data for individuals older than 75 years were not collected before 1988. (Data from National Center for Health Statistics, Health, United States, 2010: With Special Feature on Death and Dying, Hyattsville, MD: Centers for Disease Control and Prevention, 2011.)

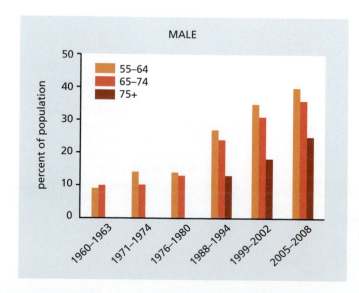

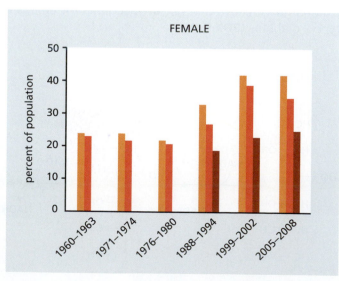

BOX 9.1 OBESITY AND TYPE 2 DIABETES: THE DIABETES PREVENTION PROGRAM RESEARCH GROUP STUDIES

Advancing age significantly increases the incidence of type 2 diabetes. The percentage of individuals 65 years and older who had type 2 diabetes was close to 8% in the United States from the time accurate records began (1975) until the early 1990s. Then the percentage of the 65+ population diagnosed with type 2 diabetes began to increase, with the current number standing at ~18%. Interestingly, the rate of obesity in the older population also began to increase at about the same time. This correlation led many researchers to suggest that obesity might have a significant role in the etiology of type 2 diabetes.

Physicians have long recognized the relationship between obesity and type 2 diabetes, but the standard dietary recommendations, such as the Food Guide Pyramid **(Figure 9.33)**, have proved ineffective in reducing the body weight of obese individuals. The rates of both obesity and type 2 diabetes continued to climb throughout the 1990s and into the new century. On the bright side, the introduction in 1994 of a new class of drugs, the biguanides, allowed the effective management of blood glucose levels without causing significant side effects. Complications due to uncontrolled blood glucose concentrations in people with type 2 diabetes started to decline. However, biguanides such as metformin (Glucophage) were being used to treat, but not prevent, type 2 diabetes. Many health professionals remained convinced that prevention was a better option than treatment for most insulin-resistant patients. Moreover, several respected nutritionists were suggesting that weight loss programs more aggressive than those recommended by the national health groups (USDA, NIH, etc.) should be used to prevent obesity and thereby reduce the incidence of type 2 diabetes.

Several small studies on human subjects have shown that type 2 diabetes could be prevented by reducing weight below the obesity standard (BMI < 30), but large clinical trials were needed to verify the data. To this end, researchers from 29 independent centers throughout the United States joined together to test whether reducing obesity in insulin-resistant individuals would prevent the development of type 2 diabetes. The research group, named the Diabetes Prevention Program Research Group, recorded the incidence of new cases of type 2 diabetes over a four-year period in male and female glucose-intolerant individuals of various ages and racial/ethnic backgrounds. The individuals were divided into groups to receive one of three treatments: (1) 850 mg doses of metformin (a biguanide) plus standard dietary recommendations as outlined in the Food Guide Pyramid; (2) standard dietary recommendations plus a placebo; or (3) an intensive weight loss program designed to reduce body weight by 7% through increased physical activity and a low-calorie diet.

The results were simply remarkable **(Figure 9.34)**. The weight lost by individuals in the intensive weight loss group was substantially greater than that of individuals who followed the standard dietary recommendations. Individuals in the intensive weight loss program lost an average of 5.6 kg of body weight in four years. Those following the standard weight loss recommendations and receiving placebo did not show any weight loss during the same period; those in the metformin group lost an average of 2.6 kg. Likewise, the incidence of type 2 diabetes in the intensive weight loss group was 58% lower than that in the group following standard dietary recommendations and taking

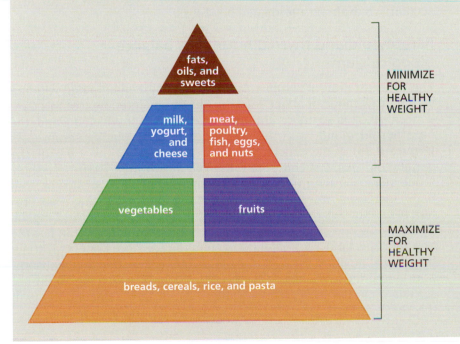

Figure 9.33 Food Guide Pyramid. The United States Department of Agriculture (USDA), the developer of this food pyramid, suggests that following these dietary recommendations will help to maintain or achieve a healthy weight.

BOX 9.1 OBESITY AND TYPE 2 DIABETES: THE DIABETES PREVENTION PROGRAM RESEARCH GROUP STUDIES

placebo. The metformin treatment prevented the development of type 2 diabetes better than did the placebo, but it was far less effective than the intensive weight loss. In fact, metformin did not reduce the incidence of type 2 diabetes in individuals aged 60 or older; only the intensive weight loss group showed a significant reduction in type 2 diabetes incidence in the 60+ group.

Figure 9.34 Four-year cumulative incidence of type 2 diabetes mellitus in patients participating in the Diabetes Prevention Program Research Group study. (Adapted from, Diabetes Prevention Program Research Group, Reduction in the incidence of type 2 diabetes with lifestyle intervention or metformin. *N Engl J Med* 326:393–403, 2002.)

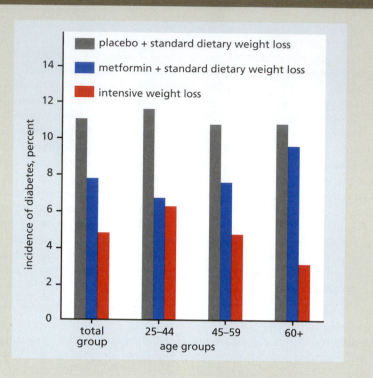

calcium concentration. Calcium is the most abundant mineral in the body. It interacts with phosphorus to form calcium phosphate, the hard, dense material that forms our bones and teeth. In fact, the human skeleton contains 99% of the body's calcium. The rest of the calcium is in the form of serum calcium ions (Ca^{2+}). Serum calcium is essential for the normal functioning of nerves and muscles and plays a role in blood coagulation and many enzymatic processes. In this section, we discuss the importance of serum calcium and calcium regulation; in the next section, we examine age-related bone loss.

Parathyroid and thyroid hormones balance blood calcium

The concentration of serum calcium needs to be maintained within a very narrow range of about 9–10 mg/dl. The maintenance of this Ca^{2+} concentration is so important that evolution has provided a gland, the **parathyroid gland**, with the sole function of regulating serum Ca^{2+} **(Figure 9.35)**. If the concentration of Ca^{2+} drops below 9 mg/dl, the parathyroid gland secretes **parathyroid hormone (PTH)**. This hormone increases serum Ca^{2+} by increasing the release of calcium from bone, increasing intestinal Ca^{2+} absorption, and increasing reabsorption of Ca^{2+} by the kidney tubule. If Ca^{2+} levels rise above 11–12 mg/dl, the thyroid gland secretes the hormone **calcitonin**. Calcitonin has the opposite effect of PTH: it inhibits calcium release from bone. Calcitonin has little to no effect on Ca^{2+} absorption by the intestine and reabsorption by the kidney tubule. Serum Ca^{2+} levels of 10–11 mg/dl do not stimulate calcitonin secretion. Rather, serum Ca^{2+} concentrations in this range are reduced to normal by increased urinary excretion.

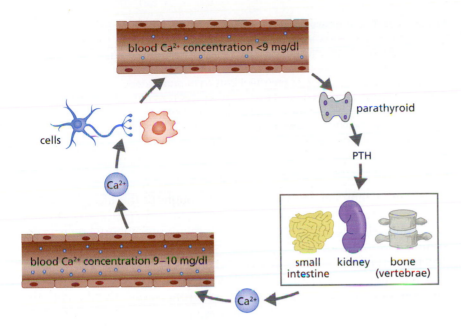

Figure 9.35 Regulation of serum calcium (Ca²⁺) by parathyroid hormone (PTH). The parathyroid gland, located just above and behind the thyroid gland in the upper neck, releases PTH in response to blood Ca^{2+} concentrations below 9 mg/dl. PTH stimulates increases in Ca^{2+} reabsorption by the kidney tubule, Ca^{2+} resorption from bone, and Ca^{2+} absorption by the small intestine. Serum Ca^{2+} is extracted from the blood by cells and used for several different physiological functions, including maintenance of neural conduction, muscle contraction, and intracellular signaling.

Hormones regulate the balance between bone mineral deposition and resorption

Hormones also help regulate the process of **bone remodeling**, the continual breakdown and renewal of adult bone tissue **(Figure 9.36)**. There are two types of bone: trabecular (or spongy) and compact. Trabecular bone is highly vascularized, and its structure creates a large surface area for bone-vessel contact. That is, Ca^{2+} can move into and out of bone more easily in trabecular bone than in compact bone. The bone cells responsible for remodeling include the **osteoblasts**, **osteoclasts**, and, to a lesser degree, the **osteocytes**. In response to PTH, osteoclasts secrete proteolytic enzymes that break down collagenous matrix containing the Ca^{2+} salt **hydroxyapatite** (Ca_{10} $(PO_4)_6(OH)_2$). Hydroxyapatite gives bone its hardness and strength. Osteoclasts also secrete acids that break apart the hydroxyapatite molecule into a variety of Ca^{2+}-linked amorphous salts, primarily calcium phosphate Ca_3 $(PO_4)_2$. Formation of these amorphous salts serves two critical functions. First, the salts provide a pool of Ca^{2+} that can be released quickly in response to PTH. Second, the salts are used by osteoblasts in the formation of new bone tissue.

The resorption of bone Ca^{2+} by osteoclast activity leaves a hole in the bone that was previously occupied by hydroxyapatite crystals. Without the formation of new bone tissue, bone mineral content would decrease and the bone's overall strength would diminish. New bone formation begins when macrophages release growth factors during their "cleanup" of the products of osteoclast activity. These growth factors—including transforming growth factor beta (TGF-β), platelet-derived growth factor (PDGF), and insulin-like growth factors 1 and 2 (IGF-1 and IGF-2)—stimulate differentiation of the preosteoblast into the mature osteoblast. The mature osteoblast secretes matrix material, primarily collagen, into the demineralized area of the bone. It also secretes **alkaline phosphatase**, an enzyme that aids in precipitation of amorphous salts into the newly formed matrix. Over the next two to three months, stochastic processes transform amorphous Ca_3PO_4 salts into hydroxyapatite crystals, ending the remodeling cycle for this section of bone.

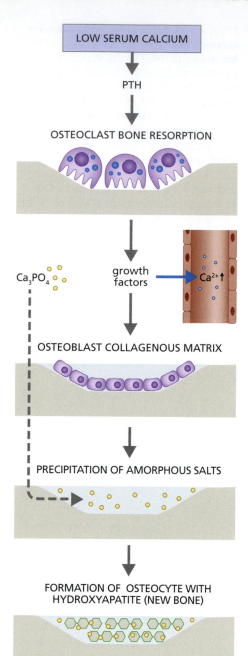

LOW SERUM CALCIUM

PTH

OSTEOCLAST BONE RESORPTION

Ca_3PO_4

growth
factors

$Ca^{2+}\uparrow$

OSTEOBLAST COLLAGENOUS MATRIX

PRECIPITATION OF AMORPHOUS SALTS

FORMATION OF OSTEOCYTE WITH
HYDROXYAPATITE (NEW BONE)

Figure 9.36 The process of bone remodeling. When serum Ca^{2+} concentration falls below 9 mg/dl, the parathyroid gland secretes parathyroid hormone (PTH), which activates osteoclast activity. Osteoclasts digest the bone matrix, releasing amorphous salts, growth factors, and Ca^{2+} ions. The growth factors stimulate conversion of preosteoblasts to mature osteoblasts, which fill in the resorbed area with collagenous matrix. Osteoblasts also release alkaline phosphatase, which aids in the precipitation of amorphous salts in the newly formed collagenous matrix. Over time, the amorphous salts form hydroxyapatite, which leads to the production of osteocytes and new bone.

Bone remodeling occurs continuously throughout the life span and is regulated by several hormones and growth factors **(TABLE 9.7)**. Greater bone mineralization during early development (pre-puberty) reflects the stimulatory influence on the osteoblast of growth-inducing hormones and growth factors such as growth hormone, insulin, IGF-1, and calcitriol (vitamin D_3). Testosterone and estrogen in both men and women promote bone growth at puberty. Testosterone seems to stimulate the differentiation of preosteoblasts into mature osteoblasts. Women with high levels of testosterone have a longer period of bone growth after the start of puberty and thus have greater peak bone mass. Estrogens promote bone growth in two ways: (1) by stimulating osteoblast activity and (2) by inhibiting osteoclast activity. While the exact mechanism by which estrogen inhibits osteoclast activity remains unknown, this hormone may act by stimulating the osteoblast's release of a protein (osteoprotegerin) that inhibits the secretion of proteolytic enzymes by osteoclasts.

AGE-RELATED DISEASES OF BONE: OSTEOPOROSIS

The average height of women and men decreases after the age of 50. The loss in height reflects a slight compression of the vertebrae caused by declining bone mineral content after peak bone mass has been achieved **(Figures 9.37** and **9.38)**. The accelerated loss of bone mass in women after menopause (~50+) reflects declining estrogen levels and places women at a significantly greater risk of developing

TABLE 9.7
EFFECT OF SELECTED HORMONES ON BONE MINERALIZATION

Hormone	Bone cell affected	Primary effect on bone mineral	Age-related increase/ decrease in secretion
Parathyroid hormone (PTH)	Osteoclast	Increased resorption	No change
Estrogen	Osteoblast Osteoclast	Increased deposition Decreased resorption	Decrease after menopause
Testosterone	Osteoblast	Increased deposition	Decrease after peak bone mass
Insulin/IGF-I	Osteoblast	Increased deposition	Decrease after puberty
Growth hormone	Osteoblast	Increased deposition	Decrease after puberty
Calcitonin	Osteoblast	Increased deposition	No change
Thyroid hormone	Osteoblast	Increased deposition	Decrease after age 50–60 years

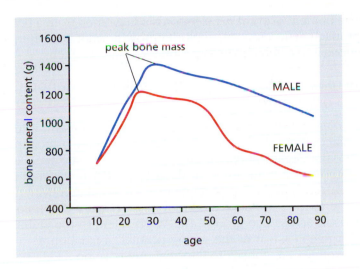

Figure 9.37 Total body bone mineral content (BMC) in males and females throughout the life span. Females tend to achieve peak bone mass at a younger age than males. The slow rate of BMC loss after peak bone mass is approximately the same for men and women, until menopause (50+ years of age). Note the accelerated rate of decline in BMC in women after the age of 50. (Data from World Health Organization, Prevention and Management of Osteoporosis, WHO Technical Report Series 921, Geneva: World Health Organization, 2003.)

osteoporosis, a bone disease characterized by low bone mass and strength, which leads to an increased risk of fracture. In fact, 80% of all cases of osteoporosis occur in women. Although men also experience age-related bone loss, they do not, in general, develop primary osteoporosis (that caused only by age-related bone loss) until the tenth or eleventh decade of life. Younger men can, however, develop **secondary osteoporosis**, caused by such things as medication, cancer, and kidney disease.

The increased rate of bone loss in women after menopause provides us with a rare opportunity to explore, in humans, the relationship between age-related loss and reproductive life span. In this section, we discuss

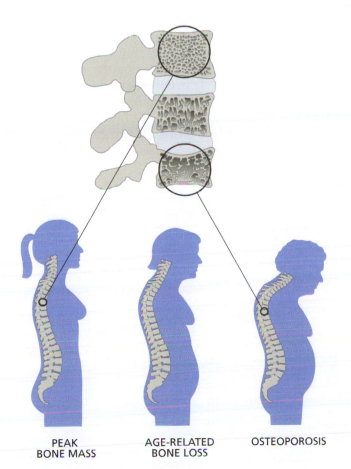

PEAK BONE MASS   AGE-RELATED BONE LOSS   OSTEOPOROSIS

Figure 9.38 Progression of vertebral bone loss in women, causing a decrease in height. The age-related loss of bone calcium results in compression of the vertebral body, which causes the entire spinal column to shrink, decreasing height. Excess bone loss in the vertebrae results in osteoporosis and a curvature of the spine sometimes known as a dowager's hump.

the development of primary osteoporosis in women from a perspective of reproductive senescence, exploring how normal bone mineral loss transitions into osteoporosis in postmenopausal women.

An increased rate of bone mineral loss at menopause can lead to osteoporosis

The increased rate at which women lose bone mineral after menopause places them at a significant risk for developing osteoporosis. The World Health Organization has suggested a diagnostic classification of osteoporosis as a **bone mineral density (BMD)** (measured in grams of mineral/cm²) at the hip or spine of ≤ 2.5 standard deviations (SDs) below the young population's normal mean **(Figure 9.39)**. For ease of comparison across age groups, the WHO refers to the statistical distribution (mean ± SD) of the young normal population as a *T*-score. *T*-scores of ≥ -1 are considered normal and pose no additional risk for developing osteoporosis. Values between –1 and –2.5 indicate a low bone mass, clinically known as **osteopenia**, and an increased risk of developing osteoporosis. Interestingly, the normal BMD for women over the age of 60 declines to the low bone mass range, suggesting that age is a major risk factor for osteoporosis. Although the WHO distribution indicates that women 90+ years of age normally have osteoporosis, limited data for this population make these values unreliable.

Environmental factors influence the risk of developing osteoporosis

More than 100 years ago, Julius Wolff, a German surgeon, found that bones become thicker—that is, BMD increases—when placed under increased load. His observation led to the development of Wolff's law: bones grow in direct proportion to the load placed upon them. For example, the BMD of the radius (the large bone in the forearm) in the dominant arm of professional tennis players is consistently greater than that of the radius in the nondominant arm. Intensive load-bearing exercise increases BMD, and disuse of bone leads to a decrease in BMD. This effect has classically been observed in patients undergoing long periods of bed rest and in astronauts after spending time free from Earth's gravitational force. Thus, lack of physical activity can increase the risk of developing osteoporosis. However, the effectiveness of moderate exercise programs in increasing BMD remains uncertain.

Figure 9.39 The WHO's diagnostic criteria for osteoporosis in women. Osteoporosis is defined as a bone mineral density (BMD, g/cm²) at the hip or spine of ≤2.5 standard deviations below the young normal mean reference (premenopausal, 30–40 years of age). This value is presented as a *T*-score (*left y-axis*). The graph lines represent the population distribution of BMD for women at various ages (*right y-axis*). (Data from World Health Organization, Prevention and Management of Osteoporosis, WHO Technical Report Series 921, Geneva: World Health Organization, 2003.)

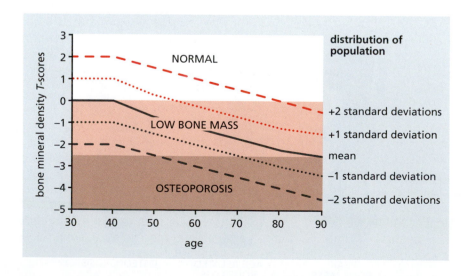

The greatest gains in BMD among postmenopausal women who start a light to moderate weight-bearing exercise program occur in previously sedentary individuals. Only small gains in BMD are observed in women who exercise regularly.

Recall that estrogen modifies the full effect of PTH-induced bone resorption by stimulating osteoclasts' deposition of bone mineral and inhibiting resorption by osteoblast. When estrogen concentrations decline during and after menopause, the bone dumps greater amounts of Ca^{2+} into the serum, and, in turn, excretion of calcium in the urine increases in order to maintain serum Ca^{2+} in the 9–10 mg/dl range. The increase in urinary calcium can often be greater than dietary intake, a phenomenon known as negative Ca^{2+} balance. Increasing the dietary intake of Ca^{2+} often eliminates a negative Ca^{2+} balance in young, premenopausal women and helps to maintain bone mass. However, similar treatment in postmenopausal women has not been as effective in maintaining premenopausal levels of BMD. The National Academy of Sciences has set the recommendation for daily Ca^{2+} intake by postmenopausal women at 1200 mg/day. This amount has proved effective in combination with medication and increased physical activity for previously sedentary individuals.

Intestinal absorption of dietary Ca^{2+} requires vitamin D. Older individuals' declining dietary intake and limited exposure to ultraviolet light (skin synthesizes vitamin D on exposure to UV light) may decrease the amount of vitamin D available for appropriate Ca^{2+} absorption. Increasing the dietary intake of vitamin D has been shown to increase the absorption of Ca^{2+} in postmenopausal women. However, the vitamin D-induced increase in Ca^{2+} absorption has not been shown to increase BMD in non-osteoporotic, postmenopausal women. Nonetheless, because the data are conflicting, clinicians often recommend increasing the intake of vitamin D along with Ca^{2+} as a precautionary measure.

Women's bone mineral content reaches its peak at the age of ~25 years. Thereafter, bone mineral content slowly declines until menopause, when the rate of mineral loss sharply increases. The rate of loss in both the pre- and postmenopausal age groups does not depend on the amount of bone mineral at the time of peak bone mass. This phenomenon is shown nicely in Figure 9.39. Women with a peak bone mineral content ≥1 SD above the mean have the same rate of loss as do other groups. However, these women do not reach the osteoporosis threshold during their life span. In fact, women with exceptionally dense bones (+2 SD above mean) may not even reach the low bone mass threshold. In other words, the more bone mineral you have at peak bone mass, the more bone mineral you will have later in life. Attaining one's genetic limit for bone mineral content at the time of peak bone mass has been widely accepted as a primary prevention strategy for reducing one's risk of developing osteoporosis.

Recommendations aimed at helping young women achieve their maximal peak bone mass have, for the most part, focused on improving dietary Ca^{2+} and vitamin D intake, as well as increasing weight-bearing exercise. The chances of achieving maximal peak bone mass are greatly enhanced if young women follow the recommendations suggested by the Food and Nutrition Board of the National Academy of Sciences. That is, girls under the age of 18 years should consume at least 1300 mg of Ca^{2+} and 5 mg of vitamin D per day. In addition, high dietary intake of specific vitamins and minerals should be accompanied by increased weight-bearing physical activity.

Drug therapies can slow bone loss in postmenopausal women

Focusing on preventive measures for young women does not imply that older women will not benefit from exercise or dietary improvement. However, increasing weight-bearing activity, dietary Ca^{2+}, and/or vitamin D levels only slows, and does not prevent, bone loss in pre- or postmenopausal women. Stopping the loss and increasing bone mineral content after peak bone mass or menopause can be achieved only through pharmacotherapy. However, increased physical activity along with increased Ca^{2+} and vitamin D intake in combination with pharmacotherapy significantly enhances bone mass and reduces the rate of fracture compared with drugs alone. Here we look at three of the more common types of drugs used to increase bone mass in postmenopausal women: estrogen, specific estrogen receptor modulators (SERMs), and bisphosphonates.

Since estrogen stimulates osteoblast activity (bone formation) and inhibits osteoclast activity (bone resorption), it should not be surprising that estrogen replacement (also known as hormone replacement therapy, or HRT) has found significant acceptance as a therapeutic agent in preventing postmenopausal bone loss. Virtually every study investigating the use of HRT as a treatment for osteoporosis has found increased bone mass and/or a decreased rate of fracture. Unfortunately, recent evidence reveals that the use of HRT at levels needed to increase bone mass also increases the risk of breast cancer. The increased risk of breast cancer has significantly reduced the use of estrogen as the first line of defense against excessive bone loss and osteoporosis. Estrogen replacement for prevention or treatment of osteoporosis is now relegated to use when other treatments fail or when the risk of health problems due to fracture outweighs the risk of cancer.

SERMs are a class of compounds that have selective actions, depending on which estrogen receptor, α or β, they bind to. For example, tamoxifen (Nolvadex) increases bone density by stimulating osteoblast activity but inhibits the growth of breast cancer tissue by blocking the binding of endogenous estrogen to its receptor. Tamoxifen's severe side effects and relatively weak action on bone have led to its use primarily as a cancer treatment. Raloxifene (Evista®) was developed specifically to treat osteoporosis, although it has recently been approved for treating women at high risk for breast cancer. Clinical trials have found that raloxifene is more effective than tamoxifen at reducing hip and vertebral factures, but is less effective than estrogen. Raloxifene does not produce the severe side effects observed with tamoxifen—namely, uterine inflammatory diseases and an increased risk of uterine cancer.

Bisphosphonates are a group of compounds that induce apoptosis in osteoclasts and thus inhibit bone resorption. Bisphosphonates disrupt normal ATP synthesis by creating a carbon-phosphate bond in place of the normal oxygen-phosphate bond. Because ATPase can break only the oxygen-phosphate bond, the carbon-phosphate ATP builds to toxic levels in the osteoclast and induces apoptosis. The three FDA-approved bisphosphonates are alendronate (Fosamax®), ibandronate (Boniva®), and risedronate (Actonel®). All three are equally effective at slowing the rate of bone mineral loss in postmenopausal women and decreasing the risk of hip and vertebral fractures. There remains some controversy as to whether bisphosphonates help increase bone mass or simply stop the loss of bone mineral content.

ESSENTIAL CONCEPTS

- The human nervous system is composed of the central nervous system (CNS), consisting of the brain and spinal cord, and the peripheral nervous system (PNS), consisting of all other nerves outside the CNS.

- Neural cells (neurons) are made up of three basic components: dendrites, the cell body, and the axon.

- Neurotransmission is propagated by establishment of an action potential, a brief depolarization of a small section of the nerve cell membrane.

- Neurotransmitters relay electrical messages from one neuron to another across synapses. A synapse is made up of the axon terminal of the presynaptic neuron, the dendrite of the postsynaptic neuron, and the synaptic cleft.

- Age-related changes in the human brain are minor and do not seem to significantly affect brain function.

- Accumulation of two protein aggregates, amyloid plaques and neurofibrillary tangles, in the human brain appears to be a normal age-related phenomenon.

- The cause of Alzheimer's disease is unknown, but individuals with Alzheimer's are found to have high concentrations of amyloid plaques and neurofibrillary tangles.

- There are three general forms of Alzheimer's disease: early-onset, late-onset, and familial. Early-onset and familial Alzheimer's are genetic in origin, whereas the late-onset form does not seem to be inherited.

- Parkinson's disease results from the loss of dopamine-producing neurons in the substantia nigra region of the basal ganglia.

- The heart consists of two separate pumps: the right side delivers oxygen-poor blood to the pulmonary system (lungs); the left side supplies oxygen-rich blood to the body.

- The muscle of the heart and arteries is excitable tissue. That is, the muscle generates action potentials in response to neural and hormonal stimulation.

- Cardiac output is determined by the amount of blood returned to the heart. Stretching of the heart muscle in response to venous return determines the force necessary to eject blood from the ventricles.

- Arterial plaques, accumulations of cholesterol-filled fatty deposits in the arteries, are present in 100% of the aged population. Excess accumulation of arterial plaques results in the disease atherosclerosis.

- Age, smoking, hypertension, and hyperlipidemia are the major risk factors for coronary artery disease.

- More than 50% of individuals 65 years and older have hypertension (high blood pressure), defined as a consistent systolic blood pressure of ≥ 140 mm Hg or diastolic pressure of ≥ 90 mm Hg.

- Heart failure can be defined as the inability of the myocardium to generate the contractile force needed to eject the blood volume required for the oxygen demands of the body—that is, an abnormal decrease in cardiac output.

- Blood glucose is maintained within a narrow range of 90–120 mg/dl by the competitive action of two hormones produced by the endocrine cells of the pancreas: insulin and glucagon. This level of blood glucose must be maintained because the brain uses only blood glucose for its energy needs.

- Insulin facilitates the uptake of glucose into cells by initiating an intracellular signaling cascade. Insulin can also regulate the metabolism of fats and proteins.

- Insulin resistance is the inability of insulin to effectively stimulate glucose uptake into cells. It is a major risk factor for type 2 diabetes.

- The major outcome of type 2 diabetes is damage to microvascular blood flow. Without proper blood flow, tissue can be damaged, leading to tissue death.

- Estimates indicate that more than 80% of new cases of type 2 diabetes are caused by obesity. Maintaining a healthy weight is considered the best preventive measure.

- Bone mineral content declines during the post-reproductive period in both men and women, but the decrease is significantly greater in women.

- Estrogen promotes bone growth by stimulating osteoblast activity and inhibiting osteoclast activity. Thus, at menopause, the loss in bone mineral content that normally occurs between the ages of 25 and 50 begins to accelerate.

- Excess bone loss can result in osteoporosis and a significantly increased risk of bone fracture; 80% of all individuals with osteoporosis are women.

- Exercise and dietary calcium, and vitamin D can affect bone mineral content. Although increasing the level of physical activity and dietary calcium after menopause has proved effective in slowing bone mineral loss, their combined effect is modest and has only a minor impact on the development of osteoporosis

- The primary preventive measure for osteoporosis is to encourage women between the ages of 14 and 30 to increase their level of exercise and their intake of calcium to achieve the greatest possible peak bone mass.

DISCUSSION QUESTIONS

Q9.1   As you learned in previous chapters, an accumulation of damaged proteins might be a symptom of aging. Apply this theory to the aging brain and the development of Alzheimer's and Parkinson's disease. Include an explanation of why the β-sheet protein structure lends itself to the formation of protein aggregates such as amyloid plaque.

Q9.2   Name the three types of Alzheimer's disease. Which type(s) has/have a strong genetic basis? Which type is the most common?

Q9.3 Consider the anatomical structure of an artery. Describe how age-related changes in the artery's architecture can result in changes in blood flow and pressure. Include fluid dynamics in your explanation.

Q9.4 Evaluate the graph of the Frank-Starling mechanism below **(Figure 9.40)**. How does understanding this mechanism provide an explanation for the cause of congestive heart failure?

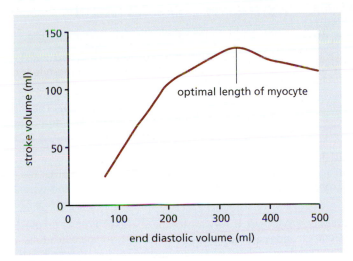

Figure 9.40

Q9.5 Describe how coronary artery disease can lead to ventricular fibrillation and death.

Q9.6 Briefly state why prevalence may be better than mortality rate as a measure of cardiovascular disease.

Q9.7 The diagnosis of type 2 diabetes is made through an oral glucose tolerance test. Briefly describe the glucose tolerance test and the criteria used to establish type 2 diabetes. What are the criteria for impaired glucose tolerance without type 2 diabetes?

Q9.8 Relate an increase in AGEs to damage caused by type 2 diabetes.

Q9.9 Explain why preventive strategies for osteoporosis target young women (14–30 years old) rather than postmenopausal women.

Q9.10 Although declining bone mineral content seems to be a normal age-related event, the rate of loss is generally greater in postmenopausal women than in men of the same age. Explain why.

FURTHER READING

THE NERVOUS SYSTEM AND NEURAL SIGNALS

Widmaier EP, Raff H & Strang KT (2004) Neuronal signaling and the structure of the nervous system. In Human Physiology: Mechanisms of Body Functions, 9th ed. Boston: McGraw Hill, pp 153–204.

AGE-RELATED DISEASES OF THE HUMAN BRAIN: ALZHEIMER'S DISEASE AND PARKINSON'S DISEASE

Dawbarn D & Allen SJ (eds) (2007) Neurobiology of Alzheimer's Disease, 3rd ed. London: Oxford University Press.

Devine MJ & Lewis PA (2008) Emerging pathways in genetic Parkinson's disease: tangles, Lewy bodies and LRRK2. FEBS J 275:5748–5757.

Mobbs C (2006) Aging of the brain. In Principles and Practice of Geriatric Medicine (Pathy MSJ, Sinclair AJ, Morley JE eds), 4th ed. Hoboken, NJ: Wiley, p 47–51.

National Institute on Aging (2008) Alzheimer's Disease: Unraveling the Mystery. www.nia.nih.gov/sites/default/files/alzheimers_disease_unraveling_the_mystery.pdf.

Nourhashemi F, Sinclair AJ & Vellas B (2006) Clinical Aspects of Alzheimer's Disease. In Principles and Practice of Geriatric

Medicine (Pathy MSJ, Sinclair AJ, Morley JE eds), 4th ed. Hoboken, NJ: Wiley, pp 1083–1093.

Parkinson J (2002) Classic articles: an essay on shaking palsy. *J Neuropsychiatry Clin Neurosci* 14:223–236.

Plassman BL, Langa KM, Fisher GG et al (2007) Prevalence of dementia in the United States: the aging, demographics, and memory study. *Neuroepidemiology* 29:125–132.

Playfer JR (2006) Parkinson's disease and parkinsonism in the elderly. In Principles and Practice of Geriatric Medicine (Pathy MSJ, Sinclair AJ, Morley JE eds), 4th ed. Hoboken, NJ: Wiley, pp 765–776.

Shenk D (2001) The Forgetting: Alzheimer's, Portrait of an Epidemic. New York: Random House.

Verhey FRJ (2009) Alois Alzheimer (1864-1915). *J Neurol* 256:502–503.

Wong PC, Price L & Tanzi RE (2008) Alzheimer's disease: genetics, pathogenesis, models, and experimental therapeutic. In Molecular Biology of Aging (Guarente LP, Partridge L, Wallace DC eds). Cold Spring Harbor, NY: Cold Spring Harbor Laboratory Press, pp 371–407.

Yoshikai S, Sasaki H, Doh-ura K et al (1990) Genomic organization of the human amyloid beta-protein precursor gene. *Gene* 87:257–263.

THE CARDIOVASCULAR SYSTEM

Widmaier EP, Raff H & Strang KT (2004) Cardiovascular physiology. In Human Physiology: Mechanisms of Body Functions, 9th ed. Boston: McGraw Hill, pp 375–466.

AGE-RELATED DISEASES OF THE CARDIOVASCULAR SYSTEM: CARDIOVASCULAR DISEASE

Lakatta EG (2003) Arterial and cardiac aging: major shareholders in cardiovascular disease enterprises. Part III: cellular and molecular clues to heart and arterial aging. *Circulation* 107:490–497.

Lakatta EG & Levy D (2003) Arterial and cardiac aging: major shareholders in cardiovascular disease enterprises. Part I: aging arteries—a "set up" for vascular disease. *Circulation* 107:139–146.

Lakatta EG & Levy D (2003) Arterial and cardiac aging: major shareholders in cardiovascular disease enterprises. Part II: the aging heart in health—links to heart disease. *Circulation* 107:346–354.

National Center for Health Statistics (2011) Health, United States, 2010: With Special Feature on Death and Dying. Hyattsville, MD: Centers for Disease Control and Prevention, p 563.

Patterson C (2006) In Handbook of Models for Human Aging (Conn PM ed). Boston: Elsevier, pp 865–871.

Stott DJ (2006) In Principles and Practice of Geriatric Medicine (Pathy MSJ, Sinclair AJ, Morley JE eds), 4th ed. Hoboken, NJ: Wiley, p 96.

THE ENDOCRINE SYSTEM AND GLUCOSE REGULATION

Widmaier EP, Raff H & Strang KT (2004) Chapters 11 and 16. In Human Physiology: The Mechanisms of Body Function, 9th ed. Boston: McGraw Hill, pp 331–374; 605–642.

AGE-RELATED DISEASES OF THE ENDOCRINE SYSTEM: TYPE 2 DIABETES MELLITUS

American Diabetes Association. www.diabetes.org.

Diabetes Prevention Program Research Group (2002) Reduction in the incidence of type 2 diabetes with lifestyle intervention or metformin. *N Engl J Med* 326:393–403.

Gregg EW, Cheng YJ, Narayan KM et al (2007) The relative contributions of different levels of overweight and obesity to the increased prevalence of diabetes in the United States: 1976–2004. *Prev Med* 45:348–352.

Magkos F, Yannakoulia M, Chan JL & Mantzoros CS (2009) Management of the metabolic syndrome and type 2 diabetes through lifestyle modification. *Annu Rev Nutr* 29:223–256.

Whitlock G, Lewington S, Sherliker P et al (2009) Body-mass index and cause-specific mortality in 900,000 adults: collaborative analyses of 57 prospective studies. *Lancet* 373:1083–1096.

THE SKELETAL SYSTEM AND BONE CALCIUM METABOLISM

Guyton AC & Hall JE (2001) Parathyroid hormone, calcitonin, calcium and phosphate metabolism, vitamin D. Bone, and teeth. In Textbook of Medical Physiology, 10th ed. Philadelphia: W. B. Saunders Company, pp 899–915.

AGE-RELATED DISEASES OF BONE: OSTEOPOROSIS

Beck BR & Snow CM (2003) Bone health across the lifespan—exercising our options. *Exerc Sport Sci Rev* 31:117–122.

Frost HM (2001) From Wolff's law to the Utah paradigm: insights about bone physiology and its clinical applications. *Anat Rec* 262:398–419.

Seeman E (2003) Invited review: Pathogenesis of osteoporosis. *J Appl Physiol* 95:2142–2151.

Standing Committee on the Scientific Evaluation of Dietary Reference Intakes, Food and Nutrition Board, Institute of Medicine (1997) Institute of Medicine Dietary Reference Intakes for Calcium, Phosphorus, Magnesium, Vitamin D, and Fluoride. Washington, DC: National Academies Press.

Suzuki A, Sekiguchi S, Asano S & Itoh M (2008) Pharmacological topics of bone metabolism: recent advances in pharmacological management of osteoporosis. *J Pharmacol Sci* 106:530–535.

Troen BR (2003) Molecular mechanisms underlying osteoclast formation and activation. *Exp Gerontol* 38:605–614.

World Health Organization (2003) Prevention and Management of Osteoporosis. WHO Technical Report Series 921. Geneva: World Health Organization.

MODULATING HUMAN AGING AND LONGEVITY

10

"THE IDEA IS TO DIE YOUNG AS LATE AS POSSIBLE."

-ASHLEY MONTAGU, ANTHROPOLOGIST (1905-1999)

Humans have searched for the fountain of youth since the beginning of recorded history. Records of the primitive cultures and the religions that have shaped the societies of today tell of individuals who stepped outside the bounds of the normal life span. Abraham—the patriarch of Muslims, Jews, and Christians—lived to 200 years of age. And Abraham was a mere teenager compared with the biblical character Methuselah, who was said to have lived for 969 years. In addition, many religions involve a core belief in a place where the pain and suffering of old age are alleviated—heaven or nirvana, for example. A belief in the possibility of a long life and an end to pain reflects our most basic instinct: survival.

Scientists are not immune to the cultural forces and instinctive behaviors that drive our desire to extend life. Some biogerontologists believe that we are living in a time when the modulation of aging and longevity will soon be commonplace. These scientists suggest, for example, that with advances in stem cell research, we will be able to grow new organs from our own DNA to replace those worn out by age. The lay press and the scientific literature abound with prognostications about living to 100 years of age and why it will become the rule rather than the exception.

Are these predictions of a longer life based on a realistic interpretation of current research? Or are they simply the hopes and longings of humans striving to extend their time on Earth? In this chapter, we discuss the current state of science on the modulation of aging and longevity. We focus first on biological aging, then on the only two interventions that are scientifically established to modulate the rate of aging or longevity: (1) reducing caloric intake and (2) maintaining physical activity throughout life. Finally, we take a look at some possible implications of halting human aging and/or extending the life span.

MODULATING BIOLOGICAL AGING

As you learned in earlier chapters, the causes of aging and longevity are now known. Aging is caused by random, stochastic damage to cellular molecules that leads to altered cell function. Longevity—here meaning the length of the life span, independent of aging—has arisen as a by-product of genes selected for reproductive success. The biochemical and physiological mechanisms that underlie aging and longevity are not yet fully understood. Nonetheless, many biogerontologists agree on one point: the rate of aging and alterations in longevity reflect the intracellular accumulation of damaged proteins.

In this section, we discuss why aging cannot be modulated, then look at what types of research might be needed to better understand the biological basis of aging.

Aging cannot be modulated

We live. We grow old. We die. Although aging is something most of us would like to stop, or at least slow down, the simple truth is that aging cannot be modified. To understand why this is so, you must accept as fact three important principles of biological systems that we discussed in previous chapters: (1) aging did not evolve; (2) biological organisms are subject to the same laws of thermodynamics as inanimate objects; and (3) the second law of thermodynamics operates constantly and randomly. Since aging did not evolve, there are no genes that regulate the process. Aging must therefore be a random event. The randomness of aging arises because of the second law of thermodynamics. Millions of reactions take place in an organism, and each one must conform to the first and second laws of thermodynamics. This fact is nonnegotiable with the universe. Forces of the universe push every chemical reaction, even the minutest, toward increasing entropy and disorder. At some point, in every organism, one of these reactions will attempt to take place in a system where the entropy exceeds the free energy. There will be a loss in molecular fidelity and an accumulation of damaged proteins. This will start a chain reaction that will end with the loss of cellular function. Over time, all the cells in the human body experience a loss of molecular fidelity. Humans are not able, and most likely never will be able, to alter this fundamental fact of the universe.

One can argue, however, that the laws of thermodynamics apply to a closed system, a system having no input from the environment. Organisms are open systems, constantly interacting with the environment, and perhaps we can intervene to offset the effect of the second law. Indeed, some suggest that, in the twentieth century, the human intervention that led to the unprecedented increase in life expectancy proves that biological systems can successfully combat the second law. A close inspection of the gains in life expectancy over the past 100 years, however, suggests only that we made headway against disease. Aging remains. As shown in **Figure 10.1**, there was indeed a rapid increase in life expectancy during the twentieth century and the first decade of the twenty-first century. But, the increased life expectancy between 1900 and 1950 was achieved by curtailing the infant mortality rate, reducing deaths due to childhood diseases, and reducing deaths due to infections that killed most people before the end of their reproductive period. The slower gains after 1960 were the result of advances in medical technology that reduced deaths from diseases that had been major killers of older people before they reached an advanced age. For example, individuals who would have died in their fifties or sixties from a heart attack are now living into their seventies and eighties because of technologies that diagnose and fix the problems before they become fatal. Thus, the increase in life span during the twentieth century was the result of modulating age-related disease, not aging.

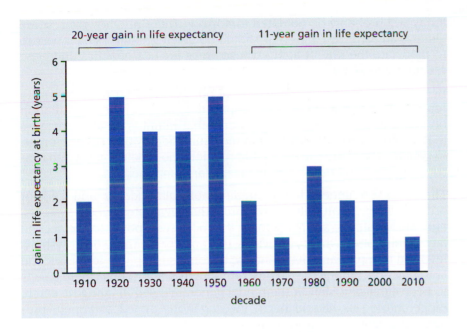

20-year gain in life expectancy · 11-year gain in life expectancy

Figure 10.1 Gain in life expectancy at birth by decade, 1910–2010. Note that 60% of the gain in life expectancy occurred during the first 50 years of the twentieth century, reflecting a decrease in infant mortality and improvements in disease control. (Data from F.C. Bell and M.L. Miller, Life Tables for the United States Social Security Area 1900–2100, Washington, DC: Social Security Administration, 2005, p 194.)

Some scientists suggest that the rapid advancements in biotechnology will lead to anti-aging therapies. These researchers point to advances in stem cell research that show great promise for restoring age-related functional loss. Others suggest that organs grown *in vitro* from our own cells will replace those damaged by age or disease. Both types of therapy will undoubtedly be realities in the future and will lead to increased life expectancy. But the question before us is, Will these interventions modulate aging? The answer is an unequivocal "no"; they will only postpone the inevitable. As soon as you fix one tissue or organ, another system will fail, and so on and so on. Knee replacement, a common surgery in the elderly population, has not stopped kidney cells from accumulating damage. Perpetual motion in humans is impossible, just as the nineteenth-century physicists who developed the laws of thermodynamics predicted.

Mechanisms that lead to loss of molecular fidelity may be modulated in the future

Since its beginning, biogerontological research has focused almost exclusively on trying to alter the process of aging by correcting damage that has already been done. This approach to modulating aging will never be successful: increasing entropy is a fundamental law of the universe, and all matter strives to reach energy equilibrium. We can, however, modulate the *rate* of aging by evaluating mechanisms that lead to the loss of molecular fidelity.

All age-related disease will be cured someday, and we will be left with nothing more than increasing entropy as the cause of death. If we are to modify the rate of aging, biogerontological research must focus on the mechanisms of aging, not on the mechanisms of age-related disease. Biogerontology should be evaluating why, given the laws of thermodynamics, organisms survive at all. In other words, biogerontologists need to stop asking "Why do we die?" and begin asking "Why do we live?"

It is not easy to predict what specific research will be undertaken when biogerontologists frame their research in the why-do-we-live context. With the rapid pace of discoveries in biology, any predictions made today will probably be outdated tomorrow. There are, however, some general areas of research that should be given greater focus if we are to modulate the rate of aging. These areas include genetics and gene regulatory systems.

Because evolution has selected all of our genes to ensure survival to reproductive age, it would seem appropriate to focus greater attention on genes and gene regulatory systems that maintain usable energy—that is, those genes that maintain molecular fidelity and cellular order. In general, these are the genes that regulate the repair of DNA and proteins and the removal of damaged cellular components.

The focus on genes that maintain molecular fidelity and cellular order will lead naturally to research that evaluates when systems, and what systems, are likely to be the most susceptible to the second law. Evolutionary theory provides the answer to the *when*: systems involved in repair and maintenance become susceptible to the second law after the organism reaches reproductive age. Thus, there needs to be a significant shift in the model generally used in aging research. Biogerontologists should begin investigations aimed at young, pre-reproductively active and reproductively active populations, without making comparisons with the aged (post-reproductive) population. Comparison with the aged population should occur only after achieving a clear understanding of genetic pathways or regulatory systems in younger populations that are most susceptible to the second law. Then, research can begin to test whether or not these systems affect the rate of aging.

Given the current state of biotechnology, answering the question of *what* system or systems are most susceptible to increasing entropy is made difficult by the randomness of the second law. Biogerontologists need significant advances in genomic research (a topic discussed in Chapter 5) before they can identify genetic pathways that are likely to be important to the rate of aging (some of this research has begun in invertebrates). This will undoubtedly require a greater use of mathematical models—a method almost nonexistent in biogerontology—to help predict outcomes of age-related functional loss resulting from gene expression (or non-expression) in young individuals.

Until general genomic research provides biogerontologists with the tools necessary for investigating the impact of increasing entropy on age-related functional loss, some areas of research might be helpful in the short term. General medical science and biogerontology have determined that the cardiovascular system tends to decay at a greater rate than other systems. Fatty streaks in arteries, precursors to an accumulation of damaged protein that leads to narrowing of the vessel, can be found in children as young as six months old. Using the cardiovascular system as a model, researchers can begin to understand how a loss in molecular fidelity begins. The development of certain types of cancer, particularly those likely to occur in younger populations, may also provide some understanding of why genetic pathways regulating damage and repair become altered.

MODULATING LONGEVITY: CALORIE RESTRICTION

The popular press is filled with stories of how specific foods and nutrients extend life span or increase health. Life spans of 130–140 years have been reported in individuals claiming that their long life can be attributed to special diets, such as diets consisting of apricots, yogurt, and "special" breads. Personal testimony on the life-extending and anti-aging benefits of vitamins E, A, B_{12}, and C appear regularly in books and popular magazines. But rigorous scientific evaluations of individual foods and nutrients reported to extend life have been unable to demonstrate any life-span extension or any specific health benefit beyond that associated with alleviating a particular deficiency (**BOX 10.1**). Only a reduction in caloric intake, commonly referred to as **calorie restriction (CR)**, or **dietary restriction (DR)**, has been shown to extend both mean and maximum life span.

The life-extending properties of calorie restriction have been effective in virtually all nonhuman species tested, including yeast, worms, flies, and rodents. Only a few genetically altered species do not respond to CR. In addition, recently completed CR investigations in nonhuman primates also find extended life span and a delay in or prevention of age-related diseases. The current data, particularly those on nonhuman primates, strongly suggest that reducing caloric intake may be one mechanism for modulating the rate of aging and/or increasing longevity in humans. In this section, we explore the use of CR as a method to modulate longevity and the possible effectiveness of this dietary intervention in extending life span and improving health in humans.

Calorie restriction increases life span and slows the rate of aging in rodents

A publication by McCay, Crowell, and Maynard in 1935 described, for the first time, an intervention that led to increased life span. As shown in **Figure 10.2**, these investigators found that reducing rats' caloric intake prolonged mean and maximum life span, when compared with rats allowed to consume as many calories as they wished (known as *ad libitum* feeding). The uniqueness of this experiment compared with previous investigations was that the investigators reduced only the caloric content of the food without altering the nutrient composition—that is, vitamins and minerals—needed to prevent deficiencies. Previous investigations had demonstrated an increased life span by food, rather than calorie, restriction. While longer life was achieved by food restriction, many animals died early in the life span due to nutrient deficiencies. Thus, McCay and colleagues showed that calorie restriction increased life span, whereas food restriction reflected only the fact that hardy animals capable of surviving nutritional deficiencies lived longer—that is, it reflected selective mortality.

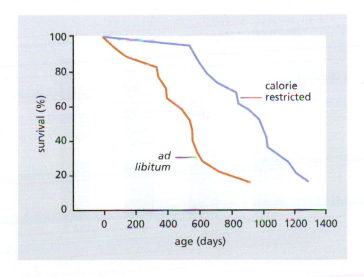

Figure 10.2 Life span of non-calorie-restricted (feeding *ad libitum*) versus 40% calorie-restricted male rats. These data, published by McCay and colleagues in 1935, were the first to show that calorie restriction without malnutrition increased life span in rodents. (Data from C.M. McCay, M.F. Crowell, and L.A. Maynard, *J. Nutr.* 10:63–79, 1935. With permission from the American Society for Nutrition.)

In the experiments by McCay and his colleagues, as well as numerous others, the CR was begun at or near weaning (about 3 weeks of age in mice and rats). As a result, calorie-restricted mice and rats were smaller in both weight and size than *ad libitum*-fed animals. This observation led most researchers to conclude that retarded growth and development was the mechanism underlying the life-extending proprieties of CR. It was not until the early 1980s that experiments showed that calorie restriction begun as late as 18 months of age in previously *ad libitum*-fed mice and rats—that is, in fully grown animals—increased mean and maximum life span.

The level of CR used in most experiments has been rather severe, at about 60–70% of *ad libitum* feeding. These values were experimentally established as the level that provided maximum life extension without early-life mortality due to starvation. The use of such levels of CR has been criticized, because it is unlikely that a 40% restriction could be achieved in humans (see the discussion on the effectiveness of CR in humans later in this section). To answer this criticism, researchers conducted life-span experiments that tested different levels of calorie restriction. The data from these studies showed that extension of life increases as caloric intake decreases **(Figure 10.3)**. Other data have shown that a reduction in calories of only 5–10% can increase life span, although only very modestly. The results from these experiments indicate that calorie restriction at levels appropriate for humans could extend the life span.

The CR procedures used today involve diets that ensure sufficient levels of vitamins and minerals to prevent nutritional deficiencies and malnutrition. Moreover, numerous experiments have shown that increasing the concentration of vitamins and minerals above the amount needed to prevent deficiencies does not increase life span. Together, the prevention of nutrient deficiency and the lack of effect with further supplementation strongly suggest that calories, rather than individual nutrients, are the underlying mechanism for the life-span extension observed under calorie restriction.

Calories are provided by the three **macronutrients**: proteins, fats (lipids), and carbohydrates (starches and sugars). To reduce the caloric content of the diet, the concentrations of these three macronutrients must be altered. Since these macronutrients have other functions in the body in addition to supplying calories, it is possible that the altered concentrations of proteins, fats, and carbohydrates, or some combination

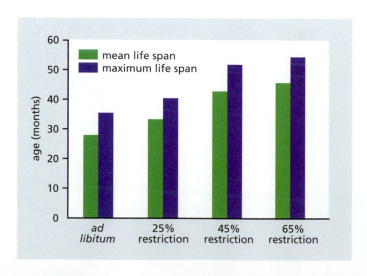

Figure 10.3 Mean and maximum life spans of male rats at different levels of calorie restriction. Note that the differences in mean and maximum life span between 25% and 45% CR are considerably larger than those between 45% and 65% CR. In fact, the researchers found that many rats died due to starvation in the 65% CR group, and they suggested that the degree of restriction should be no more than 50%. Nonetheless, these data show that life span increases as the degree of CR increases. (Data from R. Weindruch et al., *J. Nutr.* 116:641–654, 1986. With permission from the American Society for Nutrition.)

of the three, may be the factor leading to life extension in CR. However, several CR studies using various levels of macronutrients have shown that, in general, changes in calories, not in macronutrients, are responsible for the life extension in rodents.

Although the mechanism underlying the life-extending effect of CR in rodents has yet to be determined, virtually every age-related decline in a physiological system is delayed or slowed when compared with findings in *ad libitum*-fed animals. This includes metabolic rate, neuroendocrine changes, glucose regulation, thermoregulation, immune response, and circadian rhythms. The slowing of the rate of aging observed in CR can be correlated with a reduction in cellular oxidative stress and other processes that lead to damaged proteins. One of the more interesting and consistent findings from CR studies in rodents is that various chaperone proteins are up-regulated. Chaperone proteins, you'll recall, mark misfolded or damaged proteins for catabolism and removal from the cell. Thus, a reduction in the accumulation of damage may be one mechanism that slows the rate of aging during CR.

Calorie restriction also delays or prevents the appearance of many age-related diseases. In calorie-restricted rats, the incidence of glomerulonephritis, a disease leading to renal failure and the most common cause of natural death in rats, is about 50% of the rate observed in *ad libitum*-fed animals. In addition, glomerulonephritis in calorie-restricted rats occurs significantly later in the life span. The occurrence of cancer and number of tumors are lower in calorie-restricted than in *ad libitum*-fed mice **(Figure 10.4)**.

Calorie restriction in simple organisms can be used to investigate genetic and molecular mechanisms

The vast majority of investigations using CR have been carried out in mice and rats, which are, in general, poor models for the study of genetic and molecular mechanisms. Thus, biogerontologists have turned to less complex organisms such as yeast, worms, and flies to better understand the genetic and molecular mechanisms underlying life extension through CR **(Figure 10.5)**. Most results have been consistent with our current understanding of the evolutionary and genetic bases of longevity. Extension of life span by CR in *C. elegans* and *Drosophila* has

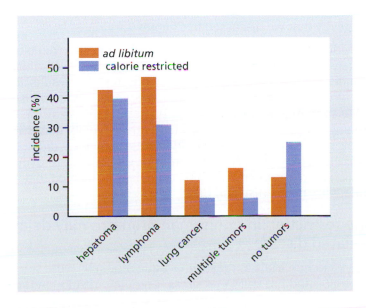

Figure 10.4 The incidence of cancer in mice at the time of death, following long-term calorie restriction versus *ad libitum* feeding. The CR was started at 12 months of age. Thus, CR can have a significant impact on age-related disease even when started in middle age. (Data from R. Weindruch and R.L.Walford, *Science* 215:1415–1418, 1982. With permission from AAAS.)

Figure 10.5 Effect of calorie restriction in *C. elegans* and *Drosophila*. Calorie restriction in *C. elegans* is achieved by reducing the concentration of *E. coli* in the culture medium, in this case to 50% of that given to the *ad libitum*-fed worms. Calorie-restricted adult *Drosophila* are fed a sucrose/yeast solution at 50% of the concentration given to *ad libitum*-fed flies.

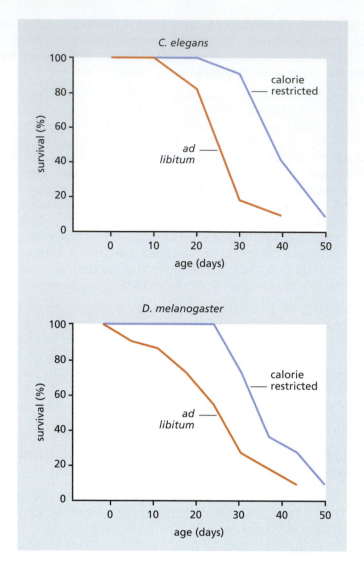

been associated with a delay in the start of reproduction. In addition, fecundity in calorie-restricted worms and flies decreases significantly compared with that in *ad libitum*-fed animals. These findings are exactly what would be predicted based on mathematical models and experimental evidence of the trade-off between reproductive success and longevity that we discussed in Chapter 3.

The genetic and molecular events responsible for life extension in CR are largely unknown, because research using simple organisms has been underway for only a few years. Thus far, the results in yeast, worms, and flies are consistent with the genetic mechanisms that affect life span that we discussed in Chapter 5. Life extension in yeast undergoing CR seems to be associated with longevity genes involved in aerobic metabolism, that is, the *SIR* genes. In fact, CR did not further increase the extension of life span in mice engineered to overexpress *SIR2*. Much more research on simple organisms is needed to determine the underlying mechanisms of CR.

Calorie restriction in nonhuman primates may delay age-related disease

Calorie-restriction trials in rhesus macaque monkeys were started during the 1980s and are still in progress. Complete results on the effects on longevity are years away, because rhesus macaques fed *ad libitum* have mean and maximum life spans of approximately 25 and 40 years, respectively, and the calorie-restricted monkeys may live as long as 45–50 years. Preliminary results suggest that the mean life span of rhesus macaques whose diet is calorie-restricted to 70% of *ad libitum* intake is slightly greater than that of monkeys fed *ad libitum*, but potential differences are small compared with those observed in other species. The scientists conducting the studies in nonhuman primates have been reluctant to make any conclusions about the effect of CR on life span.

Notwithstanding the lack of conclusion on the longevity question, CR in rhesus macaques seems to have positive effects on markers of age-related disease. As would be expected, calorie-restricted monkeys weigh significantly less than *ad libitum*-fed animals. The difference in weight reflects a decrease in both fat and lean mass. In addition, fasting blood insulin and glucose levels decrease and insulin sensitivity increases in CR. Combined with the lower total fat mass, these data indicate that calorie-restricted monkeys are at a lower risk than *ad libitum*-fed monkeys for developing type 2 diabetes. The lower serum triglycerides, LDL cholesterol, and blood pressure observed in the calorie-restricted animals are consistent with a decreased risk of cardiovascular disease.

The effectiveness of calorie restriction in humans remains unknown and controversial

The positive effect of low-calorie diets for overweight and obese individuals is well known. What is less known is the effect of reduced-calorie diets on normal-weight individuals. A few short-term CR studies in normal-weight individuals suggest a positive effect, but these investigations have been fraught with difficulties that lessen the impact of the results. High dropout rates, small sample sizes, and problems with participants adhering to the strict dietary protocol required to achieve the appropriate level of CR have led many biogerontologists to question whether CR in humans is logistically possible.

The largest human trial evaluating CR that is currently underway emphasizes that maintaining adherence to the 25% CR for a 2-year period is the most significant challenge. To address these issues, the participants receive intensive counseling from psychologists and registered dietitians. This use of intensive counseling to ensure adherence to the dietary requirements reflects the single most important question facing the use of CR in humans: Could the general population maintain a diet requiring meticulous attention to every calorie consumed and eliminating or severely restricting many foods that provide pleasure (ice cream, cake, chocolate, etc.)? We suggest the answer is "no." Evolution has provided humans with an overwhelming drive to eat by associating taste with pleasure. It is highly unlikely that a 25% reduction in calories could be sustained over a lifetime.

MODULATING THE RATE OF AGING: PHYSICAL ACTIVITY

We are all familiar with the health benefits of regular physical activity. Adults of all ages who participate in regular aerobic activity have lower resting heart rates, lower blood pressure, lower levels of circulating triglycerides and LDL cholesterol (the "bad" type), and higher levels of HDL cholesterol (the "good" type) when compared with sedentary individuals of similar age **(TABLE 10.1)**. The decrease in body fat that results

BOX 10.1 FROM MYTH TO SCIENCE: THE HEALING AND HEALTH POWERS OF FOOD

Rigorous scientific investigations have conclusively shown that low-calorie diets slow the rate of aging and decrease the risk for many age-related diseases. However, even with the overwhelming evidence of the benefits of a low-calorie diet for health and aging, we live in a time when almost 70% of the American population is overweight or obese. Clearly, the majority of Americans are having difficulty accepting the indisputable scientific evidence that low-calorie diets are a simple and low-cost way of improving health, delaying age-related diseases, and slowing the rate of aging.

On the other hand, Americans spend billions of dollars per year on individual foods and food supplements with supposed health benefits, as promoted through personal testimony, oversimplification of basic biological mechanisms, and misinterpretation of valid scientific results. A current Web-based advertisement by a food supplement company lists 40 benefits of wheatgrass, including, "Chlorophyll is the first product of light and, therefore, contains more light energy than any other element." In other words, eating wheatgrass will give you the energy of light. Wow! At present, there is no reliable or complete scientific evidence showing that any single food or nutrient will, in itself, prevent, delay, or cure disease or slow the rate of aging.

Why are we so easily persuaded by outrageous claims about the health benefits of food when the scientific evidence shows otherwise? The complexity of human survival and health psychology provides many answers. One of the more important factors would certainly be that the health and healing powers of food are deeply rooted in our cultural and religious traditions. It was not until the late nineteenth century that safe and effective medicines, as we know them today, began to appear. This means that humanity has had only a little over 100 years to separate food

from medicine, an extremely short time in which to adopt a cultural change. Here, we briefly explore the origins of naturally occurring food as medicine and how science has begun to separate the two.

Food as medicine

"Let food be thy medicine and medicine be thy food." These words were written by Hippocrates of Cos (460–370 B.C.E.), widely regarded as the father of Western medicine. Hippocrates's therapeutic method was based on a close relationship between the natural world and the human body. According to the Greeks of his day, all matter, including food and the four bodily humors (phlegm, blood, yellow bile, and black bile), was composed of four elements (earth, air, fire, and water), and an imbalance of these elements caused the abnormal functioning of matter, including bodily functions (disease). Hippocrates's method was to diagnose which element was out of balance and then prescribe a food with a high concentration of the opposite element—earth, air, fire, or water—so as to put the body back in balance.

The Hippocratic method, with its emphasis on food, established the direction of Indo-European medicine, a method that would last until the Enlightenment of the eighteenth century. The noted Roman physician Galen (129–200 C.E.) relied heavily on Hippocrates when he compiled his writing on medicines into a multi-volume set, *On the Powers of Food*. Galen described how properties in grains, fruits, vegetables, and some animal flesh have specific uses in curing disease. He suggested that food should always be the first step in treatment, followed by drugs, and surgery as the last resort. With the adoption of Galenic medicine by the Roman emperors, food as medicine became the standard medical treatment throughout the empire.

The rise of Islamic medicine in the Arab and Persian cultures came after the fall of the Roman Empire, and

this also relied heavily on food as medicine. The famous Islamic physician Ibn Sinā (Avicenna) translated the medical writings of Hippocrates and Galen into Arabic and added his own techniques—techniques that would be the basis of Islamic medicine for centuries. Ibn Sinā suggested that the healing powers of food were Allah's gift to humanity. This prophetic statement would prompt some Islamic clerics to suggest that maintenance of health through food was one way of achieving holiness.

Traditional Chinese medicine, known as *shi liao*, most likely got its start during the first millennium B.C.E. with the writing of the *Huangdi Neijing*, also known as *The Yellow Emperor's Classic of Internal Medicine*. Like Hippocrates, the *Huangdi Neijing* placed prime importance on keeping the body in balance. Food was classified into cold, wet, hot, and dry and was used to oppose disease through the concept of *yin–yang*, the belief that there are two complementary forces in the universe. The *Huangdi Neijing* stated, "If there is heat, cool it; if there is cold, warm it; if there is dryness, moisten it; if there is dampness, dry it." *Yin* foods are those that are considered cold and wet, such as fruits, seafood, beans, and vegetables; these decrease energy and increase the water content of the body. *Yang* foods are hot and dry foods, such as ginger, onions, alcohol, and most meats, which increase energy and decrease the water content of the body. The energy-enhancing effect of hot foods was based on the idea that *yang* foods contain significantly more fat and protein than *yin* foods and thus have greater energy content. Traditional Chinese medicine remains active throughout China and other parts of the world.

The Western concept of food as medicine began to change with the rise of the scientific method during the Enlightenment. For example, invention of the microscope provided evidence that microorganisms, not

BOX 10.1 FROM MYTH TO SCIENCE: THE HEALING AND HEALTH POWERS OF FOOD

food, were the cause of the diseases that killed most people at that time. Ironically, spoiled food and contaminated water were discovered to be the major vectors for the microorganisms causing diarrheal diseases, one of the major causes of death in children until 1920. During the eighteenth and nineteenth centuries, the biological and medical sciences discovered that internal therapies for disease were most effective when highly focused and specific to the chemical and biochemical nature of the cell or invading organism. The food-based medical treatments advocated by Hippocrates, Galen, and Ibn Sinā were far too general to be effective against the specificity of disease.

The rise of nutrition as a biological science and the decline of food as medicine

Western medicine's reliance on drugs as the treatment of choice for disease, which began in the late nineteenth century, brought about a shift in how scientists approached the study of food. In the early part of the twentieth century, the physical and biological rather than the medical properties of food became the focus of research, and nutrition was added to the growing list of biological sciences. With this new biological focus, it became apparent that the nutrients in foods function primarily to support metabolism for growth, reproduction, and maintenance of bodily functions. It also became clear that individual foods or nutrients do not cause or cure the infectious diseases that still caused the majority of deaths.

Not until the 1950s, when a significant number of people were living beyond 60 years of age, did the link between nutrition and disease again become an area of interest. Unlike in the preceding 5000 years, scientists now applied the scientific method to determining whether individual foods or nutrients caused disease. To this end, between 1950 and 1980, several epidemiological studies

reported that individual foods and nutrients such as dietary cholesterol, saturated fat, red meat, and so forth were associated with a higher risk of heart disease and cancer. Conversely, people consuming diets low in fat and rich in antioxidants had a lower risk of heart disease. The epidemiological evidence was often corroborated by results in laboratory animals and in small and short-term human trials. The short-term human studies did not, however, measure actual health outcomes, of cancer rates or heart disease, for example. Rather, they evaluated some biological marker that may be related to the disease outcome. For example, short-term human studies of the effect of vitamin E on cancer would often report the level of antioxidants in the blood after a few months of taking a supplement, rather than the actual rate of cancer in the population. Together, the results from epidemiological, laboratory, and small-scale human studies led to a frenzy of recommendations by governments, independent nutrition organizations, and the food industry suggesting that people should increase or decrease their intake of a specific food or individual nutrient in order to lower their risk of disease.

Antioxidants prevent cancer! No, wait, they increase the risk of cancer!

Unfortunately, conclusions about the medical and health benefits of an individual food or nutrient are often premature and reflect the failure of both scientists and the lay press to wait until completion of all the studies needed to conclude cause and effect. For example, during the 1970s and 1980s, many epidemiological studies reported that cancer rates were significantly lower in people who consumed diets with high amounts of the antioxidants vitamin E (α-tocopherol) and β-carotene (precursor of vitamin A) than in individuals with diets low in these vitamins. Studies in laboratory animals showed that these

vitamins, given in the diet at levels several times the animals' requirement, prevented the growth of tumors. Hundreds of human trials with numerically small populations corroborated the epidemiological and animal studies, describing changes in biological markers for cancer. Based on these results, food companies began to advertise the benefits of vitamin E and β-carotene on their product labels and in commercials, and sales of supplements skyrocketed. The dogma that vitamin E and β-carotene prevent cancer had entered the American culture.

The dogma crash-landed in the 1990s, with the news that β-carotene and vitamin E were not showing any effect on cancer. In 1994, a cancer-prevention study of Finnish male smokers demonstrated that β-carotene not only did not prevent lung cancer in men who smoked but may actually *increase* the risk of cancer (**Figure 10.6**). The results of the study drew substantial criticism, because it focused on smokers rather than on otherwise risk-free individuals. Subsequent trials performed in nonsmokers confirmed the initial findings and failed to demonstrate any effect of vitamin E or β-carotene on mortality rate. At about the same time, longevity studies in mice also found that neither cancer rates nor life span were altered by increasing the vitamin E content in the diet. The results of these investigations prompted many government agencies and the American Cancer Society to issue warnings over the safety of β-carotene and to unequivocally state that β-carotene does not reduce the risk of cancer or increase the length of life.

Basic nutritional advice has been right all along

Clearly, not all foods and nutrients have been tested to the degree of β-carotene, vitamin E, or the few other nutrients now shown to have no effect on the risk of disease. It would be irresponsible to suggest that no individual food or nutrient

BOX 10.1 FROM MYTH TO SCIENCE: THE HEALING AND HEALTH POWERS OF FOOD

could, at some point, be shown to have a specific effect on a disease. Nonetheless, nutritionists, public health experts, and other health professionals are starting to gain a clearer picture of the effect of food on disease. Their conclusion, after the millions of dollars spent on researching the health benefits of food, is what nutritionists have been saying for years: eat your vegetables, eat everything in moderation, and eat a wide variety of foods.

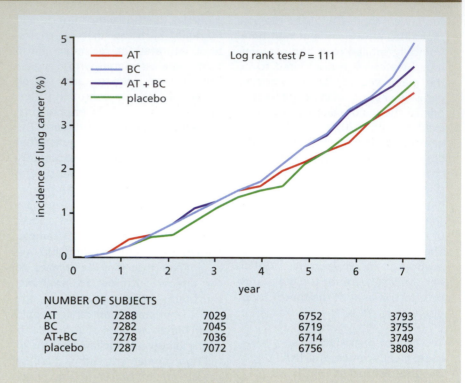

NUMBER OF SUBJECTS				
AT	7288	7029	6752	3793
BC	7282	7045	6719	3755
AT+BC	7278	7036	6714	3749
placebo	7287	7072	6756	3808

Figure 10.6 Studies of the effects of vitamin E and β-carotene on lung cancer rates. The data show the cumulative incidence of lung cancer in study participants receiving α-tocopherol, or vitamin E (AT), β-carotene (BC), α-tocopherol and β-carotene (AT + BC), or a placebo. (Data from D. Albanes et al., *J. Natl Cancer Inst.* 88:1560–1570, 1996. With permission from Oxford University Press.)

from increased aerobic activity substantially lowers the risk of type 2 diabetes and certain types of cancer. Moreover, some recent research indicates that lifelong regular physical activity may be associated with a lower incidence of Alzheimer's and Parkinson's diseases.

Regular physical activity lowers the risk of age-related disease, which, in turn, increases the mean life span. However, exercise does not increase maximum life span. This effect was shown in two classic studies published during the 1980s. In one study, the life spans of sedentary rats with different types of dietary restrictions were compared with the life spans of rats given access to running wheels. Rats with running wheels in their cages had mean (but not maximum) life spans slightly longer than those of animals without running wheels **(Figure 10.7)**. Interestingly, animals that were sedentary but calorie-restricted had the greatest mean and maximum life span. Thus, we can conclude that physical activity alters only the rate of aging, whereas calorie restriction alters both the rate of aging and longevity. Consistent with these findings are the human studies reporting that the length of the post-reproductive life span directly correlates with the amount of regular physical activity. That is, the greater the amount of regular physical activity, the longer the post-reproductive life span and the longer the mean life span. In these studies, investigators showed that the increase in post-reproductive life span due to regular physical activity was most associated with a decrease in the risk for heart disease and cancer.

TABLE 10.1
DOCUMENTED EXAMPLES OF THE BENEFITS OF REGULAR PHYSICAL ACTIVITY FOR INDIVIDUALS OVER THE AGE OF 65

Change with exercise	Benefit
Cardiovascular	
Increased stroke volume	Increased blood flow to tissue
Decreased arterial resistance	Decreased blood pressure
Decreased LDL cholesterol	Decreased risk of coronary artery disease
Increased blood volume	Increased venous return; lower risk for congestive heart failure
Muscle	
Increased mitochondrial concentration	Increased aerobic capacity for ATP synthesis
Increased capillary density	Increased capacity for oxygen delivery
Increased insulin sensitivity/glucose uptake	Decreased risk of type 2 diabetes
Increased muscle fiber recruitment/action potential threshold	Increased strength
Body composition	
Decreased body fat	Reduced risk of coronary artery disease, type 2 diabetes, and certain cancers
Increased lean muscle mass	Increased strength, stability, and resting metabolic rate
Lungs and respiration	
Increased pulmonary capillaries	Increased ventilation-perfusion
Increased strength of intercostal muscle	Easier breathing
Maintenance of lung tissue elasticity	Easier breathing
Maintenance of alveolar size	No loss in dead space; maintenance of proper ratio of surface area to diffusion
Skeletal system	
Increased bone mineral density	Increased bone strength and decreased risk of osteoporosis
Improved rate of red and white blood cell production (bone marrow)	Increased oxygen-carrying capacity of the blood; improved capacity of the innate immune system

In this section, we explore the biological basis for the increased function or maintenance of function resulting from regular physical activity. We begin with a brief introduction to the basic principles of exercise biology; adaptations that are thought to slow the rate of aging are discussed in terms of these basic principles. We end with some speculation

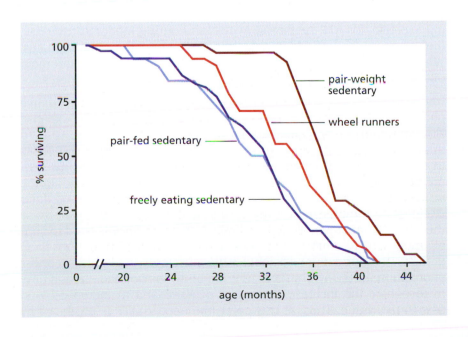

Figure 10.7 Survival curves for male rats with or without access to running wheels. "Pair-fed" sedentary rats received (daily) an amount of food equal to the amount eaten by the wheel runners. "Pair-weight" sedentary rats received an amount of food that resulted in a weight equal to that of the wheel runners; this amounted to ~30% of the CR diet. The wheel runners had a mean life span (34 months) greater than that of the pair-fed sedentary rats (31 months) and the freely eating sedentary rats (31 months) but less than that of the pair-weight sedentary rats (37 months). These data show that voluntary exercise (wheel running) significantly increases mean but not maximum life span when compared with sedentary animals. Calorie restriction increases both mean and maximum life span when compared with all other groups. (Data from J.O. Holloszy et al., *J. Appl. Physiol.* 59:826–831, 1985. With permission from the American Physiological Society.)

on a general theory explaining why an intervention such as physical activity may slow the rate of aging.

Exercise increases the muscles' demand for oxygen

Most likely the first response you notice when beginning to exercise is the increase in your breathing rate, what pulmonary physiologists refer to as **ventilation**. The increase in ventilation occurs because **chemoreceptors** (neural receptors that bind specific chemicals) in the lungs detect increasing carbon dioxide levels in the blood returning from the extremities. Nerve impulses are then sent to the brain, which, in turn, sends signals back to the lungs, causing an increased rate of breathing. At the same time, the right ventricle of the heart senses a decrease in the amount of blood returning from the extremities, which is due to more capillaries opening up in the muscles. As a result, the heart beats faster to deliver more blood. All these physiological responses occur for one reason: the muscles need more oxygen to synthesize ATP so that they can carry out the increased work (rate of muscle contraction) of exercise.

Recall from Chapter 4 that the energy for contraction of muscle fibers comes from the breaking of one phosphate bond in ATP (ATP + $H_2O \rightarrow$ ADP + P_i + energy + heat). Because muscle contains only 5–10 seconds worth of stored energy, ATP must be synthesized continually during exercise. This process requires oxygen. In addition, the increased metabolic by-products of ATP synthesis, carbon dioxide and lactic acid, must be removed from the muscle cells to prevent acidosis, a decrease in pH that disrupts normal cellular chemical reactions.

ATP can be synthesized in muscle through two separate but connected pathways. The first, glycolysis, needs no oxygen (is anaerobic) and can use only glucose as the starting substrate for the synthesis of ATP. The anaerobic metabolism of glucose limits the amount of ATP that can be synthesized. For every molecule of glucose metabolized, only two molecules of ATP are synthesized. The other pathway of ATP synthesis, oxidative phosphorylation (discussed in Chapter 4), needs oxygen (is aerobic). Oxidative phosphorylation can utilize both fat and glucose to produce 16 times more ATP than glycolysis. The use of oxygen combined with the ability to utilize both fat and carbohydrates makes oxidative phosphorylation much more metabolically efficient than glycolysis. Humans use both systems simultaneously, at rest and during exercise, for ATP synthesis. At rest and during moderate exercise, the use of free fatty acids in oxidative phosphorylation predominates. When quick energy is needed, or during periods when ATP requirement outpaces a cell's ability to take up oxygen, glycolysis predominates.

Understanding how aerobic exercise stresses or overloads the oxidative phosphorylation system is the key to understanding how such exercise slows the rate of aging and leads to the benefits listed in Table 10.1. To this end, it is helpful to look at the physiological and cellular responses to a single bout of exercise. We'll use the example of a long-time sedentary individual who sets off to run a mile at increasing speed. In the first few moments, 5–10 seconds after she starts running, the cellular stores of ATP are used for muscle contraction **(Figure 10.8)**. Because the intracellular stores of ATP are limited, the muscle cells need to supply more ATP to maintain the new level of muscle contraction. Even though it would be more efficient for the muscles to use oxidative phosphorylation for the synthesis of ATP, the body has not had enough time to recognize the increase in oxygen demand and to increase blood flow to the tissue. Thus, at this point, the needed ATP is produced by glycolysis, because this pathway does not need oxygen. However, the speed at which we can synthesize ATP in glycolysis comes at a cost. In addition to supplying only two ATP molecules for each molecule of glucose metabolized, glycolysis ends with the production of lactic acid. Humans are rather inefficient at neutralizing lactic acid, and if we continued to use only glycolysis for ATP synthesis, the muscle contraction would soon cease due to the drop in intracellular and blood pH.

After about 1–2 minutes of sustained moderate exercise (moderate meaning you can still talk while exercising), the body responds to the increase in muscle activity by increasing blood flow, which increases oxygen delivery. The increased supply of oxygen results in initiation of the oxidative phosphorylation pathway, a decrease in the reliance on glycolysis, and an increase in the use of fat as energy substrate. As long as this moderate level of exercise is maintained, ATP synthesis and exercise will continue without any negative metabolic consequences.

Now supposed this long-time sedentary individual decides to increase the speed at which she is exercising. The increase in speed increases her heart and breathing rate to meet the new demand for oxygen needed for ATP synthesis. She cannot increase her speed indefinitely, however, because the system has limits. Since she has been sedentary for a long time, her muscle cells have only enough metabolic capacity for a sedentary life. The heart and lungs are able to supply enough oxygen to the muscles, but the muscle cells do not have sufficient capacity to use the oxygen for ATP synthesis. This individual will eventually reach a point at which ATP synthesis through the oxidative pathway cannot keep up with the demand. At this point, the body starts to supply the majority of ATP through glycolysis.

Overloading cellular oxidative pathways increases the capacity for ATP synthesis

Our long time sedentary individual has only enough aerobic metabolic capacity to maintain a muscle contraction rate consistent with a sedentary lifestyle. But, if she repeats the aerobic exercise daily over an extended period, she can increase her muscles' capacity for ATP

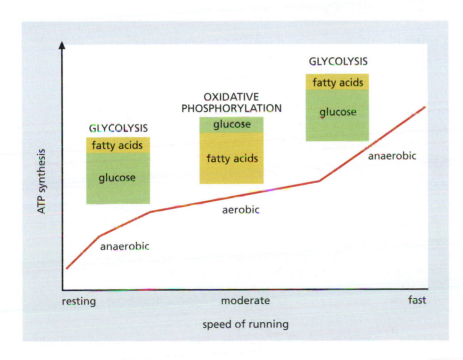

Figure 10.8 The metabolic pathways and energy substrates used for ATP synthesis during a single bout of exercise at different speeds.

synthesis. In every individual, aerobic exercise induces several physiological and cellular responses, such as increasing blood volume, red blood cell production, and expression of proteins and enzymes of the glycolytic and oxidative phosphorylation pathways **(Figure 10.9)**. Repeating the aerobic exercise daily over a few weeks will make these responses permanent, even at rest. This metabolic adaptation explains the health benefits of exercise. As long as the ATP-synthesizing pathways are overloaded by aerobic exercise, the change will remain intact.

Increasing blood volume and red blood cell number through aerobic exercise increases the amount of oxygen per unit of blood delivered to the periphery. Moreover, because the aerobic exercise has increased the metabolic capacity of the muscles, oxygen extraction from the blood by the muscle cells has become more efficient. As a result, the heart does not have to pump as much blood to the periphery as it did before regular physical activity. The cardiovascular system adapts to this new level of muscle efficiency by reducing resting heart rate and blood pressure, thereby decreasing the work of the heart. In addition, a greater capacity for oxidative metabolism means a greater use of fat as energy substrate. Since fat is stored in adipose tissue, a regular physical exercise program will reduce weight, as long as there is no increase in dietary calories. Many studies have shown that reduction in weight lowers the risk of type 2 diabetes and coronary artery disease.

Figure 10.9 The training effect. The increased need for ATP synthesis during exercise is the driving force behind the benefits of regular physical activity. Increasing the capacity for ATP synthesis leads to improvements in cardiovascular function, with decreased resting heart rate and blood pressure, as well as a reduction of fat stores. Individuals of all ages can benefit from the training effect.

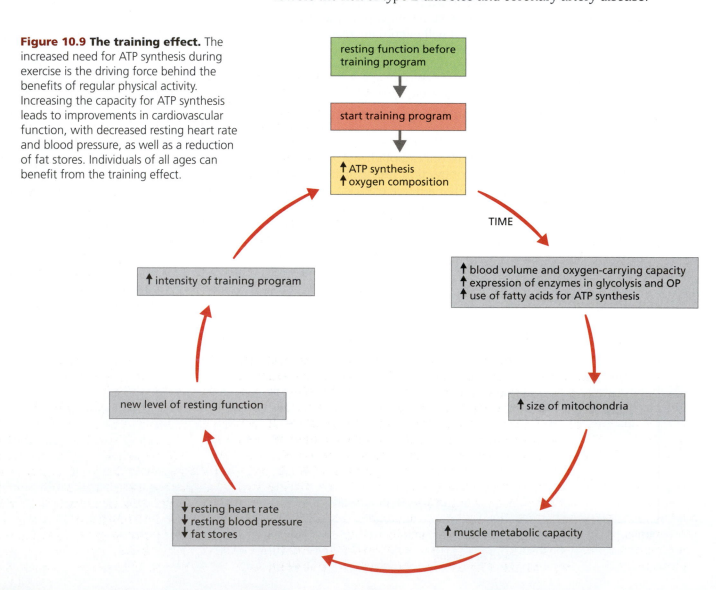

Regular physical activity prevents a decline in cellular reserve capacity

We presented the discussions on regular physical activity to illustrate that the biological process of aging can be slowed but not prevented. As you have seen, overloading a system, such as the capacity to synthesize ATP, results in adaptations that improve cardiovascular function, decrease body fat, and reduce the risk for certain diseases. In other words, a regular program of physical activity slows the rate at which systems fail due to increasing entropy—that is, it slows the rate of aging. These discussions have given you a brief introduction to how an intervention such as regular physical activity can slow the rate of aging. We now turn our attention to why: Why does physical activity slow the rate of age-related functional loss?

Regular physical activity slows the rate of aging because it helps to preserve reserve capacity in many physiological systems. Reserve, or redundant, capacity helps humans survive to reproductive age by providing a mechanism for defending against stress—such as disease, accident, or environmental insult—that exceeds the basal level of function. From the evolutionary viewpoint, the most successful hominids were those with the most reserve capacity that allowed them to reach reproductive age. But as we know, there would be no reproductive advantage to maintaining that reserve capacity after reaching reproductive age. At this age, reserve capacity would begin to decline.

The rate of decline in reserve capacity of a physiological system is determined, in large part, by the capacity that existed when the organism reached reproductive age. A higher reserve capacity results in a slower rate of aging. In Chapter 9, we discussed how, for women, reserve capacity of bone mineral content prior to menopause helps to prevent excess loss of bone mineral after menopause. However, many physiological systems retain the ability to increase reserve capacity throughout the life span. An increased reserve capacity will slow the rate of aging.

To illustrate how physical activity can slow the rate of aging, let's look at an example of reserve capacity in the human thermoregulatory system. The ability to maintain a constant internal temperature (37°C in humans) represents a survival advantage. Not having to rely on environmental conditions to raise or lower internal temperature means that humans can live virtually anywhere on the planet—a huge reproductive advantage in terms of obtaining sufficient food. Humans maintain a constant internal temperature through the heat produced in biochemical reactions. Basal levels of metabolism are sufficient to maintain the internal temperature when the environmental temperature, or ambient temperature, is between 20°C and 25°C. Ambient temperatures below or above this range require a physiological adjustment so that the internal temperature does not become too low or too high. For example, if the ambient temperature is 35°C, we maintain our internal temperature by increasing blood flow to the skin, so that heat from the body is radiated to the atmosphere, and by sweating, transferring heat to the water in the blood, which is then released onto the skin for the cooling effect of evaporation. These mechanisms for heat dissipation rely on the reserve capacity of many physiological systems: heart rate increases to increase blood flow; stress hormones are secreted to signal the vascular system of the skin that internal temperature is rising; more skin capillaries open and larger vessels dilate. Maintaining internal temperature in a cold environment also requires responses from various physiological systems so that we conserve heat. Our ability to perform thermoregulation depends directly on the extent of our reserve capacity, and this reserve capacity is critical for survival.

Exercise increases our reserve capacity for defending our internal temperature against environmental challenges. Individuals who participate in regular physical activity maintain or increase the density of capillaries in muscle and skin, the rate of blood flow to the extremities, the stroke volume of the heart, and oxidative metabolic capacity. In other words, a physically fit individual should be able to respond faster and more efficiently to temperature challenges than a sedentary individual. Numerous studies have shown that aged people who participate in regular physical activity defend against a thermoregulatory challenge better than sedentary individuals.

Maintaining a regular program of physical activity has also been shown to increase or maintain the reserve capacity of the cardiovascular system and to slow the rate of aging in various physiological systems. These adaptations taking place during physical activity are being used to investigate the mechanisms underlying a human's ability to slow the rate of aging. Results from these investigations are still preliminary but are consistent with the hypothesis relating the maintenance of molecular fidelity to the rate of aging. For example, age-related atrophy of muscle tissue appears to be related to genes involved in the repair of damaged tissues. These genes are not expressed to the same degree in aged muscle as in young muscle. Aged muscle fibers seem to accumulate damaged proteins, which, in turn, reduces molecular fidelity. Whether physical activity induces the expression of genes involved in the repair of damaged tissue remains to be seen. Nonetheless, physical activity may provide a suitable model to investigate the relationship between the maintenance of molecular fidelity and the rate of aging.

LOOKING TOWARD THE FUTURE: THE IMPLICATIONS OF MODULATING AGING AND LONGEVITY

Biogerontology underwent an important transition during the late 1990s. Identification of the fundamental causes of aging and longevity shifted the emphasis of research from observational studies focused on defining aging to experimental investigations of how the rate of aging can be slowed and longevity extended. As we have seen, low-calorie diets and increased physical activity are proven methods for slowing the rate of aging, but the population as a whole has had difficulty adopting such a strategy. This strongly suggests that more passive, medically centered interventions—taking a pill, undergoing gene therapy, and so on—will be needed to persuade people to take an active role in delaying aging.

At present, however, there is no credible scientific evidence supporting a medical intervention that slows, stops, or reverses human aging. Any suggestion to the contrary is either a misleading marketing technique used to sell a product or the claim of a fringe group expounding its own vision of reality. Nevertheless, the new biotechnologies being applied to research in biogerontology provide great hope that slowing the rate of aging and extending the average life span through medical interventions will be possible in the not too distant future. The precise number of years that will be added to current life expectancy is anybody's guess, but average life spans into the tenth, eleventh, and twelfth decades are often predicted by knowledgeable biogerontologists and demographers.

The increased life span and extended youth expected from the biotechnology revolution will take humanity into unknown territory, one that will profoundly change the basic structure of society. For the first time in history, society will have to make room for the wants, needs, and desires of a population that has moved well beyond the promises of youth on which society has traditionally been based. The outcomes from this new social order are unforeseeable. But the conversation about what our society might look like and the consequences of this new order has at least started. Here, we briefly consider that conversation and what bioethicists, sociologists, biologists, and other experts are saying about this new society. We focus primarily on what individuals' longer life will mean to the culture and structure of society.

The purpose here is to give you a frame of reference for entering this important discussion, not to provide you with answers about what society will be like. Recall what you read in the first few pages of the first chapter of this book: "Biogerontological research that leads to improved health and extended life must be reconciled with the fact that aging will occur no matter how successful the remedy for a specific age-related dysfunction and that death will be the endpoint for the individual." Biogerontologists are required not only to be experts in their particular field but also to be active participants in the discussion on the psychological, social, and economic consequences of the improved health and well-being of the older population.

Extended youth and the compression of morbidity will characterize aging in the future

The quest for a long life, and its consequences, are as old as civilization. Greek mythology tells of the god Apollo granting Cumaean Sibyl one request in exchange for her virginity. She chose eternal life. Sibyl forgot, however, to include eternal youth in her request, and she spends eternity suffering the disabilities of old age. The mythology of ancient Greece reminds us that most people are unwilling to accept a long life without good health. Significant life extension must and will come with extended youth and relatively good health into old age.

Some have predicted that life extension will be analogous to the stretching of a rubber band. That is, all phases of biological life will be extended, including the period of increased morbidity that often precedes the end of life. Most biogerontologists whose research focuses on the future of aging disagree with this rubber-band analogy. Rather, they suggest that the increased morbidity at or near the end of life will last, at most, a few months—the same as it does now—or, more likely, a significantly shorter time. Thus, the percentage of the life span spent experiencing senescence-related morbidity will decline, a phenomenon known as the **compression of morbidity**.

With the promise of extended youth and the compression of morbidity, our discussion can focus more on the ethical and cultural issues of an aging society by eliminating, to a large degree, concerns about economic catastrophes resulting from overburdened health and social security systems. Total expenditures on health care are expected to rise with extended longevity (though some suggest that curing disease will lower expenditures on costly diagnostic and therapeutic procedures). However, this rise in health care costs will result from the increase in overall population due to a larger aging cohort, rather than from the ill health of the oldest sector. The issue, then, becomes a political argument about how best to deal with health care costs, a discussion not unlike that taking place today. A similar consideration is the insolvency of the U.S. Social Security system that is likely to result if policies for participation (as outlined in Box 7.1) are not changed. Again, this is a political issue, and societies, especially democratic societies, have demonstrated time after time that these problems can be worked out—maybe not to everyone's satisfaction or without some pain, but nonetheless worked out.

Long life may modify our perception of personal achievement and a progressive society

Many view the prospect of a long life and extended youth with great optimism and see it as a possibility for new opportunities. With the expectation of a longer life, individuals are more likely to engage in projects or new ventures that are beyond reach in a shorter life that requires a more careful prioritizing of goals. Mistakes may simply encourage a person to choose a new career that fits more closely with his or her personality and aspirations. Such a move could easily be undertaken at age 70 or 80, if you have the prospect of living another 30–40 years. The expectation of a 120-year healthy life span may mean the young will be more willing to experiment with risky vocations and avocations, knowing they will have time to try again with the same venture or start a new one. Since risk is the cornerstone of discovery and progress, society will be the beneficiary here.

Other bioethicists take a more pessimistic view of what healthy longer lives might mean for personal achievement. They theorize that an extended youth could remove a sense of urgency and result in a decreased commitment to goals and aspirations. If you knew you had the chance for a "redo," you might be less inclined to be totally committed to any particular venture. The stick-to-it attitude would give way to prematurely abandoning one's ventures. The redo then becomes the way of life. Goals, aspirations, and the value of personal achievement, as we know them today, could simply lose their meaning, leading us to become a society of the status quo rather than a progressive society, as has historically been the case.

Extended longevity may change our responsibility for renewal of the species

An extended life span could also disrupt individuals' sense of responsibility for renewing the species. Proponents of this position point to the fact that the increase in life span during the twentieth century was accompanied by a decrease in birth rate and the trend toward starting reproduction later in life. The sense of urgency about having children is being replaced by the feeling that it can wait—there's plenty of time, later, to enjoy a more stable family life; youth should be spent living the "good life." However, many couples find that the "good life," without the responsibilities of children, is too difficult to give up, and so they fail to reproduce at all. In addition, those delaying reproduction may find that, later in their reproductive life span, the will to have babies now exists but the physiology of the reproductive system does not cooperate. Both men and women become less fertile as they approach the end of their reproductive period. Moreover, for women, the increase in birth defects after the age of 35 may seem too great a risk to consider having babies. A reduction in the birth rate will simply reflect biology.

Others believe that birth rate has little to do with life span. Rather, birth rates are more tied to the economic prosperity that comes with the shift from agrarian to industrial societies; longer lives are simply one of many positive outcomes of the industrial society. In an agrarian society, much like all societies before the start of the twentieth century, children are viewed as an asset because they add value to the family by providing labor for work on the land. Children in industrial societies are dependents, costing the family resources. Historically, agrarian societies could expect one in four of their children to die before reaching the age of reproduction, so couples often had many children as a hedge against the loss of labor. Today, in the developed countries, it is the rare couple that loses a child to disease or accident. Life in an industrial society, together with the gains in health care for children, means that couples have fewer children.

The declining birth rate in the twenty-first century may simply reflect the natural evolution of populations living in environmentally stable conditions, rather than a decreased desire to renew the species. As you learned in Chapter 3, the number of live births is significantly lower for populations living in environmentally stable conditions than for populations with more variable environmental conditions. Most economically developed countries are characterized by a vast majority of the population having sufficient food, housing, clothing, health care, and other basic needs of life—that is, a stable environment. There is no need to add children to the family as insurance to offset future losses of family members due to environmental hazards. Such a society becomes stable, with birth rates approaching death rates **(Figure 10.10)**. The stability of this population will most likely continue, even with extended youth and longevity.

The issue of reproductive abilities in individuals with extended youth and life span is also controversial. As you have learned throughout this text, reproduction and longevity are tightly coupled. A change in longevity implies a change in reproductive ability and reproductive life span. This is all the more possible given that extended youth and a longer life will most likely be the result of interventions taking place early in life, during development (see Chapters 3 and 5). In addition, solutions to the problems of decreased fertility and birth defects are being investigated much more vigorously than ways of extending the life span. One can expect advances in reproductive medicine to at least parallel advances in other areas of health.

Low birth rates and extended longevity may alter the current life cycle of generations

The structure of families and society is established around a life cycle in which each generation expects to play a well-defined role. For most of

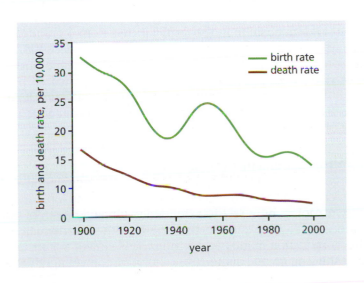

Figure 10.10 Birth and death rates in the United States, 1900–2000. (Death rates from Department of Health and Human Services, National Center for Health Statistics, *Natl Vital Stat. Rep.* 54(20), Aug. 21, 2007. Birth rates from U.S. National Center for Health Statistics, U.S. Public Health Service, Vital Statistics of the United States, Vols. I and II, 1971–2001.)

history, and for modern society, the life cycle of the family has included three generations: children, parents, and grandparents. Children, the new generation, are nurtured into adulthood by relatively young parents. Parents are expected to be the primary caregivers and to teach this new generation the basic tools needed for an independent adult life. Grandparents often are more or less extraneous to the caregiving for the new generation, playing only a supporting role in the needs of their grandchildren—babysitting, providing gifts, and so on. The important role of grandparents in the family structure centers more on imparting the wisdom of a lifetime of experience to their children, the parents of the new generation. The path for transfer of knowledge and wisdom in this three-generation life cycle, grandparents to parents to children, is short enough that family traditions are passed on without significant alteration. The family gains its own identity, and the individuals within the family feel a sense of belonging.

The three-generation structure of the family can be used as a metaphor for society in general, with each generation having a particular role. The new generation of children is expected to participate in a training and education system that prepares them for their eventual role as productive citizens of society—similar to the parent-child relationship. On the transition from training to work and inclusion as adult members of society, individuals must take their place at the bottom of the ladder and learn the "ways of the world." As in the grandparents' role in the family structure, the ways of the world are imparted by the older generation at the top of the ladder, those who have been through the process themselves. The three-generation structure of society also provides the individual with incentives to achieve and to move society forward in a positive direction. We learn early on that success in school will lead to success at work and in the community, which will lead to authority, power, and wealth.

The three-generation life cycle in the family and in society has developed based on a 60- to 80-year life span. In this model, the child can expect to progress to the parent, and the parent can expect to progress to the grandparent. The expected and orderly progression occurs because the grandparent or older generation is removed by death or infirmity, leaving room at the "top." But what if extended youth and a longer life delay death and infirmity so that the "wisdom" positions in the family and in the top tiers of society become overcrowded? Imagine, for example, a family structure in a world where average life span is 120 years. The family of today, children, parents, and grandparent, will give way to a family that includes healthy and active great-grandparents, great-great-grandparents, and possibly great-great-great-grandparents. Thus, the wisdom position in the family grows from 4 grandparents to as many as 32 great-great-great-grandparents. The succession of family tradition and identity made possible by the three-generation model weakens due to the sheer bulk of input from the wisdom positions. There are those that suggest the loss of family identity will lead to the loss of individuals' connection to the past and will challenge individuals' view of their own place in the world.

Some predict that as birth and death rates approach equilibrium (see Figure 10.10)— a trend that will undoubtedly continue in a society having a large and healthy older population—total population size will stagnate, economic growth will become sluggish, and job creation will slow down. Moreover, with the extended youth and health of the older population, fewer people in top-tier jobs or with seniority will be likely to retire. A reduction in jobs at the top means slower advancement for those on the way up and a shortage of jobs for the younger generation moving from training to work. Our concept of hard work leading to success will be challenged. Given all this, we will need to redefine what constitutes a productive member of society. Many experts suggest that this redefinition will be slow and will take a considerable toll on the first generation to deal with this new economic structure.

THE FUTURE OF BIOGERONTOLOGY

We have presented both the optimistic and the pessimistic views on what individuals might encounter in a society where the average healthy life span extends beyond the century mark. The optimists tend to see a society of great opportunity, where the more distant reality of death allows significantly more freedom to explore new possibilities without urgency and the constraints of time. Greater exploration and experimentation in the arts, sciences, medicine, and other fields will ultimately lead to an expansion of human knowledge, thereby accelerating human progress. All of humanity will benefit. The pessimists suggest that a sense of urgency arising from a not-so-distant end of life is a powerful force through which individuals feel the need to put their heart and soul into their work or ambition. It is this sense of urgency that underlies the progress of a society. A long life may remove that urgency and lead to less rather than more progress.

No one can say with certainty what society will become. Of course, that was not our point here. Rather, our hope is that you will have gained a greater appreciation for points of view that you may not have expected or experienced. Most of us may never have given thought to the idea that reducing or even halting research on how to retard aging and increase longevity might be an acceptable alternative to the current push for slowing the rate of human aging. If we accept that the pessimistic view has validity—and we suggest that it does—then we must consider how valuable research on aging retardation is to society. Placing the research emphasis on curing disease and compressing morbidity within the life span we already have might be more valuable to humanity.

The reality is, however, that research into the slowing of aging will increase rather than decrease in importance. Suffering the maladies of disease and old age will remain repugnant to most humans, even if methods are discovered to compress this morbidity into just a few months. The average life span of humans will continue to increase, and the health of the older population will improve. The resulting increase in life span and health will bring new and difficult challenges for our society. Any adverse effect on the individual can be kept at a minimum and be significantly less dramatic if we keep this discussion alive and allow room for all opinions. The science of biogerontology has a major role to play in this discussion.

ESSENTIAL CONCEPTS

- Aging cannot be modulated, because the laws of thermodynamics are universal and cannot be modified.

- Although aging cannot be modulated, the *rate* of aging can. To modulate the rate of aging, biogerontologists must begin to ask "Why do we live?" rather than "Why do we die?"

- A genomics approach is required to determine which genes are most susceptible to the second law of thermodynamics.

- Calorie restriction without malnutrition extends mean and maximum life span in every non-genetically modified species tested.

- Altering the macronutrient composition of the diet or increasing its vitamin and mineral content has little effect on life span in calorie-restricted animals.

- Biogerontologists are using simple organisms such as yeast, worms, and flies to investigate the mechanisms underlying life extension through calorie restriction.

- Preliminary results suggest that rhesus macaques that are calorie-restricted to 30% of *ad libitum* intake seem to have a mean life span slightly greater than that of monkeys fed *ad libitum*; maximum life span may not be different. Potential differences are small compared with those observed in other species.

- Calorie-restricted monkeys have lower rates of age-related diseases than *ad libitum*-fed animals.

- The effectiveness of calorie restriction in humans remains unknown and controversial.

- The lower risk of age-related disease achieved through physical activity has the potential to increase mean life span; exercise does not increase maximum life span.

- The primary effect of physical activity in slowing the rate of aging is to increase reserve capacity.

- The aging population of the future will be characterized by extended youth and the compression of morbidity.

- Extended youth and compression of morbidity will challenge our concepts of personal achievement, the importance of renewal of the species, and the generational structure of the family and society.

DISCUSSION QUESTIONS

Q10.1 The second law of thermodynamics operates constantly and randomly in biological systems. Explain why this property of the second law prevents the modulation of aging.

Q10.2 Some suggest that the gains in life expectancy observed in the past 100 years reflect successful interventions in aging. Is this correct? Discuss why or why not.

Q10.3 List the evidence showing that life extension through calorie restriction reflects reduced calories rather than the restriction (or supplementation) of other nutrients.

Q10.4 Using examples from both nonhuman primates and humans, describe why calorie restriction may not be as effective in humans as in other species.

Q10.5 Explain why exercising at an intensity above that at which you can comfortably talk results in the use of more glucose than fatty acids for ATP synthesis.

Q10.6 A regular program of physical activity decreases resting heart rate and blood pressure and increases red blood cell number. Why?

Q10.7 Discuss the relationship between rate of aging and reserve capacity. Explain how regular physical activity may increase reserve capacity.

Q10.8 Discuss how the transition from an agrarian society to an industrial society resulted in declining birth rates. Include in your answer how concepts from population biology help to explain declining birth rates.

Q10.9 How might the addition of great-grandparents to the family generational structure alter a person's identity within the family?

Q10.10 Based on current information on what our society may look like in the future, do you wish to have a significantly longer life span? Why or why not?

FURTHER READING

MODULATING BIOLOGICAL AGING

Hayflick L (2007) Entropy explains aging, genetic determinism explains longevity, and undefined terminology explains misunderstanding both. *PLoS Genet* 3:e220.

Hayflick L (2007) Biological aging is no longer an unsolved problem. *Ann N Y Acad Sci* 1100:1–13.

Lambert FL (2007) A student's approach to the second law and entropy. http//entropysite.oxy.edu/students_approach.html.

Mitteldorf J (2010) Aging is not a process of wear and tear. *Rejuvenation Res* 13:322–326.

Toussaint O, Raes M & Remacle J (1991) Aging as a multi-step process characterized by a lowering of entropy production leading the cell to a sequence of defined stages. *Mech Ageing Dev* 61:45–64.

MODULATING LONGEVITY: CALORIE RESTRICTION

Colman RJ, Anderson RM, Johnson SC et al (2009) Caloric restriction delays disease onset and mortality in rhesus monkeys. *Science* 325:201–204.

Guarente L (2005) Calorie restriction and SIR2 genes—towards a mechanism. *Mech Ageing Dev* 126:923–928.

Klass MR (1977) Aging in the nematode *Caenorhabditis elegans*: major biological and environmental factors influencing lifespan. *Mech Ageing Dev* 6:413–429.

Masoro EJ (2005) Overview of caloric restriction and ageing. *Mech Ageing Dev* 126:913–922.

Mattison JA, Black A, Huck J et al (2005) Age-related decline in caloric intake and motivation for food in rhesus monkeys. *Neurobiol Aging* 26:1117–1127.

McCay CM, Crowell MF & Maynard LA (1935) The effect of retarded growth upon the length of lifespan and upon the ultimate body size. *J Nutr* 10:63–79.

Rochon J, Bales CW, Ravussin E et al (2011) Design and conduct of the CALERIE study: Comprehensive Assessment of the Long-term Effects of Reducing Intake of Energy. *J Gerontol A Biol Sci Med Sci* 66:97–108.

Rogina B, Helfand SL & Frankel S (2002) Longevity regulation by *Drosophila* Rpd3 deacetylase and caloric restriction. *Science* 298:1745.

Weindruch R & Walford RL (1982) Dietary restriction in mice beginning at 1 year of age: effect on lifespan and spontaneous cancer incidence. *Science* 215:1415–1418.

Weindruch R, Walford RL, Fligiel S & Guthrie D (1986) The retardation of aging in mice by dietary restriction: longevity, cancer, immunity and lifetime energy intake. *J Nutr* 116:641–654.

Yu BP, Masoro EJ, Murata I et al (1982) Lifespan study of SPF Fischer 344 male rats fed ad libitum or restricted diets: longevity, growth, lean body mass and disease. *J Gerontol* 37:130–141.

MODULATING THE RATE OF AGING: PHYSICAL ACTIVITY

Brooks GA, Fahey TD & Baldwin KM (2005) Exercise Physiology: Human Bioenergetics and Its Applications, 4th ed. Boston: McGraw Hill.

Goldspink DF (2005) Ageing and activity: their effects on the functional reserve capacities of the heart and vascular smooth and skeletal muscles. *Ergonomics* 48:1334–1351.

Holloszy JO, Smith EK, Vining M & Adams S (1985) Effect of voluntary exercise on longevity of rats. *J Appl Physiol* 59:826–831.

Paffenbarger RS Jr, Hyde RT, Wing AL & Hsieh CC (1986) Physical activity, all-cause mortality, and longevity of college alumni. *N Engl J Med* 314:605–613.

LOOKING TOWARD THE FUTURE: THE IMPLICATIONS OF MODULATING AGING AND LONGEVITY

Benecke M (2002) The Dream of Eternal Life: Biomedicine, Aging and Immortality. New York: Columbia University Press.

Chen NN (2009) Food, Medicine, and the Quest for Good Health: Nutrition, Medicine, and Culture. New York: Columbia University Press.

Lesnoff-Caravaglia G ed (1987) Realistic Expectations for Long Life. New York: Human Sciences Press.

President's Council on Bioethics (2003) Beyond Therapy: Biotechnology and the Pursuit of Happiness. Washington, DC: Government Printing Office.

Solomon LD (2006) The Quest for Human Longevity: Science, Business, and Public Policy. New Brunswick, NJ: Transaction Publishers.

National Vital Statistics Reports

Volume 58, Number 21

June 28, 2010

United States Life Tables, 2006

by Elizabeth Arias, Ph.D., Division of Vital Statistics

> The data in the second panel of Table II in the Technical Notes have been corrected.

Abstract

Objectives—This report presents complete period life tables by age, race, and sex for the United States based on age-specific death rates in 2006.

Methods—Data used to prepare the 2006 life tables are 2006 final mortality statistics, July 1, 2006 population estimates based on the 2000 decennial census, and 2006 Medicare data for ages 66–100. The 2006 life tables were estimated using a recently revised methodology first applied to the final annual U.S. life tables series with the 2005 edition (1). For comparability, all life tables for the years 2000–2004 were reestimated using the revised methodology and were published in an appendix of the United States Life Tables, 2005 report (1). These revised tables replace all previously published life tables for years 2000–2004.

Results—In 2006, the overall expectation of life at birth was 77.7 years, representing an increase of 0.3 years from life expectancy in 2005. From 2005 to 2006, life expectancy at birth increased for all groups considered. It increased for males (from 74.9 to 75.1) and females (from 79.9 to 80.2), the white (from 77.9 to 78.2) and black populations (from 72.8 to 73.2), black males (from 69.3 to 69.7) and females (from 76.1 to 76.5), and white males (from 75.4 to 75.7) and females (from 80.4 to 80.6).

Keywords: life expectancy • survival • death rates • race

Introduction

There are two types of life tables—the cohort (or generation) and the period (or current). The cohort life table presents the mortality experience of a particular birth cohort, all persons born in the year 1900, for example, from the moment of birth through consecutive ages in successive calendar years. Based on age-specific death rates observed through consecutive calendar years, the cohort life table reflects the mortality experience of an actual cohort from birth until no lives remain in the group. To prepare a single complete cohort life table requires data over many years. It is usually not feasible to construct cohort life tables entirely on the basis of

observed data for real cohorts due to data unavailability or incompleteness (2). For example, a life table representation of the mortality experience of a cohort of persons born in 1970 would require the use of data projection techniques to estimate deaths into the future (3,4).

Unlike the cohort life table, the period life table does not represent the mortality experience of an actual birth cohort. Rather, the period life table presents what would happen to a hypothetical (or synthetic) cohort if it experienced throughout its entire life the mortality conditions of a particular time period. Thus, for example, a period life table for 2006 assumes a hypothetical cohort subject throughout its lifetime to the age-specific death rates prevailing for the actual population in 2006. The period life table may thus be characterized as rendering a "snapshot" of current mortality experience, and shows the long-range implications of a set of age-specific death rates that prevailed in a given year. In this report the term "life table" refers only to the period life table and not to the cohort life table.

Data and Methods

The data used to prepare the U.S. life tables for 2006 are final numbers of deaths for the year 2006, postcensal population estimates for the year 2006, and age-specific death and population counts for Medicare beneficiaries aged 66–100 for the year 2006 from the Centers for Medicare & Medicaid Services.

The populations used to estimate the life tables shown in this report were produced under a collaborative agreement with the U.S. Census Bureau and are consistent with the postcensal estimates of the 2000 census. Reflecting the guidelines issued in 1997 by the Office of Management and Budget (OMB), the 2000 census included an option for individuals to report more than one race as appropriate for themselves and household members (5). The 1997 OMB guidelines also provided for the reporting of Asian persons separately from Native Hawaiian or other Pacific Islander persons. Under the prior OMB standards (issued in 1977), data for Asian or Pacific Islander persons were collected as a single group (6). Beginning with deaths occurring in 2003, some states implemented multiple-race categories on the death certificate. Approximately one-half of the states continue to collect only one race for the decedent in the same categories as

U.S. DEPARTMENT OF HEALTH AND HUMAN SERVICES
Centers for Disease Control and Prevention
National Center for Health Statistics
National Vital Statistics System

specified in the 1977 OMB guidelines (death certificate data do not report Asian persons separately from Native Hawaiian or other Pacific Islander persons). Death certificate data by race for these states (the numerators for death rates) are thus currently incompatible with the population data collected in the 2000 census (the denominators for the rates). To produce death rates for 2006 it was necessary to "bridge" the reported population data for multiple-race persons back to single-race categories. In addition, the 2000 census counts were modified to be consistent with the 1977 OMB race categories, that is, to report the data for Asian persons and Native Hawaiian or other Pacific Islander persons as a combined category, Asian or Pacific Islander, and to reflect age as of the census reference date (7). The procedures used to produce the "bridged" populations are described in a separate publication (8). Multiple-race data for those states that implemented the 1997 OMB guidelines are bridged back to single-race categories. Once all states are collecting data on race according to the 1997 OMB guidelines, it is expected that use of the bridged populations will be discontinued.

Readers should keep in mind that the population data used to compile death rates by race are based on special estimation procedures. They are not true counts. This is the case even for the 2000 populations that are based on the 2000 census. The estimation procedures used to develop these populations contain some errors (8). Over the next several years, additional information will be incorporated in the estimation procedures, possibly resulting in further revisions of the population estimates (see the "Technical Notes" section).

Data from the Medicare program are used to supplement vital statistics and census data for ages 66 years and over. Death rates based on Medicare data for the oldest ages are considered to be more accurate than death rates based solely on vital and census data because beneficiaries must prove their date of birth in order to qualify for benefits while there is no such requirement in the census form question about a respondent's age. The prevalence of age misreporting

at the oldest ages in census data has been found to be significant enough to lead to underestimated death rates at the oldest ages (see the "Technical Notes" section).

Life tables can be classified in two ways according to the length of the age interval in which data are presented. A complete life table contains data for every year of age. An abridged life table typically contains data by 5- or 10-year age intervals. A complete life table, of course, can be easily aggregated into 5- or 10-year age groups (see the "Technical Notes" section for instructions on how to do this). Other than the decennial life tables, U.S. life tables based on data prior to 1997 are abridged life tables constructed by reference to a standard table (9). The 2006 U.S. life tables are complete life tables calculated using a revised method that blends vital statistics and Medicare data at ages 66–100 (1). See the "Technical Notes" section for more information on the method used to construct the life tables in this report.

Expectation of life—The most frequently used life table statistic is life expectancy (e_x), which is the average number of years of life remaining for persons who have attained a given age (x). Life expectancy and other life table values for each age in 2006 are shown for the total population and by race and sex in Tables 1–9. Life expectancy is summarized by age, race, and sex in Table A.

Life expectancy at birth (e_0) for 2006 for the total population was 77.7 years. This represents the average number of years that the members of the hypothetical life table cohort may expect to live at the time of birth (Table A).

Survivors to specified ages— Another way of assessing the longevity of the synthetic life table cohort is by determining the proportion who survive to specified ages. The l_x column of the life table provides the data for computing the proportion. Table B summarizes the number of survivors by age, race, and sex. To illustrate, 54,201 persons out of the original 2006 synthetic life table cohort of 100,000 (or 54.2 percent) were alive at exact age 80. In other words, the probability that a person will survive from birth to age 80, given 2006 age-specific mortality, is

Table A. Expectation of life, by age, race, and sex: United States, 2006

Age	All races			White			Black		
	Total	Male	Female	Total	Male	Female	Total	Male	Female
0	77.7	75.1	80.2	78.2	75.7	80.6	73.2	69.7	76.5
1	77.2	74.7	79.7	77.6	75.1	80.0	73.2	69.7	76.5
5	73.3	70.8	75.8	73.7	71.2	76.1	69.4	65.8	72.6
10	68.4	65.8	70.8	68.7	66.3	71.1	64.4	60.9	67.7
15	63.4	60.9	65.9	63.8	61.3	66.1	59.5	56.0	62.7
20	58.6	56.1	61.0	59.0	56.6	61.3	54.7	51.3	57.8
25	53.9	51.5	56.1	54.2	51.9	56.4	50.1	46.8	53.0
30	49.2	46.9	51.3	49.5	47.3	51.5	45.5	42.4	48.2
35	44.4	42.2	46.4	44.7	42.6	46.7	40.9	37.9	43.5
40	39.7	37.6	41.7	40.0	37.9	41.9	36.4	33.5	38.9
45	35.2	33.1	37.0	35.4	33.4	37.2	32.0	29.2	34.5
50	30.7	28.8	32.5	30.9	29.0	32.6	27.9	25.2	30.2
55	26.5	24.7	28.0	26.6	24.9	28.2	24.1	21.6	26.1
60	22.4	20.7	23.8	22.5	20.9	23.8	20.4	18.2	22.2
65	18.5	17.0	19.7	18.6	17.1	19.8	17.1	15.1	18.6
70	14.9	13.6	15.9	14.9	13.6	15.9	13.9	12.3	15.1
75	11.6	10.4	12.3	11.5	10.5	12.3	11.1	9.8	12.0
80	8.7	7.8	9.3	8.7	7.8	9.3	8.7	7.7	9.3
85	6.4	5.7	6.8	6.3	5.7	6.7	6.7	5.9	7.1
90	4.6	4.1	4.8	4.5	4.0	4.7	5.1	4.5	5.3
95	3.2	2.9	3.3	3.2	2.8	3.3	3.8	3.5	3.9
100	2.3	2.0	2.3	2.2	2.0	2.2	2.8	2.6	2.8

Table B. Number of survivors by age, out of 100,000 born alive, by race and sex: United States, 2006

Age	All races			White			Black		
	Total	Male	Female	Total	Male	Female	Total	Male	Female
0	100,000	100,000	100,000	100,000	100,000	100,000	100,000	100,000	100,000
1	99,329	99,266	99,395	99,442	99,388	99,499	98,663	98,552	98,777
5	99,216	99,144	99,291	99,341	99,279	99,406	98,492	98,367	98,622
10	99,147	99,068	99,229	99,277	99,208	99,349	98,394	98,254	98,539
15	99,065	98,972	99,164	99,200	99,117	99,288	98,285	98,125	98,451
20	98,747	98,524	98,982	98,898	98,702	99,105	97,868	97,484	98,266
25	98,253	97,797	98,739	98,430	98,017	98,874	97,174	96,435	97,940
30	97,759	97,099	98,461	97,970	97,370	98,616	96,380	95,274	97,500
35	97,213	96,371	98,105	97,466	96,697	98,292	95,452	94,001	96,892
40	96,495	95,466	97,579	96,799	95,851	97,813	94,256	92,489	95,987
45	95,397	94,112	96,740	95,771	94,569	97,048	92,515	90,398	94,564
50	93,750	92,082	95,478	94,231	92,655	95,893	89,877	87,206	92,430
55	91,352	89,083	93,681	91,992	89,850	94,231	85,930	82,211	89,426
60	88,057	85,054	91,119	88,870	86,041	91,806	80,756	75,746	85,423
65	83,251	79,346	87,200	84,216	80,526	88,012	73,917	67,414	79,910
70	76,661	71,652	81,662	77,739	72,970	82,584	65,507	57,534	72,760
75	67,331	61,057	73,449	68,440	62,425	74,416	55,000	45,743	63,292
80	54,201	46,859	61,175	55,215	48,070	62,094	42,229	32,641	50,822
85	37,805	30,371	44,685	38,526	31,170	45,373	28,469	20,043	36,141
90	20,898	15,034	26,183	21,196	15,318	26,479	15,864	9,952	21,357
95	7,991	4,895	10,685	7,979	4,873	10,656	6,716	3,675	9,558
100	1,737	850	2,460	1,672	804	2,373	1,928	905	2,845

54 percent. Probabilities of survival can be calculated at any age by simply dividing the number of survivors at the terminal age by the number at the beginning age. For example, to calculate the probability of surviving from age 20 to age 85, one would divide the number of survivors at age 85 (37,805) by the number of survivors at age 20 (98,747), which results in a 38.3 percent probability of survival.

Explanation of the columns of the life table

*Column 1—Age (x to x + 1)—*Shows the age interval between the two exact ages indicated. For instance, "20–21" means the 1-year interval between the 20th and 21st birthdays.

Column 2—Probability of dying (q_x)— Shows the probability of dying between ages x to x + 1. For example, for males in the age interval 20–21 years, the probability of dying is 0.001329 (Table 2). The "probability of dying" column forms the basis of the life table; all subsequent columns are derived from it.

Column 3—Number surviving (l_x)— Shows the number of persons from the original synthetic cohort of 100,000 live births, who survive to the beginning of each age interval. The l_x values are computed from the q_x values, which are successively applied to the remainder of the original 100,000 persons still alive at the beginning of each age interval. Thus, out of 100,000 female babies born alive, 99,395 will complete the first year of life and enter the second; 99,229 will reach age 10; 98,982 will reach age 20; and 44,685 will live to age 85 (Table 3).

Column 4—Number dying (d_x)— Shows the number dying in each successive age interval out of the original 100,000 live births. For example, out of 100,000 males born alive, 734 will die in the first year of life; 131 between ages 20 and 21; and 850 will die after reaching age 100 (Table 2). Each figure in column 4 is the difference between two successive figures in column 3.

*Column 5—Person-years lived (L_x)—*Shows the number of person-years lived by the synthetic life table cohort within an age interval x to x + 1. Each figure in column 5 represents the total time (in years) lived between two indicated birthdays by all those reaching the earlier birthday. Thus, the figure 98,459 for males in the age interval 20–21 is the total number of years lived between the 20th and 21st birthdays by the 98,524 (column 3) males who reached their 20th birthday out of 100,000 males born alive (Table 2).

*Column 6—Total number of person-years lived (T_x)—*Shows the total number of person-years that would be lived after the beginning of the age interval x to x + 1 by the synthetic life table cohort. For example, the figure 5,532,004 is the total number of years lived after attaining age 20 by the 98,524 males reaching that age (Table 2).

*Column 7—Expectation of life (e_x)—*Shows, at any given age, the average number of years remaining to be lived by those surviving to that age on the basis of a given set of age-specific rates of dying. It is derived by dividing the total person-years that would be lived above age x by the number of persons who survived to that age interval (T_x / l_x). Thus, the average remaining lifetime for males who reach age 20 is 56.1 years (5,532,004 divided by 98,524) (Table 2).

Results

Life expectancy in the United States

Tables 1–9 show complete life tables by race (white and black) and sex for 2006. Tables A and B summarize life expectancy and survival by age, race, and sex. Life expectancy at birth for 2006 represents the average number of years that a group of infants would live if the infants were to experience throughout life the age-specific death rates prevailing in 2006. In 2006, life expectancy at birth was 77.7 years, increasing by 0.3 years from 77.4 years in 2005. This increase is typical of the average annual changes that have occurred during the last 30 years. Throughout the past century, the trend in U.S. life expectancy was one of gradual improvement and this trend has continued into the new century (10).

Changes in mortality levels by age and cause of death have an important effect on changes in life expectancy. Life expectancy at birth increased from 2005 to 2006, to 75.1 for males and 80.2 for females. Increases in life expectancy for both males and females were a function of decreases in mortality from heart disease, cancer, chronic lower respiratory diseases, and stroke. The increase in life expectancy for the entire population from 2005 to 2006 could have been greater if it was not for the increase in mortality from unintentional injuries, viral hepatitis, homicide, and kidney disease (11).

The difference in life expectancy between the sexes was 5.1 years in 2006, increasing from 5.0 in 2005. From 1900 to 1975, the difference in life expectancy between the sexes increased from 2.0 years to 7.8 years. The increasing gap during these years is attributed to increases in male mortality due to ischemic heart disease and lung cancer, both of which increased largely as the result of men's early and widespread adoption of cigarette smoking (12,13). Between 1979 and 2004, the difference in life expectancy between the sexes narrowed from 7.8 years to 5.0 years and then increased slightly to 5.1 from 2005 to 2006. The general decline in the difference between males and females since 1979 reflects proportionately greater increases in lung cancer mortality for women than for men and proportionately larger decreases in heart disease mortality among men (12,13).

From 2005 to 2006, life expectancy increased by 0.4 years to 73.2 years for the black population, and by 0.3 years to 78.2 years for the white population. The difference in life expectancy between the white and black populations was 5.0 years in 2006, a historical low. The difference in life expectancy between the black and white populations narrowed from 14.6 years in 1900 to 5.7 years in 1982, but increased to 7.1 years in 1993 before beginning to decline again in 1994 (7.0 years). The increase in the gap from 1983 to 1993 was largely the result of increases in mortality among the black male population due to HIV infection and homicide (12,13).

Among the four groups shown in Figure 1, white females continued to have the highest life expectancy at birth (80.6 years), followed by black females (76.5 years), white males (75.7 years), and black males (69.7 years). From 2005 to 2006, life expectancy increased 0.4 years for black females (from 76.1 to 76.5) as well as for black males (from 69.3 to 69.7). Black males experienced an unprecedented decline in life expectancy every year for the period 1984–1989 (14), but annual increases in the years 1990–1992, 1994–2004, and 2005–2006. From 2005 to 2006, life expectancy increased by 0.3 years for white males (from 75.4 to 75.7) and by 0.2 years for white females (from 80.4 to 80.6). Overall, gains in life expectancy between 1980 and 2006 were 5.9 years for black males, 5.0 years for white males, 4.0 years for black females, and 2.5 years for white females (Table 12).

The 2006 life table may be used to compare life expectancy at any age from birth onward. On the basis of mortality experienced in 2006, a person aged 65 could expect to live an average of 18.5 more years for a total of 83.5 years, and a person aged 100 could expect to live an additional 2.3 years on average (Table A). Life expectancy at age 100, particularly for the black population, should be interpreted with caution as these figures may be affected somewhat by age misreporting (15,16,17).

Survivorship in the United States

Table B summarizes the number of survivors out of 100,000 persons born alive (l_x) by age, race, and sex. Table 10 shows trends in survivorship from 1900 through 2006. In 2006, 99.3 percent of all infants born in the United States survived the first year of life. In

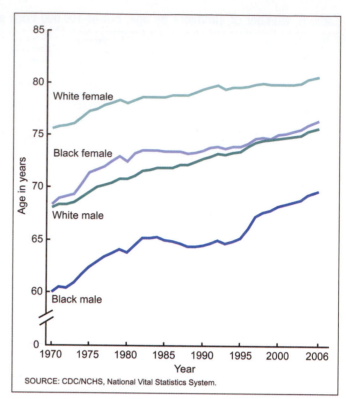

Figure 1. Life expectancy at birth, by race and sex: United States, 1970–2006

contrast, only 87.6 percent of infants born in 1900 survived the first year. Fifty-four percent of the 2006 synthetic life table cohort survived to age 80 and about 1.7 percent survived to age 100. In 1900, the median age at death was 58 and only 0.03 percent survived to age 100.

Among the four groups shown in Figure 2 and Table B, white females have the highest median age at death with about 49 percent surviving to age 84. Of the original hypothetical cohort of 100,000 infant white females, 99.1 percent survive to age 20, 88 percent to age 65, and 45.4 percent to age 85. For white males and black females, the pattern of survival by age is similar. White males have slightly higher survival rates than black females at the younger ages with 98.7 percent surviving to age 20 and 80.5 percent surviving to age 65 compared with 98.3 percent and 79.9 percent, respectively, for black females. At the older ages, in contrast, black female survival surpasses white male survival. At age 85, white male survival is 31.2 percent compared with 36.1 percent for black females. This crossover, which occurs at age 75, is clearly shown in Figure 2. The median age at death for black males is 73 years, which is 11 years less than that of white females. For black males, 97.5 percent survive to age 20, 67.4 percent to age 65, and 20 percent to age 85. By age 100, there is very little difference between the white and black populations in terms of survival. Less than 1 percent of white and black males and slightly over 2 percent of white and black females, respectively, survive to age 100.

Plotting the percentage surviving by age for the periods 1900–1902, 1949–1951, and 2006 shows an increasingly rectangular survival curve (Figure 3). That is, the survival curve has become increasingly flat in response to progressively lower mortality, particularly at the younger ages, and increasingly vertical at the older ages. The

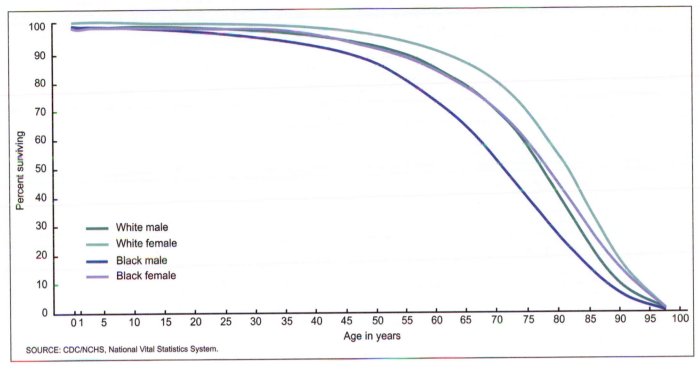

Figure 2. Percentage surviving, by age, race, and sex: United States, 2006

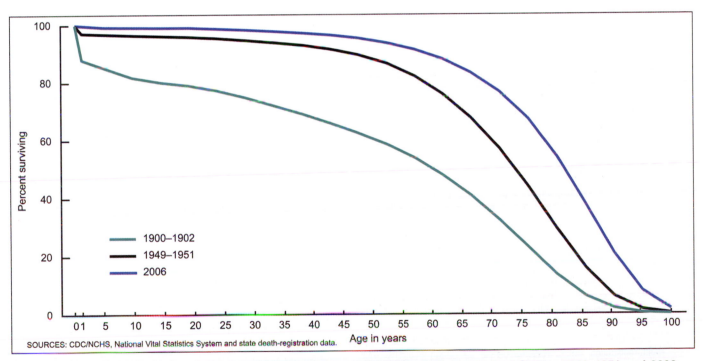

Figure 3. Percentage surviving, by age: Death-registration states, 1900–1902, and United States, 1949–1951 and 2006

survival curve for the period 1900–1902 shows a rapid decline in survival in the first few years of life and a relatively steady decline thereafter. In contrast, the survival curve for 2006 is nearly flat until about age 50 after which the decline in survival becomes more rapid. Improvements in survival between the periods 1900–1902 and 1949–1951 occurred at all ages, although the largest improvements were among the younger population. Between 1949–1951 and 2006, improvements occurred primarily for the older population.

References

1. Arias E, Rostron B, Tejada-Vera B. United States life tables, 2005. National vital statistics reports; vol 58 no 10. Hyattsville, MD: National Center for Health Statistics. 2010.

2. Shryock HS, Siegel JS, et al. The methods and materials of demography, vol 2. U.S. Bureau of the Census. Washington, DC: U.S. Government Printing Office. 1971.

3. Moriyama IM, Gustavus SO. Cohort mortality and survivorship, United States death-registration States, 1900–68. National Center for Health Statistics. Vital Health Stat 3(16). 1972.

4. Preston SM, Heuveline P, Guillot M. Demography, Measuring and Modeling Population Processes. Oxford: Blackwell Publishers. 2001.

5. Office of Management and Budget. Revisions to the standards for the classification of federal data on race and ethnicity. Federal Register 62FR58782–58790. 1997. Available from: http://www.whitehouse.gov/omb/fedreg/ombdir15.html.

6. Office of Management and Budget. Race and ethnic standards for federal statistics and administrative reporting. Statistical policy directive 15. 1977.

7. U.S. Census Bureau. Age, sex, race, and Hispanic origin information from the 1990 census: A comparison of census results with results where age and race have been modified, 1990. CPH—L—74. Washington, DC: U.S. Department of Commerce. 1991.

8. Ingram DD, Weed JA, Parker JD, et al. U.S. Census 2000 with bridged race categories. National Center for Health Statistics. Vital Health Stat 2(135). 2003.

9. Sirken MG. Comparison of two methods of constructing abridged life tables by reference to a "standard" table. National Center for Health Statistics. Vital Health Stat 2(4). 1966.

10. Arias E, Curtin LR, Wei R, Anderson RN. U.S. decennial life tables for 1999–2001, United States life tables. National vital statistics reports; vol 57 no 1. Hyattsville, MD: National Center for Health Statistics. 2008.

11. Heron M, Hoyert DL, Murphy SL, et al. Deaths: Final data for 2006. National vital statistics reports; vol 57 no 14. Hyattsville, MD: National Center for Health Statistics. 2009.

12. National Center for Health Statistics. U.S. decennial life tables for 1989–91, vol 1 no 3, some trends and comparisons of United States life table data: 1900–91. Hyattsville, MD. 1999.

13. Waldron I. Recent trends in sex mortality ratios for adults in developed countries. Soc Sci Med 36(4):451–62. 1993.

14. Kochanek KD, Maurer JD, Rosenberg HM. Causes of death contributing to changes in life expectancy: United States, 1984–89. National Center for Health Statistics. Vital Health Stat 20(23). 1994.

15. Anderson RN. A method for constructing complete annual U.S. life tables. National Center for Health Statistics. Vital Health Stat 2(129). 1999.

16. Kestenbaum B. A description of the extreme aged population based on improved Medicare enrollment data. Demography 29(4):565–80. 1992.

17. Coale AJ, Kisker EE. Defects in data on old-age mortality in the United States: New procedures for calculating mortality schedules and life tables at the highest ages. Asian and Pacific Population Forum 4:1–31. 1990.

18. Anderson RN, Arias E. The effect of revised populations on mortality statistics for the United States, 2000. National vital statistics reports; vol 51 no 9. Hyattsville, MD: National Center for Health Statistics. 2003.

19. Greville TNE, Carlson GA. Estimated average length of life in the death-registration states. National Center for Health Statistics. Vital statistics—special reports; vol 33 no 9. Washington, DC: Public Health Service. 1951.

20. Bell FC, Miller ML. Life tables for the United States Social Security area 1900–2100. Baltimore, MD: Social Security Administration, Office of the Chief Actuary. SSA Pub. No. 11–11536. 2005.

21. Research Data Assistance Center. Introduction to the use of Medicare data for research. Minneapolis, MN: University of Minnesota School of Public Health. 2004.

22. Heligman P, Pollard JH. The age pattern of mortality. J Inst Actuar 107(1):49–80. 1980.

23. Hartman M. Past and recent attempts to model mortality at all ages. Journal Off Stat 3(1):19–36. 1987.

List of Detailed Tables

Table 1. Life table for the total population: United States, 2006

Age	Probability of dying between ages x to x + 1 q_x	Number surviving to age x l_x	Number dying between ages x to x + 1 d_x	Person-years lived between ages x to x + 1 L_x	Total number of person-years lived above age x T_x	Expectation of life at age x e_x
0–1	0.006713	100,000	671	99,409	7,770,850	77.7
1–2	0.000444	99,329	44	99,307	7,671,441	77.2
2–3	0.000300	99,285	30	99,270	7,572,134	76.3
3–4	0.000216	99,255	21	99,244	7,472,864	75.3
4–5	0.000179	99,233	18	99,225	7,373,620	74.3
5–6	0.000168	99,216	17	99,207	7,274,396	73.3
6–7	0.000156	99,199	15	99,191	7,175,188	72.3
7–8	0.000143	99,184	14	99,177	7,075,997	71.3
8–9	0.000125	99,169	12	99,163	6,976,820	70.4
9–10	0.000103	99,157	10	99,152	6,877,657	69.4
10–11	0.000086	99,147	9	99,143	6,778,505	68.4
11–12	0.000088	99,138	9	99,134	6,679,363	67.4
12–13	0.000125	99,130	12	99,123	6,580,229	66.4
13–14	0.000206	99,117	20	99,107	6,481,105	65.4
14–15	0.000317	99,097	31	99,081	6,381,999	64.4
15–16	0.000438	99,065	43	99,044	6,282,918	63.4
16–17	0.000552	99,022	55	98,995	6,183,874	62.4
17–18	0.000657	98,967	65	98,935	6,084,879	61.5
18–19	0.000747	98,902	74	98,865	5,985,945	60.5
19–20	0.000825	98,828	82	98,788	5,887,079	59.6
20–21	0.000905	98,747	89	98,702	5,788,291	58.6
21–22	0.000983	98,658	97	98,609	5,689,589	57.7
22–23	0.001033	98,561	102	98,510	5,590,980	56.7
23–24	0.001049	98,459	103	98,407	5,492,471	55.8
24–25	0.001038	98,355	102	98,304	5,394,063	54.8
25–26	0.001019	98,253	100	98,203	5,295,759	53.9
26–27	0.001006	98,153	99	98,104	5,197,556	53.0
27–28	0.000998	98,055	98	98,006	5,099,452	52.0
28–29	0.001002	97,957	98	97,908	5,001,446	51.1
29–30	0.001018	97,859	100	97,809	4,903,539	50.1
30–31	0.001042	97,759	102	97,708	4,805,730	49.2
31–32	0.001072	97,657	105	97,605	4,708,022	48.2
32–33	0.001113	97,552	109	97,498	4,610,417	47.3
33–34	0.001156	97,444	113	97,387	4,512,919	46.3
34–35	0.001212	97,331	118	97,272	4,415,532	45.4
35–36	0.001276	97,213	124	97,151	4,318,260	44.4
36–37	0.001355	97,089	132	97,023	4,221,109	43.5
37–38	0.001456	96,958	141	96,887	4,124,085	42.5
38–39	0.001585	96,816	153	96,740	4,027,198	41.6
39–40	0.001739	96,663	168	96,579	3,930,459	40.7
40–41	0.001903	96,495	184	96,403	3,833,880	39.7
41–42	0.002077	96,311	200	96,211	3,737,477	38.8
42–43	0.002268	96,111	218	96,002	3,641,266	37.9
43–44	0.002479	95,893	238	95,774	3,545,264	37.0
44–45	0.002706	95,655	259	95,526	3,449,490	36.1
45–46	0.002943	95,397	281	95,256	3,353,964	35.2
46–47	0.003190	95,116	303	94,964	3,258,707	34.3
47–48	0.003453	94,812	327	94,649	3,163,743	33.4
48–49	0.003741	94,485	353	94,308	3,069,095	32.5
49–50	0.004057	94,132	382	93,941	2,974,786	31.6
50–51	0.004405	93,750	413	93,543	2,880,846	30.7
51–52	0.004778	93,337	446	93,114	2,787,302	29.9
52–53	0.005166	92,891	480	92,651	2,694,189	29.0
53–54	0.005554	92,411	513	92,154	2,601,538	28.2
54–55	0.005939	91,898	546	91,625	2,509,383	27.3
55–56	0.006335	91,352	579	91,063	2,417,759	26.5
56–57	0.006760	90,773	614	90,466	2,326,696	25.6
57–58	0.007234	90,160	652	89,834	2,236,230	24.8
58–59	0.007796	89,507	698	89,158	2,146,396	24.0
59–60	0.008470	88,810	752	88,433	2,057,238	23.2
60–61	0.009282	88,057	817	87,649	1,968,804	22.4
61–62	0.010204	87,240	890	86,795	1,881,155	21.6
62–63	0.011178	86,350	965	85,867	1,794,360	20.8
63–64	0.012118	85,385	1,035	84,867	1,708,493	20.0
64–65	0.013024	84,350	1,099	83,801	1,623,626	19.2
65–66	0.013999	83,251	1,165	82,669	1,539,825	18.5
66–67	0.014995	82,086	1,231	81,471	1,457,156	17.8

Table 1. Life table for the total population: United States, 2006—Con.

Age	Probability of dying between ages x to $x+1$ q_x	Number surviving to age x l_x	Number dying between ages x to $x+1$ d_x	Person-years lived between ages x to $x+1$ L_x	Total number of person-years lived above age x T_x	Expectation of life at age x e_x
67–68	0.016161	80,855	1,307	80,202	1,375,686	17.0
68–69	0.017527	79,548	1,394	78,851	1,295,484	16.3
69–70	0.019109	78,154	1,493	77,408	1,216,633	15.6
70–71	0.020890	76,661	1,601	75,860	1,139,225	14.9
71–72	0.022925	75,059	1,721	74,199	1,063,365	14.2
72–73	0.025280	73,339	1,854	72,412	989,166	13.5
73–74	0.027972	71,485	2,000	70,485	916,755	12.8
74–75	0.030997	69,485	2,154	68,408	846,270	12.2
75–76	0.034386	67,331	2,315	66,174	777,862	11.6
76–77	0.038027	65,016	2,472	63,780	711,688	10.9
77–78	0.042036	62,544	2,629	61,229	647,908	10.4
78–79	0.046447	59,915	2,783	58,523	586,679	9.8
79–80	0.051297	57,132	2,931	55,666	528,156	9.2
80–81	0.056623	54,201	3,069	52,667	472,489	8.7
81–82	0.062465	51,132	3,194	49,535	419,823	8.2
82–83	0.068867	47,938	3,301	46,287	370,288	7.7
83–84	0.075871	44,637	3,387	42,943	324,000	7.3
84–85	0.083524	41,250	3,445	39,527	281,057	6.8
85–86	0.091872	37,805	3,473	36,068	241,530	6.4
86–87	0.100962	34,332	3,466	32,598	205,461	6.0
87–88	0.110842	30,865	3,421	29,155	172,863	5.6
88–89	0.121558	27,444	3,336	25,776	143,708	5.2
89–90	0.133155	24,108	3,210	22,503	117,932	4.9
90–91	0.145675	20,898	3,044	19,376	95,429	4.6
91–92	0.159156	17,854	2,842	16,433	76,053	4.3
92–93	0.173631	15,012	2,607	13,709	59,620	4.0
93–94	0.189127	12,406	2,346	11,232	45,911	3.7
94–95	0.205661	10,059	2,069	9,025	34,679	3.4
95–96	0.223242	7,991	1,784	7,099	25,654	3.2
96–97	0.241869	6,207	1,501	5,456	18,555	3.0
97–98	0.261527	4,706	1,231	4,090	13,099	2.8
98–99	0.282188	3,475	981	2,985	9,009	2.6
99–100.	0.303810	2,494	758	2,115	6,024	2.4
100 and over	1.00000	1,737	1,737	3,909	3,909	2.3

GLOSSARY

Aβ protein
A subunit of amyloid precursor protein (a transmembrane protein), the protein forming amyloid plaques.

abiotic
Nonliving; describing chemical and physical factors in the environment.

abscisic acid
A plant hormone that slows growth, often antagonizing the actions of growth hormones. Two of its many effects are to promote seed dormancy and facilitate drought tolerance.

abscission
The organized, programmed, or natural loss of a part of a plant, usually a leaf, fruit, or flower; a critical component of leaf senescence.

accommodation (optical)
The degree to which the lens assists the cornea in focusing on an object.

acetylcholine (Ach)
A neurotransmitter in both the peripheral nervous system (PNS) and the central nervous system (CNS); causes excitatory action in the CNS; activates muscle in the PNS.

acetyl-CoA
An important metabolic intermediate derived from various pathways, including glycolysis, fatty acid oxidation, and degradation of some amino acids; the entry compound for the tricarboxylic acid cycle in cellular respiration.

acquired (adaptive) immunity
Immunity that develops after exposure to a particular antigen or after formation of antibodies; primarily comprised of B cells and T cells and substances secreted by these cells.

actin
One of two proteins of muscle that are responsible for muscle contraction; forms the thin filament of the myofibril; also found in microtubules of the cytoskeleton. *See also* myosin.

action potential
A wave of electrical discharge that travels along the membrane of a cell; usually associated with nerve transmission and muscle contraction.

activators
DNA-binding proteins that increase the rate of transcription.

active transport
A type of facilitated diffusion across a biological membrane, using membrane proteins to transport large molecules in a two-step process: first, the molecule binds to a transport protein in the membrane and associates with an ion that moves down its concentration gradient; second, energy is used to move the ion out of the cell against its concentration gradient.

adenine
One of the nucleotide bases of DNA and RNA; a purine. *See also* nucleotide bases.

adenosine 5′-diphosphate (ADP)
A nucleoside composed of adenine bonded to two phosphate groups. ADP is produced by breakage of one high-energy phosphate bond of ATP (ATP → ADP + P_i + energy).

adenosine 5′-triphosphatase (ATPase)
An enzyme that catalyzes the hydrolysis of ATP to form ADP and P_i (inorganic phosphate), providing energy to drive reactions.

adenosine 5′-triphosphate (ATP)
A nucleoside composed of adenine and three phosphate groups; the primary source of chemical energy used by the cell.

adipose tissue
The anatomical location of fat storage, storing energy in the form of triglycerides; also provides insulation and cushioning for organs. Most adipose tissue in humans is found under the skin (as subcutaneous fat).

ad libitum
In biology, describes an animal's eating protocol—animals are given access to food at all times. Latin for "at [one's] pleasure."

adrenergic neurons
Neurons that release the neurotransmitters norepinephrine and epinephrine.

adrenergic receptors
Neurotransmitter receptors that bind norepinephrine and epinephrine.

advanced glycation end product (AGE)
A product of the nonenzymatic linking of two proteins by Amadori products. AGEs are insoluble and do not degrade easily; they accumulate in some cells over time and are thought to be among the many compounds that cause cellular aging.

aerobic metabolism
Metabolism occurring in or requiring the presence of oxygen.

age-1 gene
A gene identified in *C. elegans* as important to the regulation of longevity; expresses protein of the highly conserved phosphatidylinositol-3-kinase (PIK-3) family.

age-dependent mortality
Death due to natural causes; the end result of biological aging.

age-independent mortality
Death that is not the result of biological aging.

age-specific mortality rate
A measure of the chance (probability) of dying within a specific age range.

age-structure analysis
A mathematical analysis used to determine the rate

at which alleles are fixed in complex eukaryotes; often used to describe a population's reproductive contribution to future generations at any given time during the life span.

alkaline phosphatase
An enzyme found primarily in liver and bone that removes phosphate groups under alkaline conditions.

allele
One of two (or more) forms of a gene at a given locus on a chromosome.

allometric scaling
The mathematical relationship between a biological property, growth, and a biological process, such as between surface area and metabolic rate in warm-blooded animals. Allometry is the study of the relationship between physiological size and anatomical shape and its effect on behavior.

α-amylase
An enzyme in saliva and pancreatic fluid that breaks down starch and glycogen into simple sugars.

α-synuclein
A small protein of unknown function, primarily associated with synaptic vesicles of dopamine-producing neurons; found in high concentrations in Lewy bodies. Mutation in its gene is associated with early-onset Parkinson's disease.

alternative RNA splicing
Synthesis of different RNAs from the same gene. Exons of the RNA transcript are connected in multiple ways during RNA splicing.

Alzheimer's disease
A type of age-related dementia that causes problems with memory, thinking, and behavior; named after Alois Alzheimer. There are three primary types: early-onset, late-onset, and familial.

Amadori product
A product formed in the second stage of the Maillard reaction. A Schiff base undergoes nonenzymatic rearrangement to form the Amadori product, which is much more stable than the Schiff base and can accumulate in the cell.

aminoacyl-tRNA synthetases
Enzymes that catalyze the attachment of a specific tRNA to its corresponding amino acid.

amplitude
Magnitude of an oscillating quantity such as sound pressure; the distance between compression and rarefaction.

amyloid plaque
A highly insoluble fibrous aggregate of Aβ protein; a pathological marker of Alzheimer's disease.

amyloid precursor protein (APP)
A large, transmembrane protein that supports dendrite outgrowth, synaptogenesis, and inhibition of platelet activation. A subunit of this protein is the Aβ protein.

anaerobic metabolism
Metabolism occurring in the absence of molecular oxygen.

anaphase
The stage of mitosis in which replicated chromosomes separate.

anaphase-promoting complex (APC)
A member of the ubiquitin ligase family of enzymes; marks M- and S-cyclins for degradation, thus initiating the separation of sister chromatids in mitosis.

andropause
Biological change experienced by men after their mid-life, characterized by a gradual decline in testosterone level.

angiotensin-converting enzyme inhibitors (ACEs)
A class of drugs that are used to treat hypertension; prevent the conversion of angiotensin I to angiotensin II, which is a vasoconstrictor.

anorexia of aging
A clinical condition of the elderly in which loss of appetite leads to excessive loss of body weight; one of the four symptoms associated with geriatric failure to thrive. The cause is multifactorial and includes disease, depression, and social isolation.

antagonistic pleiotropy
A theory originally postulated by G. C. Williams suggesting that aging occurs because genes that have a beneficial effect in early life have a harmful effect in late life.

antidiuretic hormone (vasopressin)
A hormone released from the pituitary gland that increases the water permeability of the collecting ducts in the kidney, resulting in increased arterial blood pressure.

antigen
A foreign particle, molecule, or organism.

aortic valve
One-way valve that regulates blood flow out of the left ventricle; positioned between the left ventricle and the aorta.

apolipoprotein E (ApoE)
A lipid-binding protein.

apoplast
In plant cells, the space between the cell wall and the plasma membrane through which water and soluble nutrients are transported across a tissue or organ.

apoplastic uploading
In plants, the process by which sucrose is broken down into glucose and fructose in the apoplast and the two sugars are transported into the cell.

apoptosis
A genetically determined destruction of cells from within; results from activation of a stimulus (or removal of a suppressing agent), postulated to exist to explain the orderly elimination of superfluous cells. Also called programmed cell death.

arterial plaques
Fatty deposits inside an arterial wall, made up of immune cells (macrophages), fatty acids, calcium, and cholesterol; characteristic of atherosclerosis.

arteries
Blood vessels that leave the heart; generally carry oxygenated blood (exception is the pulmonary artery).

artificial selection
A laboratory method for modeling natural selection in which eggs (or offspring) are collected from short-lived, rapidly reproducing species that have a particular trait.

The process continues for several generations until the trait has become dominant in the population.

ascospores
Spores produced in the saclike cells of the sexual stage of a fungus, such as yeast.

atherosclerosis
A pathological process in large and medium-size arteries, where deposits of fatty substances, cholesterol, calcium, and/or fibrin build up in the inner lining, forming plaques.

atria (heart) (*single*, **atrium**)
Two small, pouch-like projections from the upper right ventricle and left ventricle that act as primers for the ventricles so that blood flow is smooth and uninterrupted.

atrioventricular node (AV node)
A specialized mass of cardiac muscle fibers that picks up a signal from the sinus (sinoatrial) node and causes the ventricles to contract, pumping blood out of the heart.

atrophic gastritis
Inflammation of the stomach mucosa that causes loss of chief and parietal cells, resulting in decreased secretion of HCl and various enzymes. Cells are replaced by fibrous tissue.

autophagy (autophagic system)
The degradation of a cell's own components; a regulated process involving the lysosomal mechanism that helps maintain normal cell growth, development, and homeostasis. In plants, it is an important process supporting senescence.

autosome
Chromosome that is not associated with the sex of the individual.

autotrophy
The ability to be self-sustaining by producing food from inorganic compounds. Plants and some algae are autotrophic.

auxins
A class of plant hormones that stimulate cell division, cell enlargement, apical dominance, root initiation, and flowering.

axon
The long, tubular extension of a neuron that carries signals from the cell body to the axon terminals.

B cells
Lymphocytes (white blood cells) responsible for antibody-mediated immunity. After maturing in the bone marrow, the cells circulate in the circulatory and lymph systems, where they can differentiate into antibody-producing plasma cells when exposed to antigens.

beta-blockers
A class of drugs used to treat hypertension; prevent the binding of catecholamines to adrenergic receptors in the heart.

β-sheet
A type of secondary protein structure consisting of sequences of 5–10 amino acid residues (a β-strand) connected laterally by three or more hydrogen bonds; often the basic structure in protein aggregates, such as amyloid plaques in the brain in Alzheimer's disease.

bile salts
Alkaline salts in bile, necessary for emulsification and digestion of fats. Bile is produced in the liver and stored in the gallbladder.

biodemography
A science dealing with the integration of biology and demography.

biogerontology
A branch of gerontology that deals with the biological basis of aging.

biomarker
An observation, substance, or molecule that can be used as an indicator of a biological state. It can be objectively measured and used as a predictor of future outcomes.

biotic
Living; pertaining to all living species on Earth.

bladder (urinary)
Sac composed of smooth muscle that stores urine prior to urination.

body mass index (BMI)
A body composition index relating weight to height, calculated as weight in kilograms divided by height in meters squared; often used as a marker of obesity.

bone mineral density (BMD)
A measure of the amount of matter per square centimeter of bone.

bone remodeling
The continuous turnover of bone tissue involving resorption (osteoclastic activity) and deposition (osteoblastic activity) of minerals.

Bowman's capsule
In the kidney, the structure at the beginning of the tubular component of a nephron that performs the first step in the filtration of blood to form urine, in a process called glomerular filtration. Also called the glomerular capsule.

bradykinesia
The slowing of motor movements due to dysfunction of the basal ganglia; often seen in Parkinson's disease or in the syndrome "parkinsonism."

breeding season–specific reproduction rate
The net reproduction rate of breeding-season animals, indicating which groups have the most influence on growth and fitness.

brush border
The microvilli forming a border on the intestinal side of the epithelial cells of the small intestine.

bud scar
A circular chitin residue left on a mother cell of budding yeast after a daughter cell separates.

calcitonin
A hormone synthesized and secreted by the thyroid gland in response to high serum calcium; increases bone deposition of calcium and urinary calcium concentration.

calorie (cal)
A unit of heat energy: the energy required to heat 1 g of water 1°C.

calorie restriction (dietary restriction)
An experimental paradigm used in biogerontology to extend the life span of species maintained under laboratory conditions. Calories are restricted without altering the micronutrient composition.

Calvin cycle (dark phase of photosynthesis)
Chemical reactions taking place in the chloroplast that synthesize sugars from CO_2, using energy from the light reactions.

carbidopa
A DOPA decarboxylase inhibitor that prevents conversion of L-dopa (levodopa) to dopamine in peripheral nerves; used pharmacologically in conjunction with levodopa.

carbohydrate
A compound containing carbon, hydrogen, and oxygen with the general composition $(CH_2O)_n$. Food carbohydrates include monosaccharides (simple sugars), disaccharides (complex sugars), and polysaccharides (starches).

carboxylation
A chemical reaction in which a carboxylic acid group (–COOH) is introduced into a substrate molecule.

cardiac output
The total amount of blood pumped out of the heart over a given time period, as volume per minute; equal to stroke volume (ml) × heart rate (beats/min).

cardiovascular system
The heart and blood vessels; a closed system of fluid transport with a central pump (the heart) and conduits (arteries and veins) that carry fluid to target structures (cells) and back to the pump.

carotenoids
Organic pigments found in the chromoplasts of plants, such as in fruits and petals; absorb blue light and reflect green and red light, thus appearing yellow and orange. More than 600 carotenoids have been identified.

carrying capacity of a population, K
The theoretical population size at equilibrium; the size at which a particular population in a particular environment stabilizes when its supply of resources remains constant. This is the maximum sustainable population size, the maximum size that can be supported indefinitely into the future without degrading the environment for future generations.

catabolism
Destructive metabolism, breaking down complex materials into simpler compounds with the release of energy (e.g., digestion of starch to glucose).

catabolites
Any products of catabolism.

catalase
An enzyme that reduces hydrogen peroxide to water and oxygen; often associated with superoxide dismutase in the conversion of superoxide ions to water in cellular respiration; a potent antioxidant.

cataracts
Opacities (cloudiness) in the lens of the eye due to formation of denatured proteins; interfere with passage of light to the retina and cause blurred vision.

catecholamines
A group of neurotransmitters consisting of aromatic amines, including epinephrine, norepinephrine, and dopamine.

cell body
The part of a neuron (nerve cell) that contains the nucleus and other cellular organelles that carry out the normal functions of the cell.

cell cycle
The orderly and regulated series of events through which a single cell duplicates its contents and divides into two daughter cells. The cell cycle is divided into four phases: G_1 (gap 1) phase: the cell increases in size, duplicating organelles; S (synthesis) phase: DNA is replicated; G_2 (gap 2) phase: a checkpoint or control mechanism; and M (mitosis) phase: the cell divides into two daughter cells. In the G_0 phase, the cell has exited the cycle. G_1, S, and G_2 are collectively known as interphase.

cell lines
Cells maintained in culture that do not have a finite life span. The first cell line was established from an ovarian tumor removed from Henrietta Lacks in 1956; these cells (HeLa cells) are still available from biotech companies.

cellulose
A fibrous carbohydrate in the cell walls of plants, consisting of a linear chain of several hundred to more than 10,000 D-glucose units linked by $\beta(1{\to}4)$ bonds; an indigestible carbohydrate for humans.

cell wall
A layer of polysaccharides surrounding the plasma membrane of plant cells. There are two general types: a primary wall and a secondary wall, deposited after the primary wall has stopped growing. *See also* primary cell wall; secondary cell wall.

centenarian
A person who is 100 or more years of age.

central nervous system (CNS)
The brain and spinal cord.

central vacuole
A membrane-enclosed organelle of plant cells that contains fluid and a variety of molecules; can constitute as much as 95% of the cell volume.

centromere
A region of sister chromatids that helps hold them together and contains the proteins that will become the kinetochore during mitosis.

centrosome
A microtubule-organizing center of animal cells that divides prior to cell division; grows out of the centromere during prophase and becomes the mitotic spindle pole for each set of sister chromatids.

cerebrovascular incident (stroke)
A loss of brain function due to disturbance in the blood supply to the brain.

chaperone proteins
Proteins that assist other proteins in achieving or maintaining proper folding.

charged tRNA
A tRNA coupled to its corresponding amino acid. Also called an aminoacyl-tRNA.

chemokines
Proteins produced and released by a wide variety of cell

types during the initial phase of the body's response to injury; induce chemotaxis, attracting white blood cells to the location of an injury or infection.

chemoreceptor
A sensory receptor that is sensitive to chemical stimuli.

chemotaxis
The movement of bacteria, individual body cells, or multicellular organisms in response to certain chemicals in their environment.

chitin
A polymer of nitrogen-containing polysaccharide; the principal component of yeast bud scars and an organic component of the exoskeleton of insects.

chloroplasts
One of several types of plastids, membrane-surrounded organelles of plant cells; the site of photosynthesis and storage of the green pigment chlorophyll.

chromatin
The material of which eukaryotic chromosomes are composed; consists of protein, DNA, and RNA.

chromoplasts
One of several types of plastids in plant cells; the site of pigment storage (other than chlorophyll).

chromosomes
Threadlike structures made up of nucleic acids and proteins, found in the nucleus of most living cells; carry genetic information in the form of genes.

chyme
The semi-solid mass consisting of partially digested food, water, and gastric juices that enters the intestine from the stomach.

ciliary body
A circular muscle of the eye that attaches to the lens by zonular fibers; contraction or relaxation changes the shape of the lens.

circadian rhythm
Physiological changes that repeat at about 24-hour intervals, often synchronized with changes in the external environment, such as day-night cycles.

***clk-1*, *clk-2*, *clk-3* genes—clock genes**
A group of genes associated with longevity; seem to express proteins important to mitochondrial function.

cloning (genetic)
The creation of several identical copies of a gene.

cochlea
A spiral tube forming part of the inner ear, containing the structures that transform sound wave vibrations into neural impulses.

codon
A sequence of three nucleotides (or bases) in RNA that codes for a specific amino acid.

cohesins
Proteins that hold sister chromatids together.

cohort
A group of individuals with similar life experiences, usually having birth dates within 5–10 years of each other.

cohort effect
The confounding effect on data when comparing different cohorts.

cohort life table
One of two types of life table; follows the death characteristics of a population in a single birth cohort through its entire life span. *See also* current life table.

collagen
A structural protein of animal connective tissue and bone; in the skin, found in the dermis.

comparative biogerontology
The observational study of longevity in wild animals that have long life spans. It identifies wild species that show resistance to aging and have extended longevity in environments that are otherwise conducive to short life spans.

complete life table
A life table in which the age interval is one year.

compliance (physiological)
A measure of the ease with which a hollow structure (such as heart ventricles, lungs) expands under pressure.

compression (acoustic)
The portion of a sound wave in which molecules are pushed together, forming a region with higher than normal atmospheric pressure. *See also* rarefaction.

compression of morbidity
The decreasing percentage of the life span spent experiencing senescence-related morbidity.

condensins
Proteins synthesized during the G_2 phase of the cell cycle that condense sister chromatids so that mitosis can occur more efficiently.

confluence (in cell culture)
In replicating cell cultures, the maximum capacity within the confines of the culture dish.

congestive heart failure
The loss of ability of the heart to pump enough blood to the tissues for sufficient oxygenation; characterized by fluid buildup in the extremities and lungs.

consensus sequences
Sequences that most often occur in DNA, RNA, or protein.

contractility
The ability of muscle fibers to contract and develop tension in response to resistance.

control element
A region of noncoding DNA that helps regulate the transcription of genes by binding transcription factors.

cornea
The transparent front part of the eye; accounts for approximately 66% of the eye's total optical power.

corpus luteum
The endocrine tissue of a ruptured ovarian follicle; produces progesterone.

cross-sectional studies
Studies that compare two or more groups of separate individuals at a single point in time.

crude mortality
The death rate in a total population without regard to age.

crystallin
A water-soluble structural protein of the lens and

cornea; accounts for the transparency of these structures.

current life table
One of two types of life table; describes the death rate of a hypothetical population as established from age-specific death rates in the current population; assumes that the age-specific deaths at this time represent the death rates for the whole generation. Current life tables are used when a cohort life table is not possible, such as in human populations. Also called period, cross-sectional, or time-specific life table. *See also* cohort life table.

cyclin-dependent kinase (Cdk)
A protein kinase involved in regulation of the cell cycle.

cyclins
Nuclear proteins that form complexes with cyclin-dependent kinases to control various steps in the cell cycle.

cytogerontology
The study of the aging cell and its mechanisms.

cytokines
A family of proteins, secreted by a variety of cells, that regulate the behavior of cells by binding to receptors on their surfaces. Binding triggers a variety of responses (autocrine or paracrine), depending on the nature of the cytokine and the target cell.

cytokinesis
The process in which the cytoplasm of a single eukaryotic cell divides to form two daughter cells.

cytokinins
A class of plant hormones that promote and control growth responses; they seem to delay senescence.

cytosine
One of the nucleotide bases of DNA and RNA; a pyrimidine. *See also* nucleotide bases.

cytoskeleton
The network of microtubules and other structural elements, such as actin, that create the molecular scaffold of the cell. It keeps all the organelles in place, cushions the cell against damage, and maintains basic cell shape.

cytotoxic T cells
A class of T cells (T lymphocytes) that kill damaged cells, such as cells infected with viruses.

daf-2 gene—*see* dauer formation gene

dauer
A sexually immature *C. elegans* that forms at larval stage 3 in response to environmental conditions that would not support offspring; metabolically active but reproductively silent and can survive without food for months.

dauer formation gene (*daf-2*)
A gene that encodes a transmembrane receptor protein that has strong homology with insulin/insulin-like growth factor receptors; a highly conserved gene that has links to normal growth, reproduction, and longevity.

decibel
A measure of loudness; 1 decibel (dB) is the logarithmic increase in loudness above a sound that is barely audible to the human ear.

demography
The statistical study of populations, especially with reference to size and density, distribution, and vital statistics.

denaturation
A disruption in the secondary, tertiary, and quaternary structures of proteins resulting in unfolding and loss of function; similar disruption in the structure of DNA.

dendrites
Branchlike structures that extend from the cell body of a neuron, increasing its surface area.

dendritic cells
White blood cells with the primary function of presenting antigens to T cells (antigen-presenting cells); generally considered part of the innate immune system; act as messengers to the adaptive immune system.

deoxyribonucleic acid (DNA)
Nucleic acid composed of two polynucleotide strands wound around a central axis to form a double helix; the repository of genetic information.

depolarization (neural)
In a biological membrane, a change in electrical charge that causes the membrane to become less polarized.

dermis
The layer of skin tissue below the epidermis, containing blood capillaries, nerve endings, sweat glands, hair follicles, and other structures.

development (biological)
The period of the life span in which growth takes place; usually the period of life before reproductive capability.

diabetes mellitus
A disease caused by the inability of cells to take up glucose, resulting in high blood glucose levels. *See also* type 1 diabetes mellitus; type 2 diabetes mellitus.

diapause
A period of suspended development in an insect, other invertebrate, or mammalian embryo, especially during unfavorable environmental conditions.

diastole
The period of time when the heart muscle is at rest and fills with blood; the period in which blood pressure is lowest.

diastolic pressure
The pressure in the arteries (in mm Hg) when the heart is at rest; the lower number in the measurement of blood pressure (e.g., the "60" in 120/60).

dietary restriction—*see* calorie restriction

diet-induced thermogenesis (DIT)
The amount of energy required for digestion, absorption, and nutrient storage; accounts for 10–20% of total energy expenditure (TEE).

differentiated cells
Cells of a multicellular organism that have a specialized function.

diploid
Containing two identical (homologous) copies of each chromosome and thus two copies (alleles) of each gene.

disaccharide
Sugar consisting of two monosaccharides bonded together. Common disaccharides are sucrose (glucose

+ fructose), maltose (glucose + glucose), and lactose (galactose + glucose).

disposable soma theory
T. B. Kirkwood's evolutionary theory of senescence, predicting that the mortality of somatic cells arises as a "cost" involved in the preferential distribution of resources to the maintenance and repair needed to ensure an error-free and immortal germ line.

diuretics
A class of drugs used to treat hypertension; increase urine output and lower water retention.

DNA ligase
An enzyme that joins the ends of two strands of DNA.

DNA microarray
A technique for analyzing the simultaneous expression of large numbers of genes in cells; isolated cellular RNA is hybridized to a large array of short DNA probes immobilized on a glass slide.

DNA polymerase
An enzyme that catalyzes the synthesis of double-stranded DNA, using a single strand of DNA as a template.

dopamine
A neurotransmitter in regions of the brain that regulate movement and emotion. A decrease in dopamine-producing neurons in the substantia nigra produces the initial symptoms of Parkinson's disease.

eclosion
Emergence from the pupal casing.

edema
Fluid buildup in tissue.

effector cells
A class of neurons that receive the motor output from the central nervous system; muscle cells or gland cells.

effectors
Agents (e.g., a protein) that produce an effect (e.g., on another protein).

electrolytes
Substances that dissociate into ions in aqueous solution; ions that conduct an electrical current in the body—calcium, sodium, potassium, magnesium.

electron transfer system (ETS)
The final stage of cellular respiration in which electrons generated by the tricarboxylic acid cycle are "passed" between intermediates, and the energy generated by differences in redox potential drives ATP synthesis.

embolus
An abnormal particle circulating in the blood.

end diastolic volume
The amount of blood in each ventricle at the end of diastole (the relaxed phase).

endocrine system
The system of glands that secrete hormones directly into the blood.

energy balance
The difference between energy intake and energy expenditure.

enhancer
A site on the protein-coding region of DNA that binds activator proteins and "enhances" the transcription rate of a specific gene; may be located thousands of nucleotides away from the coding region of that gene.

enthalpy, H
A measure of the heat energy in a system.

entropy, S
A measure of the degradation of matter and energy in the universe to an ultimate state of inert uniformity; a process of degradation or running down or a trend to disorder.

epidermis
The outer layer of skin cells, overlying the dermis.

epigenetic trait
A phenotype resulting from changes in a chromosome without alteration in the DNA sequence.

epigenome
A second level of trait development resulting from chemical changes in the DNA and histone proteins. These traits can be transmitted to future generations.

epinephrine
A hormone secreted by the adrenal glands that increases heart rate, constricts blood vessels, and opens airways; most often associated with the fight-or-flight response. Also called adrenaline.

erectile dysfunction—*see* **impotence**

esophagus
The part of the digestive system that connects the pharynx (throat) to the stomach.

estrogen
A hormone secreted by the ovaries that promotes the development and maintenance of female characteristics.

ethylene
A plant hormone that stimulates fruit ripening and leaf abscission.

eukaryotic
Describing organisms and cells that contain their genetic material inside a membrane-enclosed nucleus.

Euler-Lotka equation of population growth
An equation describing population growth in continuous-breeding populations, proposed by the statistician Alfred Lotka, who built on the work of eighteenth-century Swiss mathematician Leonhard Euler. The equation is the integration of the net reproduction rate.

eusociality
Social structure characterized by the specialization of tasks and cooperative care of the young; most often found in social insects such as ants, wasps, termites, and bees.

eustachian tube
The narrow channel connecting the middle ear and nasopharynx; equalizes pressure between the outer and inner ear.

eutelic
Describing animals that have a fixed number of cells when they reach maturity.

eutherians
Mammals with a placenta.

exocrine hormones
Hormones synthesized by exocrine glands or cells and

secreted directly into the target tissue through a duct (as opposed to hormones secreted directly into the blood by endocrine glands); generally affect only one organ or one type of cell.

exons
The protein-coding regions of a gene; mRNA contains only exons. *See also* introns.

extracapsular cataract surgery
Cataract surgery in which the front of the lens is removed.

extrachromosomal rDNA circles (ERCs)
Nonchromosomal DNA molecules in *S. cerevisiae* that are created through homologous recombination of rDNA. Accumulation of ERCs in *S. cerevisiae* is associated with replicative senescence.

extrinsic rate of aging
The rate of aging in a population due to environmental hazards; often associated with the phenotype.

facilitated diffusion
Transport of a molecule across a biological membrane from a region of higher concentration to a region of lower concentration. The molecule binds to a specific transport protein in the cell membrane that facilitates its transport into the cell.

facultative
Functioning under varying environmental conditions; capable of occurring by various pathways or under various conditions; in biodemography, influenced by environmental factors that cause mortality rate or the trajectory of mortality rate to have significant plasticity.

fatty acids
Carboxylic acids consisting of a hydrocarbon chain and a terminal carboxyl group; major components of fats that are used by the body for energy and tissue development.

fecundity
The ability to produce offspring; the ability to cause growth; the number of offspring or the rate of or capacity for producing offspring; the rate of production of young by a female.

Fenton reaction
The reaction between hydrogen peroxide and an iron catalyst to produce a reactive oxygen species.

fibrillation
Uncontrolled contractions of muscle fibers; when occurring in heart muscle, significantly decreases the efficiency of the pumping of blood.

fibrointimal hyperplasia
Abnormal growth of the intimal layer of the arterial wall.

first law of thermodynamics
One of the principles governing the relationship between work, energy, and heat; states that energy is neither created nor destroyed, it only changes form.

fitness
The capacity of an organism to survive and transmit its genotype to its offspring, compared with that of competing organisms; the relative ability of an individual (or population) to survive, reproduce, and propagate genes in a particular environment.

follicle-stimulating hormone (FSH)
Pituitary hormone that causes ovarian follicles to increase in size and, together with luteinizing hormone, causes them to produce estrogen and bring about ovulation; helps regulate sperm production and synthesis of testosterone.

forkhead box transcription factor family (FOXO)
A family of highly conserved transcription factors that regulate the expression of genes involved in cell growth, proliferation, differentiation, and longevity. The name is derived from its structure, which has a "forked" or winged helix motif that binds to DNA.

fovea centralis
A structure of the retina that is responsible for sharp central vision; necessary for any activity for which visual detail is of primary importance.

free energy, G
In thermodynamics, the amount of work that can be extracted from a system; a quantity of energy that interrelates entropy (S) and the system's total energy (H). The change in free energy of a system is calculated by the equation $\Delta G = \Delta H - T\Delta S$, where T is absolute temperature.

free radicals—*see* oxygen-centered free radicals

frequency (acoustic)
Measure of the rapidity of alteration of a periodic signal, expressed in cycles per second or hertz (Hz); the distance between the peaks of either compression or rarefaction.

gametes
Mature haploid germ cells, male or female, that can unite with another germ cell of the opposite sex in sexual reproduction to form a zygote.

gastric pits
Structures of the stomach wall, formed by the folds and grooves of the stomach lining.

gel electrophoresis
The separation and identification of molecules based on their movement through a gel with an electrically charged field.

gene homologs
Genes with the same sequence.

gene homology
The degree to which the DNA sequence of one gene matches the DNA sequence of a second gene; may apply to genes of the same or different species.

gene knockout
A mutant organism in which one or more genes have been removed by genetic engineering; used to eliminate the expression of a protein.

gene orthologs
Genes of different species that retain the same function in the course of evolution.

general recombination—*see* homologous recombination

general transcription factor—*see* transcription factor.

gene silencing
The repression of gene transcription.

genetic code
The set of 64 triplet codons that determine the 20 amino acids used to construct proteins and code for translation initiation and termination.

genetic determinism
The concept that all genes are selected for a particular purpose and only for that purpose.

genetic drift
The process in which genes can be fixed in a small population as a result of the random sorting of alleles during meiosis.

genetic engineering—*see* **recombinant DNA technology**

genetic screening
The systematic search for individuals in a population who have a certain genotype.

genome
All of an organism's genetic information, usually in the form of DNA.

genotype
All or part of the genetic constitution of an individual or group.

geriatric failure to thrive
A state of decline in the elderly that is multifactorial and may be caused by chronic concurrent diseases and functional impairments. The name comes from the observation that the individual does not respond to treatment.

geriatrics
A branch of medicine that deals with the problems and diseases of old age and aging people.

germ cells (germ plasm)
The cells produced by the sex organs or tissues of multicellular organisms that transmit the hereditary information to offspring. Transmission can occur asexually or sexually (by means of gametes). *See also* **somatic cells**.

gerontological biodemography
A science integrating biological knowledge with demographic research on human longevity and survival.

gerontology
The comprehensive study of aging and the problems of the aged.

gibberellins
A class of plant hormones that stimulate leaf and stem growth, bring buds out of dormancy, and cause seed germination.

glial cells (glia)
Non-neuronal cells of the brain that support and maintain neurons; classified as neuroglia and microglia. They outnumber neurons 10:1.

glomerular filtrate
In the kidney, the fluid filtered from the glomerulus into Bowman's capsule.

glomerular filtration
The process whereby fluid and low-molecular-weight molecules in the blood are filtered across the capillaries of the glomerulus and into Bowman's capsule.

glomerular filtration rate (GFR)
The rate of glomerular filtration, in volume per unit of time.

glomerulosclerosis
A kidney disease in which glomerular function (blood filtration) is lost as fibrous scar tissue replaces the glomeruli; can be associated with hypertension, infection, or atherosclerosis.

glomerulus
In the kidney, capillaries at the beginning of the nephron that begin the process of filtering the blood.

glucagon
An endocrine hormone secreted by alpha cells in the islets of Langerhans of the pancreas; causes blood glucose levels to rise.

gluconeogenesis
The synthesis of glucose from non-carbohydrate substrates: pyruvate, lactate, amino acids (primarily alanine and glutamine), and fatty acids.

glucose intolerance—*see* **insulin resistance**

glutathione peroxidase
A cytosolic enzyme that is a powerful scavenger of oxygen-centered free radicals; a cellular antioxidant.

glycogen
The storage form of carbohydrate in animals and some fungi (not found in plants); stored in muscles and liver. The glycogen molecule is a highly branched structure containing up to 3000 glucose units.

glycogenesis
The synthesis of glycogen, in which glucose molecules are added to the glycogen chains.

glycolysis
The metabolic pathway in which six-carbon sugars are split to form two molecules of three-carbon pyruvate; pathway for the anaerobic breakdown of glucose; takes place in the cytosol.

glycosylation
The addition of sugar units to a molecule. Enzymatic glycosylation is a regulated process used in cellular signaling. Nonenzymatic glycosylation is an uncontrolled, nonregulated process that increases intracellular and extracellular damage; tends to increase with age.

Gompertz mortality function
The function that describes the rate of mortality in a population.

G$_0$ phase (gap)—*see* **cell cycle**

G$_1$ phase (gap 1)—*see* **cell cycle**

G$_2$ phase (gap 2)—*see* **cell cycle**

G-protein-coupled receptors
A family of transmembrane receptor proteins that transduce an extracellular signal into an intracellular signal. These receptors traverse the membrane seven times and are often associated with signal transduction through cyclic AMP (cAMP).

grandmother hypothesis
The hypothesis that aging women gain an inclusive fitness advantage from investing in their grandchildren. In practical terms, grandmothers aid in the rearing of children, allowing the mother to have more children and thus increasing fitness.

group selection
Selection occurring as a result of competition between groups or populations, as opposed to between individuals. A concept most often associated with V. C. Wynne-Edwards's book *Animal Dispersion in Relation to Social Behaviour*.

guanine
One of the nucleotide bases of DNA and RNA; a purine. *See also* nucleotide bases.

Haber-Weiss reaction
The generation of hydroxyl radicals (•OH) from hydrogen peroxide (H_2O_2) and the superoxide radical (•O_2^-); uses iron and sometimes copper as the catalyst.

haploid
Containing only one set of chromosomes and thus only one copy (allele) of each gene.

Hayflick limit
The maximum number of times a cell will divide during serial cell culture. For normal human fibroblasts, for example, the Hayflick limit is 50 (±10) population doublings.

heart rate
The number of contractions of the heart ventricles in a fixed time, usually a minute (beats/min).

helicase
An enzyme that separates double-stranded DNA into two single strands.

***Helicobacter pylori* (*H. pylori*)**
A bacterium resistant to the normal effects of stomach acid. *H. pylori* infection causes 90% of gastritis cases.

helper T cells
T lymphocytes (T cells) that "help" the immune system respond adequately to an invader; do not attack antigens directly, but secrete lymphokines (interleukins and interferon) that stimulate the proliferation and differentiation of other cells that do attack the antigen.

hemoglobin
An iron-containing protein in red blood cells, responsible for the transport of oxygen.

hemoglobin A1c
The glycosylated form of hemoglobin; used as a diagnostic criterion.

hexokinase
An enzyme that phosphorylates glucose, a six-carbon sugar (hexose), to produce glucose 6-phosphate; the first step of glycolysis. Phosphorylation of glucose may be important as a signal during plant senescence.

high-density lipoproteins (HDLs)
A class of lipoproteins, consisting of protein and lipids, that transport water-insoluble lipid in the bloodstream; transport cholesterol to the liver for catabolism.

histone acetylation
The modification of chromatin structure by attachment of acetyl groups to core histones; one method of regulating DNA transcription.

histone octamer
A histone protein complex at the center of a nucleosome that binds to DNA; consists of two copies each of four core histone proteins (H2A, H2B, H3, and H4).

histones
Proteins associated with DNA in chromosomes; involved in the packaging and regulation of gene expression.

homeostasis
The tendency of an organism or cell to regulate its internal conditions so as to stabilize its functioning, regardless of the changing environment.

homologous recombination
A process in which two chromosomes or two highly homologous DNA repeats within the same chromosome break and reconnect, exchanging sections from the starting pair. Also called general recombination.

hormone-sensitive lipase
A lipase that is stimulated by hormones; found in adipose tissue.

hydrophilic
Having an affinity for water; describing a substance that absorbs, dissolves in, or is attracted to water. *See also* hydrophobic.

hydrophobic
Water-repellent; most often describing a molecule or region of a molecule having water-insoluble group(s) or surfaces. *See also* hydrophilic.

hydroxyapatite
A crystalized structure of calcium and phosphorus, as calcium phosphate, that is a primary contributor to the strength of bone; $Ca_{10}(PO_4)_6(OH)_2$.

hyperlipidemia
A high serum cholesterol level; a major risk factor for development of atherosclerosis.

hyperpolarization
In a biological membrane, a brief period occurring after the action potential has passed; the transmembrane potential is slightly greater than at rest.

hypertension
High blood pressure.

impotence (erectile dysfunction)
Male sexual dysfunction characterized by an inability to develop or maintain an erection of the penis sufficient for satisfactory sexual performance.

incus
One of the bones of the middle ear that amplify and transfer sound wave vibrations from the tympanic membrane to structures of the inner ear.

indirect calorimetry
A method of estimating energy expenditure by measuring the amount of oxygen and carbon dioxide in respiratory gases.

indoleacetic acid
A water-insoluble compound that stimulates growth and root formation in plants; auxins are variants of indoleacetic acid. *See also* auxins.

infant mortality rate
In a human population, the death rate of individuals from birth to 1 year of age.

innate immunity
Immunity to disease that is part of an individual's natural biological makeup; includes defenses such as skin, mucus, stomach acid, and various types of white blood cells.

***in situ* hybridization**
A technique that localizes specific nucleic acid sequences in intact chromosomes, eukaryotic cells, or bacterial cells through the use of specific nucleic acid-labeled probes.

insudation
The accumulation in a vessel wall of substances

derived from the blood; often seen in vessels of the kidney.

insulin
An endocrine hormone secreted by beta cells of the islets of Langerhans in the pancreas; induces the cellular uptake of glucose and amino acids and synthesis of glycogen in the liver.

insulin/insulin-like growth factor (IGF-1) receptor
A class of receptors that bind insulin or molecules with a structure similar to insulin; involved in nutrient metabolism and growth.

insulin resistance (glucose intolerance)
A physiological condition in which the ability of insulin to stimulate glucose uptake by cells is reduced, often a precursor to type 2 diabetes.

interneurons
One of the three main types of neurons; integrate sensory input and motor output.

interphase
The G_1, S, and G_2 phases, collectively, of the cell cycle. *See also* cell cycle.

intrinsic rate of aging
The rate of biological aging attributable to the genotype and not affected by external influences (e.g., accidents).

intrinsic rate of natural increase, *r*
The rate of increase in a population whose growth is not impaired by environmental constraints. Growth rate is determined by the biological makeup of the individuals in the population.

introns
Noncoding regions of a eukaryotic gene; removed from the RNA transcript by splicing to form the messenger RNA (mRNA). *See also* exons.

ischemia
Inadequate blood supply to tissue.

ischemic heart disease
Disease resulting when a large artery of the heart, one of four coronary arteries, is blocked by an atherosclerotic lesion and blood flow to the heart tissue decreases.

islet of Langerhans
A grouping of endocrine cells in the pancreas; secretes insulin and glucagon.

iteroparous
Capable of reproduction across multiple seasons; humans are iteroparous.

jasmonic acid
A plant hormone with a role in regulating responses to abiotic and biotic stresses; also important in the formation of tubers such as potatoes, yams, and onions.

joule (J)
A unit of energy: the energy expended when 1 kg is moved a distance of 1 m by the force of 1 newton.

juvenile hormone
An insect hormone important in development and in egg production by ovaries; also seems to stimulate recovery from diapause.

keratinocytes
The predominant cell type in the epidermis, the outermost layer of human skin, constituting 95% of epidermal cells; produce keratin.

kilocalorie (kcal)
1000 calories. *See also* calorie.

kilojoule (kJ)
1000 joules. *See also* joule.

kinetochore
A protein of the centromere that links the chromosome to the mitotic spindle.

lagging strand
In DNA replication, one of two newly synthesized strands of DNA; made up of discontinuous lengths called Okazaki fragments that are later joined together.

lagging strand template
The DNA strand that acts as template for the synthesis of Okazaki fragments.

Langerhans cells
Phagocytic immune cells in the epidermis.

laws of thermodynamics
Three principles that govern the relationship between work, energy, and heat. The first law: energy is neither created nor destroyed; it only changes form. The second law: in the conversion of energy from one form to another, some energy becomes unusable; the unusable energy is known as entropy. The third law: the first and second laws apply to all reactions above absolute zero (–279 C). Since absolute zero cannot be reached, the first and second laws always apply.

leading strand
In DNA replication, one of two newly synthesized strands of DNA; made by continuous synthesis in the $5' \rightarrow 3'$ direction.

leading strand template
The DNA strand that acts as template for the new, continuously synthesized strand.

lens (of the eye)
The clear structure directly under the cornea that increases the optical power of the eye; involved in optical accommodation.

leucoplast
One of several types of plastids in plant cells; colorless plastids used primarily for storage of starch and oils.

levodopa (L-dopa)
The primary medication used to treat Parkinson's disease; in the body. converted to dopamine by DOPA decarboxylase.

Lewy bodies
Protein aggregates that accumulate in the cytoplasm of neurons during aging; composed primarily of α-synuclein and ubiquitin; the histological hallmark of Parkinson's disease.

Leydig cells
Cells of the testes that produce testosterone.

life expectancy
The amount of life remaining after a specific age; a value generated by a life table.

life history
The changes an organism undergoes from conception to death, focusing particularly on the schedule of reproduction and survival.

life span
The length of life of an individual cell, organ, or organism.

life table
A table describing the mortality (death) characteristics of a population for specific ages or age intervals. Analysis of mortality begins with construction of a life table. *See also* cohort life table; current life table.

ligand
A substance capable of binding specifically and reversibly to another substance; in biochemistry, often designates the molecule that binds to a receptor.

light reactions
In photosynthesis, the chemical and physical reactions, taking place in chloroplasts, in which light energy is converted to chemical energy with the aid of chlorophyll.

limbic system
A system of interconnected brain structures that control emotions and memory; includes the hypothalamus, thalamus, hippocampus, and amygdala.

lipases
Enzymes that catalyze the hydrolysis of triglycerides to form glycerol and free fatty acids.

lipid bilayer
The core structure of biological membranes, consisting of two layers of lipid molecules; contains phospholipids arranged so that the hydrophobic regions face each other in the center of the bilayer and the hydrophilic regions face outward—in the case of plasma membranes, facing the extracellular and intracellular spaces.

lipids
Nonpolar molecules containing carbon and hydrogen with small amounts of oxygen; soluble only in organic solvents such as ether or benzene. In biological systems, lipids include fatty acids and compounds made from fatty acids: monoglycerides, diglycerides, triglycerides, phospholipids, and sterols (e.g., cholesterol).

lipoproteins—*see* high-density lipoproteins; low-density lipoproteins

longevity
The evolved length or duration of life for a species.

longitudinal studies
Correlational studies that involve repeated observations of the same characteristics in the same individuals over long periods of time.

low-density lipoproteins (LDLs)
A class of lipoproteins, consisting of protein and lipids, that transport cholesterol from the liver to body cells.

luteinizing hormone (LH)
A hormone secreted by the pituitary gland (in the brain) that stimulates the growth and maturation of eggs in females and sperm in males.

lymphocytes
White blood cells that mediate the immune response. *See also* B cells; T cells.

lymphokines
Compounds released by T cells that activate macrophages and stimulate the production of antibodies by B cells.

lysozyme
An enzyme that catalyzes destruction of the cell walls of certain bacteria; found in saliva.

macronutrients
Proteins, fats (lipids), and carbohydrates; nutrients required in much larger quantities than micronutrients (vitamins and minerals).

macrophages
Phagocytic immune cells, found in tissues, that destroy invading bacteria and other pathogens; function in both innate and adaptive immunity. Macrophages attract other immune cells by presenting them with small pieces of the invaders.

Maillard reaction
A nonenzymatic reaction between sugars and proteins that occurs on heating and may produce browning of foods. A variant of this reaction occurs in animals and may cause age-related cellular damage.

malleus
One of the bones of the middle ear that amplify and transfer sound wave vibrations from the tympanic membrane to structures of the inner ear.

maltose
A disaccharide consisting of two molecules of glucose.

marsupials
Nonplacental mammals.

maturity
A period of the life span of a molecule, cell, or organism during which function remains at optimal levels or slowly declines. The end of maturity occurs when the molecule, cell, or organism no longer has the capacity to resist the force of entropy.

maximum life span
The length of life of the longest-lived individual member of a species or a population of a species.

mean life span
The average of the individual life spans of members of a group (cohort) having the same birth date.

mediated transport
The movement of molecules or ions across a membrane with the help of transport proteins and, in some cases, energy. *See also* active transport; facilitated diffusion.

mediator
In DNA transcription, a complex of proteins attached to general transcription factors that facilitates binding of the activator/enhancer site to the transcription initiation complex.

meiosis
The process of two consecutive cell divisions in the diploid progenitors of sex cells, resulting in four daughter cells (rather than two, as in mitosis), each with a haploid set of chromosomes.

melanocytes
Skin cells that produce the protein pigment melanin.

melanoma
A tumor of melanin-forming cells (melanocytes), typically a malignant tumor associated with skin cancer.

membrane potential
The difference in electrical charge across a membrane.

See also depolarization; hyperpolarization; resting membrane potential.

memory cells
A subset of T cells and B cells that are capable of responding to a particular antigen on its reintroduction, long after the initial exposure that prompted production of the T and B cells.

menopause
The cessation of menstruation.

menstrual cycle
The process of ovulation and menstruation in women and other female primates; in humans, occurring in a 28-day cycle.

Merkel cells
Receptor cells of the epidermis that have synaptic connections with sensory neurons.

messenger RNA (mRNA)
A molecule that carries the genetic information for protein synthesis on ribosomes. An RNA transcript is synthesized on a DNA template by RNA polymerase, then the transcript undergoes RNA splicing to remove introns, forming mRNA.

metaphase
The stage of mitosis in which condensed chromosomes become aligned, before separating in anaphase.

metazoa
Multicellular organisms, with cells having distinct functions.

Methuselah gene (mth)
A gene of *Drosophila melanogaster* that seems to extend life; its function has not been identified.

microarray chips—*see* **DNA microarray**

microglia
Neuromacrophages; immune cells in the brain that can differentiate into macrophages.

micronutrients
Essential vitamins and minerals needed in only very small quantities for proper growth and metabolism.

microtubules
Fibrous, hollow rods that function primarily to support and shape the cell; a component of the cytoskeleton; also function as routes along which organelles can move; typically found in all eukaryotic cells.

mitogen
An agent that stimulates mitosis.

mitosis (M phase)
Cell division; in eukaryotes, a process that takes place in the nucleus of a dividing cell, usually involving a series of steps—prophase, metaphase, anaphase, and telophase—and resulting in the formation of two daughter nuclei, each with the same number of chromosomes as the parent nucleus.

mitotic clock theory
A theory of cellular replicative senescence predicting that old cells sense short telomeres and that this causes arrest of the cell cycle.

mitotic spindle
In cell division, cytoskeletal structure to which chromosomes attach before separating, moving toward opposite poles of the cell.

mitral valve
One-way valve that regulates blood flow from the left atrium; positioned between the left atrium and the left ventricle.

molecular brakes
A class of proteins that inhibit progression of the cell cycle at the transition from G_1 to S phase, and from G_2 to M phase, generally inactivated by phosphorylation.

molecular fidelity
The degree to which the amino acids of a protein or the nucleotides of DNA are in the proper order or sequence; high fidelity implies high functionality; low fidelity implies low functionality.

monocarpic
In plants, bearing fruit or seed only once and then dying.

monophyletic group
In phylogenetics, a group that contains all the descendants of a common ancestor.

monosaccharide
The simplest form of sugar that cannot be broken down to form other sugars; usually colorless, water-soluble, crystalline solids. The three most common monosaccharides found naturally in foods are glucose, fructose, and galactose.

morbidity
A state of illness or disease.

morphology
A branch of biology that deals with the form of living organisms and with relationships among their structures.

mortality
The state of being certain to die.

mortality rate
The number of deaths that occur at a given time, in a given group, and/or from a given cause, usually expressed as number of deaths per 100 or 1000 or 10,000.

mortality-rate doubling time
The time required for the mortality rate of a population to double.

motor end plate
A complex structure in which the axon of a motor neuron establishes synaptic contact with a striated muscle fiber. Also called a neuromuscular junction.

motor neurons
One of the three main types of neurons; convey motor output from the central nervous system to **effector cells** (muscle or gland cells).

M phase (mitosis)
Period of the eukaryotic cell cycle during which the nucleus and cytoplasm divide. *See also* cell cycle; mitosis.

multipotent stem cells
Adult stem cells that form the tissue in which they are found: liver stem cells produce liver cells; muscle stem cells produce muscle cells; and so on.

muscarinic receptor
A type of acetylcholine receptor that uses a G-coupled mechanism to propagate the signal; found primarily in the peripheral nervous system.

muscle fiber
An elongated contractile cell of muscle tissue; generally of two types: type I and types II. *See also* type I muscle fibers; type II muscle fibers.

mutation accumulation theory of senescence
An evolutionary theory of senescence, first proposed by Sir Peter Medawar in 1952, suggesting that a decline in the force of natural selection with age allows the fixing of an accumulation of late-acting, deleterious genes in the genome. In other words, at older ages, the force of natural selection is too low to eliminate deleterious mutations.

myelin sheath
An insulating layer surrounding axons, formed by supporting cells; in the peripheral nervous system, formed from Schwann cells; in the central nervous system, produced by oligodendrocytes.

myocardial infarction
Tissue death of the myocardium (heart muscle) due to an interruption of blood flow; commonly known as a heart attack.

myosin
One of two proteins in a muscle cell that are responsible for muscle contraction; forms the thick filament of the myofibril; also found in microtubules of the cytoskeleton. *See also* actin.

naive T cells
Small white blood cells (lymphocytes) that will be transformed into immune cells that attach to and kill invaders.

natural killer (NK) cells
White blood cells (lymphocytes) of the innate immune system that attach to cells that lack surface proteins of the major histocompatibility complex; destroy microbial and cancer cells by injecting proteases that cause apoptosis. Also called large granular lymphocytes.

natural selection
A theory on the mechanism of evolution predicting that, in a population, heritable characteristics that improve survival and reproduction (fitness) are favored.

necropsy
Examination of an animal after death. Also called autopsy.

nephron
A microscopic structure in the kidney that filters blood and forms urine; consists of a glomerulus and tubules.

net reproduction rate
The average number of offspring an individual in a population will produce in its lifetime.

neurofibrillary tangles
Aggregates of paired helical fibers made up of tau proteins; a pathological marker of Alzheimer's disease.

neuroglia
A class of cells in the central nervous system that provide support and maintenance for neurons. They include astrocytes (matrix-secreting cells) and oligodendrocytes (myelin-secreting cells).

neurons
Nerve cells; the cells of the nervous system that send and receive electrical signals over long distances. Each neuron has three general parts: a cell body (soma), dendrites (branched projections that conduct incoming impulses from other neurons), and an axon (slender projection that conducts impulses away from the cell body).

neuroplasticity
The brain's ability to reorganize itself by forming new neural connections as a result of changing experiences or learning.

neutrophils
White blood cells (leukocytes) that attack and destroy invading bacteria, other foreign matter, and some cancer cells through phagocytosis; part of the innate immune system.

newton
A unit of force: the force required to cause a mass of 1 kilogram (1 kg) to accelerate at a rate of 1 meter per second per second (1 m s^{-1} s^{-1}) in the absence of other force-producing effects.

nicotinamide adenine dinucleotide (NAD)
A coenzyme with a niacin active site that shuttles electrons between metabolic pathways. When reduced, NAD$^+$ becomes NADH + H$^+$; often referred to as a reducing equivalent.

nicotinamide adenine dinucleotide phosphate (NADP)
A coenzyme with a niacin active site that shuttles electrons between metabolic pathways. When reduced, NADP$^+$ becomes NADPH + H$^+$; often referred to as a reducing equivalent.

nicotinic receptor
A type of acetylcholine receptor; propagates a signal by opening Na$^+$ channels on the postsynaptic neuron. Nicotinic receptors in the brain are important in functions related to attention, learning, and memory.

non-adaptive traits (non-adaptive aging)
In a theory of aging, a trait that becomes useless to an individual; natural selection no longer acts to either remove or maintain the trait. Since most physical problems of aging occur after reproduction, the traits of aging neither increase nor decrease fitness, so aging and/or senescence are neutral to the forces of natural selection.

norepinephrine
A neurotransmitter found mainly in areas of the brain that are involved in governing autonomic nervous system activity; involved in the formation and function of dopamine and serotonin; also secreted by the adrenal gland in the "fight-or-flight" response, raising blood pressure and stimulating muscle contraction.

nuclear pore complexes
Channels in the nuclear envelope that allow molecules to move between the nucleus and the cytoplasm.

nucleosome
The primary structure of chromatin; small lengths of DNA wrapped around beadlike histone proteins.

nucleotide
Any of several compounds that consist of deoxyribose or ribose (a sugar) joined to a purine or pyrimidine base and to a phosphate group; the basic structural unit of nucleic acids (DNA and RNA).

nucleotide bases
The purine or pyrimidine bases that are part of nucleotides and make up the genetic code; in DNA:

adenine (A), thymine (T), guanine (G) , and cytosine (C); in RNA: adenine, uracil (U), guanine, and cytosine.

Okazaki fragments
Short lengths of DNA produced on the lagging strand template during DNA replication.

olfactory bulb
The terminus of the olfactory nerve, located just above the nasal cavities; integrates the nerve impulses from the olfactory nerve into a signal that is sent to the limbic system for decoding into a sense of taste and smell.

olfactory nerve
A nerve of the upper nasal cavity that detects aromatic compounds; leads to the olfactory bulb.

oligodendrocytes
The supporting cells of the central nervous system that produce the myelin sheath.

oocyte
An undeveloped egg in the ovarian follicle.

optimality theory
A general theory of evolution, proposed by John Maynard Smith, predicting that an individual will optimize a behavior so that the cost associated with the behavior is minimized in accordance with the local environment.

oral glucose tolerance test (OGTT)
A test for glucose tolerance (insulin resistance); measures blood glucose concentration in a fasted state, administers an oral dose of glucose (75 g), and makes 6 to 10 additional blood glucose measurements over the next two hours.

organ of Corti
A structure of the cochlea of the inner ear; contains sensory receptors for hearing.

origin recognition complex (ORC)
A multisubunit complex that binds to origins of replication in all eukaryotes; initiates DNA synthesis.

osteoblasts
Bone cells responsible for bone formation; secrete the proteins, primarily type I collagen, that make up the bone matrix into which Ca^{2+} salts are precipitated to form crystals.

osteoclasts
Bone cells that secrete proteolytic enzymes and acids in order to remove Ca^{2+} from bone tissue during the remodeling process.

osteocytes
Osteoblasts after they have finished secreting matrix material and have become trapped in the new calcified bone. Osteocytes link together to form canals called canaliculi, which allow the bone to exchange nutrients and waste.

osteopenia
Low bone density, not reaching the diagnostic criteria for osteoporosis.

osteoporosis
A condition characterized by a decrease in bone mineral content, causing bones to become porous and fragile and increasing the risk of facture. Primary osteoporosis is caused by age-related bone mineral loss and occurs primarily in women. Secondary osteoporosis is caused by medication and/or a primary disease, such as cancer or kidney disease, and can occur in men and women.

ovarian follicle
A cavity in the ovary that contains a single egg; the structure in which eggs are nurtured to maturation.

ovulation
Release of an egg from the ovary into one of the fallopian tubes.

ovum
An unfertilized, mature female reproductive cell; the human egg.

oxidation
A decrease in electron density of an atom. In biochemical reactions, oxidation is generally the loss of a hydrogen atom from or addition of an oxygen atom to a carbon-containing molecule.

oxidative phosphorylation
Coupling of the electron transport chain to ATP synthesis via a transmembrane proton gradient and an ATP synthase; occurs in mitochondria and primarily in the presence of oxygen.

oxidative stress theory
A theory of cellular senescence predicting that a random accumulation of damage caused by reactive oxygen species leads to widespread alterations in biomolecules important to cell replication.

oxygen-centered free radicals
Free radicals that arise from the reduction of oxygen. They include the superoxide radical ($\bullet O_2^-$), hydrogen peroxide (H_2O_2), and the hydroxyl radical ($\bullet OH$). Also called reactive oxygen species (ROS).

paired helical fibrils (PHFs)
Protein aggregates formed from hyperphosphorylated tau proteins; the precursors of neural fibrillary tangles.

parathyroid gland
A group of four endocrine glands located in the neck, behind the thyroid gland; secretes parathyroid hormone.

parathyroid hormone (PTH)
A hormone synthesized and secreted by the parathyroid gland in response to low serum calcium. It increases calcium absorption in the intestine, increases resorption of bone tissue, and increases reabsorption of calcium in the renal tubules.

passive diffusion
Movement of molecules down a concentration gradient, from high concentration to low concentration, without help from other molecules; occurs directly through the membrane (for nonpolar molecules) or through ion channels.

peripheral nervous system (PNS)
The system that includes all nerves outside the central nervous system (brain and spinal cord).

peristaltic contractions
Waves of involuntary smooth muscle contraction that transport food, waste matter, or other contents through a tube-shaped organ such as the esophagus and intestine.

peritubular capillaries
Capillaries surrounding the renal tubules.

personal genomics
A branch of genomics that focuses on individual genotypes and epigenetic mechanisms, using bioinformatics techniques.

p53 pathway
A pathway involving the protein p53, a transcriptional activator of genes. High levels of p53 are associated with arrest of the cell cycle and apoptosis.

phacoemulsification
A method of cataract surgery in which the lens is emulsified and aspirated from the eye.

phagocytosis
The process of engulfing and ingesting foreign particles, such as bacteria; carried out by white blood cells, typically neutrophils and macrophages.

pharynx
The hollow tube that starts behind the nose and ends at the top of the trachea and esophagus; contains the voice box.

phenotype
The traits of an organism that are produced by interaction between the genotype and the environment.

phloem
A vascular system of plants, made up of living cells, that carries organic nutrients (mostly sugars) to all parts of the plant body; the innermost layer of bark. *See also* **xylem**.

phospholipids
Compounds composed of fatty acids and phosphate; important constituents of biological membranes.

phosphorylation
The process of adding phosphorus (as a phosphate group) to a compound, often carried out by enzymes called kinases; used in cell signaling to turn molecules on (active) or off (inactive).

photoaging
Long-term changes in the skin due to exposure to the sun; known clinically as solar elastosis.

photosynthesis
The synthesis of glucose in a series of reactions involving CO_2, water, and light, taking place in chloroplasts.

phototropins
Photoreceptors that mediate phototropic responses (phototropism) in higher plants, allowing plants to alter their growth in response to light in the environment.

phototropism
The movement (bending) of a plant toward light, induced primarily by the absorption of blue light; hormonally activated (by auxins).

phylogenetics
A field of study that describes the relatedness among organisms based on genetic similarities.

phylogenetic tree
A branching diagram showing the deduced evolutionary relationships among various species.

phylogeny
The evolutionary sequence of events involved in the development of a species or group of organisms.

pinna
The external part of the ear in humans and other mammals. Also called the auricle.

pitch (acoustic)
A subjective term for the perceived frequency of a tone; fast frequency = high pitch; slow frequency = low pitch.

plasma (blood plasma)
The liquid portion of the blood (contains no proteins, fats, or carbohydrates); also part of the extracellular fluid.

plasma cells
Antibody-producing B cells Also called effector B cells.

plasmodesma (*plural,* plasmodesmata)
In plants, cytoplasm that extends through a small pore in the cell wall and into an adjacent plant cell; site of cell-to-cell communication.

plastids
Membrane-surrounded organelles of plant cells; sites of the manufacture and storage of important chemical compounds. *See also* chloroplasts; chromoplasts; leucoplasts.

pleiotropy
The production of more than one phenotype by a single gene.

pluripotent stem cells
Stem cells that differentiate into various types of specialized tissue in the body; capable of generating the three basic germ layers: endoderm, ectoderm, and mesoderm; arise from totipotent stem cells during embryonic cell division.

poly-A-binding protein
A protein that binds to the poly-A tail of mRNA so that completed mRNA can be distinguished from other RNA fragments.

poly-A tail
A string of nucleotides containing adenine (A) at the 3' end of mRNA.

polymerase chain reaction (PCR)
A technique for selective amplification of a defined nucleic acid region in a DNA mixture by copying the complementary strand of a target DNA or mRNA molecule for a series of cycles until the desired amount is obtained.

polymorphism
The existence of two or more clearly different phenotypes in the same population of a species or two or more alleles of the same gene.

population doubling (in cell culture)
The process in which some cells are removed from a culture that has reached confluence, placed into a new flask, and allowed to grow to confluence again.

population genetics
A branch of genetics that studies the distribution of alleles and changes in allele frequencies in a population, founded in the 1920s and 1930s mainly by R. A. Fisher, J. B. S. Haldane, and S. Wright; primarily a mathematical discipline making wide use of statistical probabilities.

porphyrins
Organic pigments with four pyrrole rings and a metal cofactor bound to nitrogen. Hemoglobin and chlorophyll have a porphyrin moiety.

postmaturation
The period of the life span after growth has ceased. In mammals, this typically occurs when the growth plates

of bones have calcified. The last bone to completely calcify in humans is the femur, at about 27–29 years of age.

postsynaptic neuron
At a synapse, the neuron to which neurotransmitter binds, thus receiving the neural signal. *See also* synapse.

presbycusis
Age-related hearing loss; normally associated with loss in high-pitched sounds. The Greek word *presbys* means "old person."

presbyopia
Loss of ability of the eye to focus on near objects, occurring with advancing age; primarily caused by increased stiffness of the lens. The Greek word *presbys* means "old person."

presynaptic neuron
At a synapse, the neuron that releases neurotransmitter into the synaptic cleft, thus passing on the neural signal. *See also* synapse.

prevalence (of disease)
The total number of individuals in a population who have a particular disease at a specific point in time.

primary cell culture
A tissue culture started from cells taken directly from an organism; in biogerontology, a culture of post-mitotic cells and cells with limited proliferative capacity. *See also* population doubling (in cell culture); replicative senescence.

primary cell wall
In plants, a thin, flexible, extensible layer outside the cell membrane that is composed of cellulose, pectin, and hemicellulose. *See also* cell wall.

primase
An enzyme that creates an RNA primer for the initiation of DNA replication; part of the primosome.

primosome
A protein consisting of two enzymes, helicase and primase, that separate DNA into single strands and lay down the RNA primer.

progeria
A rare genetic disorder characterized by physical signs suggestive of premature aging.

progesterone
A hormone produced in the ovaries that prepares and maintains the uterus for pregnancy.

programmed senescence (in plants)
The purposeful, highly regulated, highly ordered process in plants that results in the dismantling of post-mitotic cells and the recycling/remobilization of nutrients.

prokaryotes
Single-cell organisms that lack a membrane-surrounded nucleus; bacteria and archaea.

prometaphase
The stage of mitosis that follows prophase and precedes metaphase in eukaryotic somatic cells. The nuclear envelope breaks into fragments and disappears; microtubules emerge from the centrosomes at the poles (ends) of the spindle.

promoter region
A nucleotide sequence in DNA that binds RNA polymerase; the region of DNA that initiates transcription.

prophase
The first stage of mitosis, during which chromosomes condense and become visible and the mitotic spindle forms.

proteases
Enzymes that catalyze the hydrolytic breakdown of proteins (proteolysis) into peptides or amino acids.

proteasome
A protein complex in the cytoplasm that degrades proteins marked for removal by ubiquitin.

protein kinase A (PKA)
A highly conserved kinase responsive to nutrient signaling; involved in longevity regulation in *S. cerevisiae*, *C. elegans*, and *Drosophila*.

protein kinases
Enzymes that catalyze the addition of a phosphate group to a protein.

protein structure
A classification system with four levels of protein organization: primary structure: the amino acid sequence; secondary structure: arrangement of the polypeptide chain into a helix or sheet configuration; tertiary structure: the three-dimensional structure of a single polypeptide chain, with the final folding that imparts functionality; quaternary structure: association of two or more polypeptide chains that make up the protein, folded into its final, functional configuration.

proteolysis (proteolytic)
The directed degradation of proteins by cellular enzymes called proteases.

protoporphyrins
The precursor compounds of porphyrins, composed of four modified subunits interconnected at their α-carbon atoms; require a metal cofactor to become porphyrins, such as heme.

protozoa
Single-cell eukaryotic organisms.

pulmonary circulation
The circulatory system limited to the heart and lungs; responsible for gas exchange in blood in the lungs, with deoxygenated blood leaving the right ventricle and oxygenated blood returning to the left atrium.

pulmonary valve
One-way valve that regulates blood flow from the right ventricle; positioned between the right ventricle and the pulmonary artery.

pulmonary veins
The blood vessels carrying oxygenated blood from the lungs to the heart.

rarefaction (acoustic)
The portion of a sound wave in which molecules are spread apart, forming a region with lower than normal pressure. *See also* compression (acoustic).

reactive oxygen species (ROS)—*see* oxygen-centered free radicals

recombinant DNA technology
Techniques used to make a new segment of DNA from different DNA sources. Also called genetic engineering.

reduction
An increase in electron density of an atom; in biochemical reactions, the addition of a hydrogen atom, proton, or electron to or the removal of oxygen from a carbon-containing molecule.

refractive power
The degree to which a lens causes the convergence or divergence of light rays. Also called optical power.

regeneration (in the Calvin cycle)
The conversion of triose phosphate (glyceraldehyde 3-phosphate) to dihydroxyacetone phosphate and then to ribulose bisphosphate, the starting compound of the Calvin cycle.

renal tubules
The microscopic tubules in the nephron that manufacture urine from filtered blood serum and conserve essential nutrients and other substances required by the body.

replicating cell cultures
Nondifferentiated mitotic cells (such as fibroblasts) that have been removed from tissue and allowed to divide until they reach confluence.

replication origin
The site on a chromosome where DNA replication begins.

replicative senescence
The period of time when a mitotic cell can no longer divide.

repressor
A protein that binds to a regulatory DNA sequence, preventing transcription of a gene.

reproduction potential
A species' relative capacity to reproduce itself under optimal conditions.

reproductive value, v_x
A value, proposed by R. A. Fisher, that can be used to predict the future reproductive contributions of an individual relative to the reproductive output of the total population.

resting energy expenditure (REE)
The amount of energy required to maintain the essential functions of life—heart rate, body temperature, brain function, and so on; makes up 60–70% of total energy expenditure (TEE).

resting membrane potential
The transmembrane voltage when a neuron or muscle cell is not producing an action potential.

retina
The light-sensitive tissue lining the inner surface of the eye.

reverse transcriptase
An enzyme that transcribes single-stranded DNA into single-stranded RNA.

ribonucleic acid (RNA)
A class of nucleic acids synthesized from a template of chromosomal DNA and involved in protein synthesis; composed of chains of sugar molecules (ribose), phosphate, and purines (adenine, guanine) and pyrimidines (cytosine, uracil).

ribosomal DNA (rDNA)
DNA in the nucleolus that codes for ribosomal RNA (rRNA) and the nucleolus itself.

ribosomal RNA (rRNA)
A type of RNA found in ribosomes that interacts with messenger RNA (mRNA) and transfer RNA (tRNA) during the translation of mRNA to form a protein.

ribosomes
The sites in the cytoplasm where protein synthesis takes place; contain enzymes, regulatory proteins, rRNA, and the binding sites for mRNA and tRNA.

risk factor
A characteristic that is statistically demonstrated to be associated with (although not necessarily the direct cause of) a particular injury or disease. Risk factors can be used for targeting preventive efforts at groups that may be particularly susceptible to the injury or illness.

RNA polymerases
Enzymes that catalyze the polymerization of RNA, using single-stranded DNA as a template.

RNA splicing
A molecular process that splices intron sequences from an RNA transcript and joins exon sequences together to form messenger RNA (mRNA).

RNA transcript
The RNA molecule synthesized on a DNA template, prior to RNA splicing; still contains introns.

salicylic acid
A plant hormone that aids in resistance to pathogens; the primary ingredient of aspirin.

saliva
A secretion of the salivary glands containing substances that lubricate food, form chewed food into a bolus, and begin the process of chemical digestion.

salivary glands
Exocrine glands in the oral cavity that produce saliva.

sarcopenia
An age-related decrease in muscle mass that occurs regardless of factors known to increase muscle mass, such as increased physical activity; reflects a decrease in muscle cell number and size; greater in men than in women.

saturated fatty acid
A fatty acid in which all carbons in the hydrocarbon chain are connected by single bonds, thus maximizing the number of hydrogen atoms attached to the carbon backbone.

Schiff base
A functional group that contains a carbon-nitrogen double bond, with the nitrogen atom connected to an aryl or alkyl group; an important intermediate in the glycosylation of proteins. The nonenzymatic rearrangement of a Schiff base produces an Amadori product, a precursor to advanced glycation end products.

Schwann cells
Supporting cells of the peripheral nervous system that produce myelin.

S-cyclins—*see* cyclins

secondary cell wall
In plants, a cell wall deposited after the primary wall has stopped growing; provides support, strength, and protection. Wood is composed primarily of secondary cell wall. *See also* cell wall.

secondary osteoporosis—*see* **osteoporosis**

securin
A protein the holds sister chromatids together in early mitosis; destroyed at the transition from metaphase to anaphase if all DNA checkpoints have been cleared.

selection pressure
The events altering the genetic composition of individuals.

selective mortality
In analyses of mortality, inclusion of only those individuals who survive to a particular age. The older the cohort, the smaller the percentage of the original group still alive.

seminiferous tubules
A network of tubes in the testicles in which sperm are formed, mature, and move toward the epididymis.

senescence
Age-related changes at the end of an organism's life span that affect vitality and function, increase the likelihood of death, and are not directly related to disease.

sensory neurons
Neurons that convey sensory information about the external and internal environments to the central nervous system.

Sertoli cells
Cells in the seminiferous tubules that support the growth and maturation of sperm in the testicles.

sex chromosome
Either of a pair of chromosomes, usually designated X or Y, in the germ cells of most animals and some plants that combine to determine the sex and sex-linked characteristics of an individual.

silencer
A sequence of DNA that inhibits transcription.

sinus node (sinoatrial node)
Neural tissue in the right atrium of the heart that generates an action potential that causes the heart to contract.

sister chromatids
Two copies of a chromatid that are connected by a centromere and cohesin proteins; the end product of the S phase of the cell cycle.

small nuclear RNAs (snRNAs)
RNA molecules of about 200 nucleotides that are involved in RNA splicing.

sodium/potassium ATP pump
A complex in the cell membrane that maintains electrolyte balance by exchanging sodium ions for potassium ions, using energy from the hydrolysis of ATP to ADP.

solar elastosis—*see* **photoaging**

somatic cells
Cells of the body that are not involved in sexual reproduction. *See also* germ cells.

S phase (synthesis phase)
The stage of the eukaryotic cell cycle in which DNA synthesis occurs. *See also* cell cycle.

spindle poles
Microtubule-organizing centers, functionally equivalent to the centrosome; the mitotic spindle grows out of these structures.

stapes
One of the bones of the middle ear that amplify and transfer sound wave vibrations from the tympanic membrane to structures of the inner ear.

starch
The storage form of carbohydrate in plants; not found in animals. Starch occurs in two forms: amylose, a straight-chain structure, and α-amylopectin, a branched-chain structure.

statins
A class of drugs that prevent the synthesis of low-density lipoproteins (LDLs) in the liver; effective in lowering the risk for atherosclerosis. Statins work by inhibiting an enzyme involved in the synthesis of cholesterol. The most commonly prescribed statin is Lipitor.

stem cells
Undifferentiated cells having the ability to renew themselves indefinitely; the first few cells after fertilization, dividing to create differentiated cells and further stem cells. *See also* multipotent stem cells; pluripotent stem cells; totipotent stem cells.

stereocilia
In the inner ear, mechano-sensing hairlike projections on the cells of the organ of Corti. Their movement is proportional to the vibration of the fluid in the cochlea; part of the transformation of mechanical movement (vibrations) into neural impulses.

stochastic (biological)
Describing a biological or chemical process having a component of change or probability.

stochastic senescence (in plants)
Random degradation of a plant cell following programmed senescence; often associated with breakdown of the cell, nuclear, and vacuolar (tonoplast) membranes.

stop codons
Three codons, UAA, UAG, UGA, in mRNA that are not recognized by a tRNA and signal the ribosome to stop translation.

stroke volume
The amount of blood ejected from the heart ventricles during systole (the contractile cycle of the heart).

subcutaneous fat tissue
The fat layer beneath the skin.

superior vena cava
The major vessel carry deoxygenated blood to the heart (right atrium).

superoxide dismutase (SOD)
An enzyme in the cytosol and mitochondria that reduces superoxide ions to hydrogen peroxide; functions as an antioxidant. The cytosolic form of the enzyme contains a copper/zinc active site; the mitochondrial form contains a manganese active site.

superoxide radical—*see* **oxygen-centered free radicals**

survival curve
A representation of the percentage of survivors in a given population. The *y*-axis normally denotes the

percentage of the population (100% = alive at birth) alive at a specific chronological age (the *x*-axis).

synapse
The junction where two nerves meet for neural transmission; composed of the presynaptic neuron transmitting the signal; the synaptic cleft or space between the two neurons; and the postsynaptic neuron that receives the signal.

synaptic cleft
The space between the presynaptic and postsynaptic neurons, into which neurotransmitters are released.

synaptic terminals
The endings of branches of an axon.

systemic circulation
The part of the circulatory system that includes all vessels except the pulmonary circulation.

systole
The contraction of heart muscle that drives blood out of the ventricles; the point at which blood pressure is highest.

systolic pressure
The pressure (in mm Hg) in the arteries when the heart beats and ejects blood from the ventricles; the higher number in measurements of blood pressure (e.g., the "120" in 120/60).

target of rapamycin (TOR)
A highly conserved nutrient-sensing protein kinase that regulates growth and metabolism in all eukaryotic cells; may be involved with regulating longevity. (Rapamycin is an immunosuppressant drug used to prevent rejection of transplanted organs; it blocks a protein involved in cell division and inhibits the growth and function of certain T cells.)

taste buds
Clusters of nerve endings on the tongue and the lining of the mouth that provide the sense of taste; the five general categories of taste buds are salty, sweet, bitter, sour, and umami.

TATA box
A nucleotide sequence consisting of TATAAAA, located in the promoter region of a eukaryotic gene, about 25 nucleotides from the transcription start site. Binding of a general transcription factor to the TATA box initiates formation of the transcription initiation complex.

tau protein
A microtubule-associated protein (MAP) that helps to stabilize axonal microtubules. Its activity depends on its degree of phosphorylation; hyperphosphorylation results in aggregation of tau into insoluble paired helical fibrils, the precursors of neurofibrillary tangles.

taxonomy
The science of classifying plants, animals, and microorganisms into increasingly broader categories based on shared features.

T cells
Lymphocytes (white blood cells) produced in the bone marrow and processed in the thymus that participate in cell-mediated immune defenses. *See also* cytotoxic T cells; helper T cells; naïve T cells; natural killer (NK) cells.

telangiectasia
Appearance on the skin surface of small dilated blood vessels; most often associated with photoaging.

telomerase
An enzyme that adds telomere sequences to the ends of eukaryotic chromosomes. It has two subunits: a catalytic center and an RNA template.

telomeres
Non-gene, highly repeated DNA sequences at the ends of chromosomes that protect the chromosomes from degradation; shorten after each round of DNA duplication.

telomere-shortening theory
A theory of cell senescence predicting that short telomeres cause the cell to halt the replication process. Telomeres shorten during the normal cell cycle as a result of the end replication problem in the DNA lagging strand.

telophase
The stage of mitosis in which the chromosomes arrive at the poles, the microtubules disappear, and the nuclear envelopes form around the two daughter nuclei.

terminally differentiated
Describing cells or groups of cells that have become incapable of mitosis and are not replaced by new cells when they die. Examples are neurons (nerve cells), myocytes (heart cells), and cells in the lens of the eye.

termination site
A location on DNA or RNA that ends transcription or translation.

tertiary structure
The three-dimensional structure of a polypeptide chain, with the final folding that imparts functionality. *See also* protein structure.

testosterone
An androgenic hormone produced chiefly by the testes; responsible for the development of male secondary sex characteristics.

thrombosis
A solid mass (formed from constituents of blood and cells) in a blood vessel or organ that can block blood flow.

thymine
One of the nucleotide bases of DNA; a pyrimidine. *See also* nucleotide bases.

thymus gland
A gland in the upper chest, behind the sternum (breastbone); the site where T cells mature and multiply after they leave the bone marrow. The gland grows throughout childhood until puberty and then gradually decreases in size.

tonoplast
The semi-permeable membrane surrounding the central vacuole of a plant cell.

total energy expenditure (TEE)
The total amount of energy expended by an organism in a given period, usually calculated in humans over 24 hours. TEE = resting energy expenditure (REE; the energy required to maintain normal body functions when not active and not absorbing food) + physical activity + diet-induced thermogenesis (DIT; the energy required for digestion, absorption, and nutrient storage).

totipotent stem cells
Stem cells that can reproduce every type of cell and tissue in the body, including the placenta. Cells from

the first few divisions of an embryo are totipotent stem cells. Totipotent stem cells give rise to pluripotent stem cells.

trachea
The tube connecting the lungs to the mouth and nose.

trade-off hypothesis (of aging and longevity)
The hypothesis that successful reproduction has to be traded for mortality: the more resources given to successful reproduction, the less resources available for survival after reproduction.

transcription
The process of constructing an RNA molecule by using a DNA molecule as template, resulting in the transfer of genetic information from DNA to RNA.

transcription factor
Any protein or other molecule that initiates or regulates the transcription of DNA. These factors bind to the promoter region of a gene and change the shape of the DNA so that DNA polymerase can recognize the start of the gene.

transcription initiation complex
A composite of several general transcription factors and RNA polymerase II bound to the promoter region of a eukaryotic gene; necessary but not sufficient for initiation of transcription.

transcriptome
All of the RNA transcripts present in a cell, tissue, or organism under specified conditions.

transfer RNA (tRNA)
A class of small RNA molecules involved in translation of mRNA into the amino acid sequence of a protein. The mRNA codon links to the corresponding tRNA, which carries a specific amino acid.

transgenic organism
A mutant organism in which an extra gene copy has been inserted into the genome by genetic engineering; used to increase the expression of a protein.

translation
The use of information in messenger RNA (mRNA) to construct a protein from amino acids; carried out on ribosomes.

translational initiation
A type of post-transcriptional control of gene expression.

tricarboxylic acid (TCA) cycle
Part of the oxidative cellular respiration process that reduces carbon-containing intermediates to CO_2. In the process of oxidizing TCA intermediates, electrons are generated and shuttled to the electron transfer system. Also called the Krebs cycle or the citric acid cycle.

tricuspid valve
One-way valve that regulates blood flow from the right atrium; positioned between the right atrium and the right ventricle.

triglyceride
A molecule containing three fatty acids bound to a glycerol, the primary form of lipid in adipose tissue and in foods.

triose phosphate
The common name for glyceraldehyde 3-phosphate, a compound that serves as an intermediate in metabolic pathways in all organisms; formed in glycolysis and in

the dark reactions of photosynthesis.

trypsin
An enzyme secreted by the pancreas that digests proteins. Bovine trypsin is used in cell culture methods to degrade the connective tissue that holds cells together.

tubular reabsorption
The movement of substances from a kidney (renal) tubule into the peritubular capillaries.

tubular secretion
The movement of substances from the kidney's peritubular capillaries into a renal tubule.

tubule (renal)—see renal tubule

tympanic membrane
The eardrum; the membrane that vibrates in response to sound waves; located between the outer and middle ear.

type 1 diabetes mellitus
A disease caused by the inability of beta cells of the pancreas to secrete insulin; characterized by high blood glucose levels, excessive thirst, and frequency of urination; arises during early childhood; requires daily injections of insulin. Also called juvenile or early-onset diabetes mellitus.

type 2 diabetes mellitus
A mild form of diabetes occurring in adulthood and characterized by the decreased ability of insulin to stimulate glucose uptake by cells; usually controlled by diet and exercise, without injections of insulin. Also called adult-onset or late-onset diabetes mellitus.

type I muscle fibers
Muscle cells with high concentrations of the myosin isoform that results in slow contraction speeds; sometimes called slow-twitch fibers. They have large numbers of oxidative enzymes and are highly fatigue-resistant.

type II muscle fiber
Muscle cells with high concentrations of the myosin isoform that results in fast contraction speeds; sometimes called fast-twitch fibers. They rely more on glycolytic pathways than on oxidative pathways for energy and fatigue quickly.

type IIX muscle fibers
Muscle cells with characteristics of both type I and type II fibers; generally found only in older, inactive humans.

ubiquitin
A highly conserved small protein (79 amino acids) that attaches to lysine residues of damaged proteins to mark them for degradation.

ubiquitin ligase
An enzyme that binds ubiquitin to the lysine residues of damaged proteins to mark them for degradation.

umami
One of the five categories of taste detected by the taste buds; associated with salts of glutamic acid and other amino acids. *See also* taste buds.

unsaturated fatty acids
Fatty acids with a hydrocarbon chain having at least one double bond.

uracil
One of the nucleotide bases of RNA; a pyrimidine. *See also* nucleotide bases.

urea
A waste product, found in the blood, that is formed from the normal breakdown of protein; normally removed from the blood by the kidneys and excreted in the urine.

ureter
The tube that connects the kidney to the bladder.

urethra
The tube that connects the bladder to the outside of the body.

urinary incontinence
The inability to control urination.

vasoconstriction
Contraction of the smooth muscle surrounding arteries that causes the arterial lumen to become smaller, decreasing blood flow.

vasodilation
Relaxation of the smooth muscle surrounding arteries that causes the arterial lumen to expand, increasing blood flow.

vasopressin—*see* **antidiuretic hormone**

veins
Blood vessels that return blood to the heart.

venous return
The amount of blood returned to the right atrium.

ventilation
Breathing rate.

ventricles
The two chambers of the heart that pump blood to the lungs and to the rest of the body. The right ventricle pumps deoxygenated blood to the lungs; the left ventricle pumps oxygenated blood to the body.

Verhulst-Pearl logistic equation
An equation describing population growth for any population, particularly those that are constrained by lack of mobility and/or are maintained under highly controlled conditions. $\Delta N = rN[(K - N)/K]$, where N = population size, r = intrinsic rate of natural increase, and K = carrying capacity of the population.

villi
Fingerlike projections of the lining of the small intestine; aid in absorption by increasing the surface area.

vital statistics
The statistics of a population that relate to major life events such as births, deaths, marriages, health, and disease.

voltage-gated ion channels
Membrane channels that allow the movement of ions into and out of the cell, opening and closing in response to electrical signals, thus allowing the cell to change its membrane potential in response to stimuli.

weak mutations
Alterations in a gene that reduce rather than eliminate expression.

wild type
A strain, gene, or characteristic that prevails among individuals in natural condition, as distinct from an atypical or mutant type.

X-ray crystallography
A method of determining the arrangement of atoms within molecules in a crystal, in which a beam of X-rays strikes the crystal and diffracts in many directions; used to determine the three-dimensional structure of molecules.

xylem
The woody part of a plant that transports water and minerals; consists mostly of dead cells; can also provide structural support for the plant.

zonular fibers
In the human eye, fibers that attach the lens to the ciliary body.

INDEX